T0299296

ISO 9001:2015— A Complete Guide to Quality Management Systems

ISO 9001:2015— A Complete Guide to Quality Management Systems

Itay Abuhav

CRC Press
Taylor & Francis Group
Boca Raton London New York

CRC Press is an imprint of the
Taylor & Francis Group, an **informa** business

CRC Press
Taylor & Francis Group
6000 Broken Sound Parkway NW, Suite 300
Boca Raton, FL 33487-2742

First issued in paperback 2021

ISBN 13: 978-1-03-224042-8 (pbk)
ISBN 13: 978-1-4987-3321-2 (hbk)

DOI: 10.4324/9781315369808

Publisher's Note
The publisher has gone to great lengths to ensure the quality of this reprint but points out that some imperfections in the original copies may be apparent.

Library of Congress Cataloging-in-Publication Data

[LoC Data here]

Visit the Taylor & Francis Web site at
http://www.taylorandfrancis.com

and the CRC Press Web site at
http://www.crcpress.com

Contents

Preface

The quality management world is about to go through a big change represented by the publication of the new revision of the ISO 9001 Standard: the 2015 revision. The new revision of the standard introduces new challenges to the organization but also a revision to old challenges. This book is a complete guide to implementing all the requirements of the standard. In order to present the reader with a practical and useful guide, I have provided a definition of my quality policy and objectives.

My Quality Policy

Presenting and reviewing the ISO 9001:2015 standard requirements through analysis, interpretation, and demonstration, with explanations, insightful examples, and events from various industries and sectors.

My Quality Objectives

- Commitment to the highest level of consulting regarding the ISO 9001:2015 standard
- Reviewing all the topics and issues related to the realization of a product or service with reference to various types of processes and products
- Providing support in the implementation of an effective quality management system
- Facilitating the documentation of processes
- Providing references to the new challenges presented in the ISO 9001:2015 standard

However, a policy and related objectives are ineffective without also having in place designed and structured tactics and methods to achieve them:

- The book is designed and structured to mirror the standard's table of contents in order to simplify navigation and use.
- Each clause and sub-clause of the standard are discussed and analyzed through quality perspectives, such as the implications for an organization—its processes, risk-based thinking, resources, infrastructures, process environment, control and effectiveness, and documented information.

- The ISO 9001:2015 standard acts like a complicated web of prerequisites with relations between them. A full and comprehensive reference to the interrelations between the different clauses and sub-clauses has been included.
- Putting words into actions—the book will assist in translating the requirements and objectives into feasible activities and tasks. It visualizes situations with everyday events from the different sectors, branches, and products or services.

List of Exclusions

I decided to exclude chapter 0 of the standard from this book since it mainly provides explanations regarding the ISO 9001:2015 that are already covered elsewhere in this book.

My biggest wish is that you, as a reader, will refer to this book as a consulting session; that you will read and explore it, draw information and knowledge that suit you and your organization; and that you will introduce this information to your quality management system and processes.

Acknowledgments

I wish to thank all the people—consultants, coworkers, auditors, mentors, bosses, and friends—that introduced me to the quality world and have aided, supported, taught, lectured, consulted, and provided valuable knowledge and information during the undertaking of this book and also in my professional career. You have helped to give an edge to this book. The list of names is too extensive to include here, but you know who you are.

I wish to thank my dear family for their warm support throughout the years.

I also wish to thank my wife Angela and daughter No'omi for understanding, pushing, believing, and supporting me throughout this project.

Thank you.

Author

Itay Abuhav has been acting as a consultant for many organizations in various areas for the past 10 years, specializing in the quality management/control industry, consulting with a number of small- to medium-sized firms dealing with implementing quality management systems, with a major focus on the European medical device industry. He is highly qualified to certify organizations to various standards: ISO 9001 and ISO 13485. The author has two active websites related to these topics, namely, www.9001quality.com and www.13485quality.com.

1 Scope

Clause 1 of the ISO 9001 Standard is used to present the purposes and concepts of the standard and define the scope of application of the standard to your quality management system. The following aspects are covered in this clause:

- The goals and purposes of the standard
- The approach and reference to customer requirements
- The approach and reference to regulatory or statutory requirements
- Applicability of the standard requirements

Before we start to understand the requirements of clause 1—Scope, let us review them first:

- The ISO 9001 Standard is an international standard for the establishment, design, and implementation of a quality management system (QMS) in an organization.
- Applying the ISO 9001 Standard requirements enables an organization to demonstrate its ability to consistently provide products or services that meet customer requirements.
- Applying the ISO 9001 Standard requirements enables an organization to demonstrate its ability to consistently provide products or services that meet applicable statutory or regulatory requirements.
- Applying the ISO 9001 Standard requirements enables an organization to enhance customer satisfaction through the use of quality management instruments that include methods for planning and improvements of processes and ensuring conformity to customer and applicable statutory and regulatory requirements.
- The requirements of this standard are generic and apply to any sector and area of business and may be implemented in any organization regardless of its size or the type of its products or services.
- Note 1—The use of the words "product" or "service" in this standard refers to a product or a service as intended or required by the customer.
- Note 2—Legal requirements may be regarded as statutory and regulatory requirements.

For me, clause 1 is an important introduction to the ISO 9001 Standard and I consider it crucial for

- The understanding of the expectations of the standard from an organization: What is the purpose of a QMS?
- Future definitions of a QMS: How may the QMS reach its purpose?

The Principles of the ISO 9001 Standard

Clause 1—Scope presents us with the principles and concepts of the standard. The ISO 9001 Standard is aimed to satisfy customers by fulfilling its requirements along with applicable international and national regulations. This is expressed through four principles:

- The goal and objective of the standard is to initiate a QMS that acts to consistently meet customer requirements as well as applicable regulatory requirements.
- The requirements of this standard initiate harmonization between a QMS of an organization and applicable regulatory requirements.
- The requirements suggested in the ISO 9001 Standard facilitate an improvement of processes included in the QMS and assurance of conformity to customer or regulatory requirements.
- The requirements suggested in the ISO 9001 Standard are applicable to all sorts of organizations regardless of their size or type, the type of their customers, and the type of products or services that they are providing.

Harmonization between a QMS of an organization and applicable regulatory requirements means brining the applicable and relevant international, national, local, and regulatory requirements into consensus with QMS aspects, for example, the planning and implementation of processes and activities, the design of documentation, and the training qualification of human resources.

The Ability to Provide Appropriate Product

What are actually the requirements here? When an organization decides to implement the ISO 9001 Standard, it is required, through the use and application of the methods and instruments presented in this standard, to prove its ability to identify customer requirements and provide products and services according to these requirements. And while identifying customer requirements, the organization must consider the application of relevant applicable regulatory, statutory, or other requirements, for example, codes or practices. In other words, the purposes presented in this clause must be reflected through the QMS of the organization. How? Through applying the quality management tools and instruments that are suggested in the standard, such as setting defining quality policy, quality objectives, planning processes, and much more.

2 Normative References

The meaning and purpose of normative references is the indication that terminology and nomenclature specified in this standard is not open for debate or an interpretive discussion. A normative reference refers to a document that includes terms, fundamental concepts, principles, and vocabulary that are essential for the application of the ISO 9001 Standard. The ISO 9001 Standard requirements are as follows:

- When dated normative references are used, only the edition cited applies (the ISO 9000:2015 Standard).
- When undated normative references are used, the latest edition of this referenced document (including any amendments) applies.
- The document ISO 9000:2015, Quality management systems—Fundamentals and vocabulary are to be normatively referred to while establishing a quality management system (QMS) according to the ISO 9001 Standard requirements.

A normative reference lists other ISO or IEC documents or standards that are necessary for the application of the standard. In other words, which documentation may assist you in how to comply with the requirements stated in the ISO 9001 Standard. The objective of a normative reference is to relate to a standard that is applicable to the implementation of the ISO 9001:2015 Standard and to relate to directives, definitions, or understanding of the ISO 9001:2015 Standard.

The ISO 9001 Standard refers us to a specific document, "ISO 9000:2015: Quality management systems—Fundamentals and vocabulary." In case questions or misunderstanding regarding the definitions or the requirements of the ISO 9001 Standard arises during the implementation and application of the standard requirements, you can turn to this document. For example, when you are discussing and planning activities related to customer focus and you are not sure what the definition of customer focus is, you may turn to the ISO 9000 Standard and understand how the ISO 9001 Standard interprets the issue of customer focus.

3 Terms and Definitions

Clause 3 of the standard—Term and Definitions is necessary to clarify matters and disputes regarding the terms and definitions mentioned in the ISO 9001 Standard. The terms and definitions mentioned and discussed in the ISO 9000 Standard—Quality management systems, Fundamentals and vocabulary—are applicable to this standard (the ISO 9001 Standard). The ISO 9001 Standard requirements are as follows:

- The terms and definitions given in ISO 9000:2015 apply for the purposes of the ISO 9001 Standard.

ISO 9000:2015 Standard

The ISO 9000 Standard provides the fundamental concepts, principles, and vocabulary that are to be used when establishing a quality management system according to the requirements stated in the ISO 9001 Standard. In other words, when you stumble upon a term in the ISO 9001 Standard that are unfamiliar to you, you cannot interpret it correctly or you have a dispute with another on the meaning, you may turn to the ISO 9000 Standard and resolve the conflict. A good example may be the term "risk"; you may consider it as one thing but your auditor may see it as another. In order to avoid unnecessary disputes in advance, I advise you to review this standard and learn the terms and definitions that are mentioned there.

4 Context of the Organization

4.1 Understanding the Organization and Its Context

Understanding the context of an organization is now standard requirement. It is not a new concept but it is officially adopted by the ISO 9001:2015 Standard, and when developing a quality management system (QMS), it is required to identify, analyze, and understand the business environment in which the organization conducts its business and realizes its product.

Any organization is a combination of different business entities (functions or systems—internal or external) that relate and interact with each other and exchange information and materials. The context of the organization is an outline of the interaction and integration of these business entities. It is the set of functions, factors, processes, inputs and outputs, and conditions and limitations that create the business environment of an organization—these encompass both internal and external issues. These issues have an impact on the ability of the organization to provide the product and thus affect the goals and objectives of the organization. Thus the deduction: when determining your business strategy and setting the objectives of your QMS, you should understand, consider, and refer to these aspects. Understanding the context of the organization is key to a correct business strategy, let alone a correct quality strategy.

Before we plunge into this topic, let us review what the ISO 9001:2015 Standard requirements are:

- The organization must determine the organizational context in which it is active. The organizational context includes issues that
  - Are relevant to its purpose
  - Are relevant to the scope of the QMS
  - Are relevant to its strategy (goals and objectives)
  - Affect the organization's ability to achieve the intended results and customer expectations
- Internal and external information and issues related to the context of the organization shall be reviewed in order to evaluate changes that might affect the objectives of the organization.

- Note 1—The review may refer to positive and negative factors or conditions that may affect the QMS and its context of the organization.
- Note 2—Issues arising from legal, technological, competitive, market, cultural, social, and economic environments conditions, whether international, national, regional, or local, may be considered for the definition of the context of the organization.
- Note 3—Organizational values, perceptions, and cultural environment may help understand and form the context of the organization.

The Principle of the Context of the Organization

The ISO 9001 Standard recognizes the importance and role of the context of the organization and requires its definition when establishing a QMS. The analysis regarding the context of the organization is a natural process that each organization must follow at some point in its life. The ISO 9001:2015 aspires to institutionalize the context as the foundation for the QMS.

Organizations usually have several strategies or concepts in various fields and areas that dictate the direction of the organization: financial strategy, sales and marketing strategy, purchase strategy, IT concept, and so on. One clear objective of determining the context of the organization is to harmonize these various strategies and concepts. By determining the context of the organization, the organization identifies the scope and boundaries of its activities that will be reflected in the QMS. From the context of the organization we derive the goals and objectives of the QMS. After attending to these points, you should have a clear picture of the business environment in which your organization is active and you would be in the right position to plan and define your QMS, its goals, and objectives.

External and Internal Issues Relevant to the Organization

The main goal of understanding the context of the organization is the identification of the external and internal issues that are relevant to the purpose of the organization. Organizational issues are significant factors, conditions, influences, situations, or events that have an effect on the QMS and will affect the ability of the organization in achieving its objectives and desired results. The issues may be affected or directed by relevant parties that have an effect on the organization, such as investors, employees, competitors, suppliers, regulators, and customers.

Every area of expertise, sector, market, product family, or other business segmentation has its own relevant issues that affect the organizational context. External or internal issues may include

- The expectations of interested parties
- The main products that provide the most value for its interested parties
- Processes and activities needed to meet specifications and expectations of its interested parties

- The influence of the business environment in which the organization is active
- Availability of resources needed for the realization
- Competence of the human resource
- Statutory and regulatory requirements

In practice, the standard expects awareness of these issues, and when issues are identified you must categorize them. I suggest two integrated methods for the effective identification of external and internal issues and their influence on the organization.

Understanding and Defining the Organizational Context

As mentioned earlier, the context of the organization consists of the environment in which it operates and refers to internal and external issues relevant to the activity of the organization. As a first step in defining the organizational context, I suggest following a course that will put you in a position to better define your quality policy, to identify the business environment in which the organization is active, to identify which internal and external issues in this environment influence the organization, and to understand who the relevant interested parties are. The following is the suggested process for an effective definition of the context of the organization (the different stages are elaborated throughout Section 4.1) (Figure 4.1):

- The definition of the context of the organization will begin with PEST (political, economic, social, and technological factors) analysis, which will provide inputs for the strengths, weaknesses, opportunities, and threats (SWOT) analysis.
- SWOT analysis will define inputs for the determination of the internal and external issues.
- The determination of the internal and the external issues will help identifying relevant interested parties.
- Identifying interested parties will assist in defining the scope of the QMS.
- The scope of the QMS will provide the foundation of the QMS.

PEST Analysis

PEST stands for political, economic, social, and technological factors that may affect the strategy of the organization. In order to understand better the context of the organization, it is recommended to conduct a PEST analysis—analysis of different conditions that affect the organization in the business environment in which it is active. This analysis reviews the macro environment of an organization

Figure 4.1   Stages in understanding the organizational context.

and is a useful tool for understanding how a business environment behaves and affects the QMS. A business environment will be analyzed by what it is addressing: the products, the customers' expectations, the activities of the organization, and so on. And these are reviewed through four different perspectives: political, economic, social, and technological.

PEST analysis identifies external factors that might change and while changing will affect or influence the organization and its operations. Each of the factors (political, economic, social, and technological) is used to assess the extent of influence of the business environment on the ability of the organization to deliver a product according to its customers' expectations. The organization, on the other hand, cannot influence these factors but must try to adapt to them. The outputs of PEST analysis will be used as opportunities and threats for the later SWOT analysis, which will be discussed later. PEST analysis may be also extended to seven or even more factors: ecological (or environmental), legislative (or legal), and industry analysis, which uses the name PESTELI analysis. But I will concentrate on the four main factors.

The four parameters for analysis vary in significance and influence depending on the type of organization, its business environment, and activities. For example, a software development company should give more scale to the technology while a voluntary association in a developing country should pay more attention to the political conditions.

Political Factors
Political factors consist of government regulations and legal factors that can affect the business environment. Issues such as political stability, trade regulations, manufacturing regulations, safety regulations, and employment laws will be assessed for their influence on the QMS. More example factors are

- Ecological/environmental issues
- Current legislation
- Anticipated future legislation
- International legislation (global influences)
- Regulatory bodies and processes
- Government policies, terms, and change
- Funding, grants, and initiatives
- Market lobbying groups
- Wars and conflicts

These factors may have a significant influence on defining the processes, their resources, and their required controls. These factors may dictate the implementation of processes and activities or may demand product specifications. Other good examples of a factor are work grants and permissions between different countries in different areas of the world that affect the availability or costs of human resources and the medical industry, when a manufacturer must implement local or regional regulations in each country or region where it is marketing its devices.

Economic Factors

The economic factors evaluate issues that are bound to impact on the cash flow, the business cycle, and any critical decisions regarding the direction of the organization. Here we can find factors such as economic growth, unemployment rate, inflation, and interest rates. Some more examples are

- National economies and trends
- General taxation issues
- Taxation to activities, products, services
- Seasonality
- Weather issues
- Market and trade cycles
- Specific sector factors
- Customer/end user drivers
- Interest and exchange rates
- International trade and monetary issues

Another good example is the big economic crisis of 2008, when many companies experienced a decline in their business. This led to the cancelation or delay of many plans and developments for new products.

Social Factors

Social factors represent the socioeconomic environment of the markets in which the organization is active. Social factors affect how one organization perceives its customers and the interested parties, their needs, and expectations (clause 4.2 of the ISO 9001 Standard), and understanding the social factors help the organization ensure meeting those needs and expectations by supplying the appropriate products and services. Examples of social factors are

- Lifestyle trends
- Demographic trends
- Consumer attitudes and opinions
- Media views
- Changes in laws that affect social behaviors
- Image of the organization
- Consumer buying patterns
- Fashion and role models
- Major events and influences
- Buying access and trends
- Ethnic/religious factors
- Advertising and publicity
- Ethical issues

Lifestyle trends are a good example of how a change of lifestyle can change realization processes of a product—today people shop online so that products are delivered directly to their home instead of going out to the stores. As a result the delivery businesses are booming. This fact impels many organizations to change their realization processes and adopt new distribution channels.

Technological Factors

Technological factors may have a significant effect on the expectations of customers, on the products or services offered, and thus on the processes and how products are realized. And in today's world technological factors are the ones that are changing most rapidly. An organization that has the ability to detect and identify and react to these changes faster has an advantage in its market. Examples for technological factors are

- Competing technology development
- Associated/dependent technologies
- Replacement technology/solutions
- Maturity of technology
- Information and communications
- Consumer buying mechanisms
- Technology legislation
- Innovation potential
- Technology access, licensing, patents
- Intellectual property issues
- Global communication
- Social media use
- Maturity of organizations

A good example that describes such significant technological changes is taken from the music recording industry in the last 20 years. In the early 2000s, the industry went from analog technology to digital technology leaving thousands of studios with irrelevant equipment. In a very short time, all studios needed to adapt to the new technologies and purchase new digital equipment because all its customers adapted it. That means adapting their substructures, their work environment, their purchase, the competence of the human resources, and so on. But guess what? After 20 years of digital sound, a social trend (social factor) has changed and today people are looking again for the warm analog sound! So studios that were smart and kept the old technology can satisfy today's customers who will pay a bit more for the vintage technology. How the world is changing and then changing back again ....

SWOT Analysis

After understanding the business environment in which the organization is active and the factors that influence this environment, the organization must evaluate itself. SWOT analysis (or SWOT model) was originally developed as a corporate planning tool. The analysis uses comparative data that enables the organization to evaluate itself and to help develop and shape a strategy—in our case, the QMS. SWOT evaluates the internal business environment of an organization through four axes: Strengths, Weaknesses, Opportunities, and Threats. Most of us know or use SWOT as part of our marketing strategy. But in fact the conclusions of this analysis may assist organizations in identifying the relevant quality elements that affect the

QMS: products and/or services, processes, and resources. The principle of SWOT analysis is easy to understand:

- What is good in the present and in the past is a strength
- What is bad in the present and in the past is a weakness
- What is good in the future is an opportunity
- What is bad in the future is a threat

The Strengths, Weaknesses, Opportunities, and Threats are axes of the analysis. Each of these axes will be analyzed according to the next index:

- Product (what are we realizing?)
- Process (how do we realize the product?)
- Customer (to whom are we delivering it?)
- Administration (and how do we manage these activities?)

What is the difference between PEST and SWOT analysis? PEST analysis measures the market and external business environment that a business unit is active in. SWOT analysis measures the internal factors of the business unit. The factors of PEST analysis cannot be influenced; the factors of SWOT can. The most effective (and easy) way to perform a SWOT analysis is to begin with answering question regarding each of the parameters mentioned earlier.

Strengths

The strengths are the internal attributes and resources that support the realization of the product and the establishment, implementation, maintenance, and improvement of the QMS. In other words, the advantages the organization has to realize the product or service. The appropriate questions in this case will be

- What are the reasons that on their behalf we achieve our objectives (process, administration)?
- Which resources in the organization are at our advantage and support the realization of the product (product, process)?
- Which knowledge in our organization is at our advantage (product, process)?
- Which processes do we perform better than others (process)?
- Which customer expectations and requirements do we meet the best (customer)?
- What are the reasons our customers are satisfied (customer)?

Weaknesses

Weaknesses are internal attributes that may interrupt the realization of the product and may produce a nonconformity in the QMS. When analyzing the weaknesses it is highly recommended to consider it from an external (as well as internal) point of view. With external point of view I mean—do other parties perceive a weakness that you are unaware of? Another external point of view is seeing the strengths of your competitors as your weaknesses. The appropriate questions in this case will be

- What are the reasons we do not achieve our objectives (process, administration)?
- Which resources in our organization we need to improve or supplement (process, administration)?

- What knowledge are we lacking in order to achieve our objectives (process, product)?
- Which processes we need to improve (process)?
- Which customer expectations do we fail to understand or answer (product, customer)?
- What are the reasons our customers are not satisfied (product, customer)?

Opportunities

Opportunities are factors the organization can draw advantages from in the future. These are the chances for improvements that the ISO 9001 is so fond of. A lot is written on this topic in this standard—many of the standard's clauses refer directly to the opportunities. In fact the standard requires an assessment of external as well as internal issues that may serve as opportunities that may assure that the QMS can achieve its objectives.

The outputs of the analysis of opportunities may serve as inputs to some of the standard's clauses (e.g., 6.1—Actions to address risks and opportunities, 6.3—Planning of changes). In these clauses, you will need to demonstrate methods and manners for identifying opportunities (aside from SWOT analysis). So the advantage of performing a clear and accurate analysis of opportunities will serve you with other standard requirements. The appropriate questions in this case will be

- Which product features, customer expectations, or trends can we develop or improve (product, customer)?
- Which processes can we do better (product, customer, admin)?
- Which outputs of PEST analysis can we view as opportunities for chances improvement (product, process, customer, administration)?

Threats

Threats may be considered as the risks that the organization faces while managing the QMS and realizing the products or services. Threats are the risks that may cause nonconformities in the QMS or in the realization processes, which the organization must address and eliminate. In fact the standard requires that the organization address risks that may affect the ability of the organization to provide a product according to the expectations of the interested parties. While conducting SWOT analysis, you need to assess the factors that may influence this ability. The appropriate questions in this case will be

- Which regulatory issues or standard specifications may affect the realization processes or the products (process, product)?
- Which technological issues may affect the product, its realization, or its distribution (process, product)?
- Which economic issues may affect the product, its realization, or its distribution (process, product)?
- Which processes, product characteristics, and/or customer satisfaction factors may be affected by upcoming changes or events (process, product, and customer)?

*Determining and Documenting the Internal and External Issues
and the Context of the Organization*

After completing the analyses discussed earlier, gathering all the information and identifying the issues that may influence the organizational context and affect the QMS, you are in a position to define what internal and external issues are relevant to the organization and its QMS. We reviewed the business environment through PEST analysis. We analyzed our SWOT. Now we should have enough inputs and should know which issues are affecting our activities and what the extent of each one is.

I highly recommend documenting this analysis for the following reasons:

- In clause 4.1 (Understanding the organization and its context) there is no explicit requirement for documented information, but in clause 4.3 (Determining the scope of the quality management system) it is required to maintain as documented information the scope of the organization's QMS. The analyses and its conclusion are a great part of the scope.
- In clause 4.2 (Understanding the needs and expectations of interested parties), the organization is required to monitor and review information about the interested parties and their relevant requirements. This review is done through update of the analysis mentioned in this clause.
- Audit—During an audit you will probably be asked to explain and demonstrate how you reached the conclusion of your analysis. When the day comes and the question will pop "please enlighten me how you came to all these conclusions?" you then will refer to the quality manual (if you decide to maintain one) or any other document and will have all the answers at hand.

In case you decide to still maintain your quality manual, you may include the conclusions there; just add chapters bearing the information and conclusions you came to during your analyses.

4.2 Understanding the Needs and Expectations of Interested Parties

Understanding the needs and expectations of the interested parties is the first step in developing the QMS and the foundation for determining its scope, defining relevant quality objectives, and developing quality activities for achieving them. After understanding the context of the organization, it must be determined who the interested parties are and what their needs and expectations are. Interested parties are individuals and other business entities that may affect the organization's ability to provide a product according to the specifications. Understanding their needs and expectations will create the organizational quality management conditions for the fulfillment of those needs and expectations over the long term and will enable the organization to achieve its objectives.

Interested parties differ from one another and thus have different needs and expectations. For example, investors want security for their investment and a good return, while customers expect to get a product that meets their needs in the most effective way. Identifying these expectations may be crucial for determining the scope of the QMS and setting the quality objectives. What does the ISO 9001 Standard require?

- Interested parties have a potential effect on the organization's ability to consistently provide products and services that meet customer and applicable statutory and regulatory requirements.
- The organization shall identify and determine the interested parties that are relevant to the QMS.
- The organization shall determine the requirements of these interested parties that are relevant to the QMS.
- The organization shall monitor and review information about these interested parties and their relevant requirements.

Identification of Interested Parties and Their Effect on the Organization

Interested parties are entities that perform and are active in the business environment, in which the organization is active that has an effect on the QMS. Interested parties may be a person or another organization that can affect, be affected by, or perceive themselves to be affected by the organization or its activities. Interested parties can be investors of the organization; suppliers of materials, products, or service; technological providers; financial institutions; and governmental and nongovernmental organizations. The first step in determining the needs and expectations of interested parties is identifying what area in the QMS they relate to. The following will help this identification process:

Interested Party	Influence
Customers	Expectations of the product, its intended use, and its characteristics. Changes in those expectations may influence the QMS and its activities.
Employees	Employees directly affect the QMS and its processes and, thus, the quality of the product.
Suppliers and external providers	The quality of materials, components, or parts that must be integrated in the end product and their delivery from suppliers and external providers may directly influence the quality and conformity of the product.
Regulators	Regulations and statutory requirements may dictate processes and activities of the organization and thus may affect the characteristics of a product.
Investors	The investors of the organization determine the availability of resources and thus may influence the quality and conformity of the product.
Competitors	Competitors may affect the profitability of the organization by delivering a product that meets customer needs and expectations better.
Other interested parties relevant to the area of expertise or activity of the organization	Other interested parties and their influence on the quality and conformity of the product identified through PEST analysis.

It is important to stay focused on interested parties that are relevant to the organization, its area of activities, and its QMS.

Identification of the Needs and Expectations of the Interested Parties

After listing the interested parties, we must identify their needs and expectation with reference to the QMS. I suggest here a very simple table that specifies the needs and expectations of each interested party.

Interested Party	Needs and Expectations
Customers	• Enhancement of customer satisfaction • Delivery of quality, price, and performance of products according to specifications • Appropriate communication channels with the organization • Handling of property belonging to customers
Owners/shareholders	• Increased revenue and market share • Sustained profitability • Flexible and fast responses to market opportunities • Evaluation of risks and opportunities
Employees of the organization	• Clear vision of the organization's future • Appropriate integration in the organization • A proper job definition • Good work environment • Work opportunities • Job security • Social conditions • Recognition and reward • Facilities for personnel in the organization • Developing and improving the competence of personnel (training)
Suppliers and partners	• Return on supply of goods or provision of services; appropriate communication channels between the organization and the supplier • Provision of resources such as information, knowledge, expertise, technology, processes, and training • Mutual benefits and continuity • Sticking to payment conditions • Handling of property belonging to external providers
Society in which the organization is active	• Environmental protection • Ethical behavior • Delivering product in accordance to the social, economic, ecological trends, and local cultural aspects
Regulators	• Compliance with statutory and regulatory requirements • Anticipating and reacting to expected changes in statutory and regulatory requirements • Understanding the labor market and its effect on the loyalty of people in the organization

After the identification of needs and expectations of the interested parties, you are in a position to develop and plan your QMS effectively. The ISO 9001 Standard refers through the following issues to the needs and expectations of interested parties:

- Defining the scope of the QMS according to their expectations
- Communicating the quality policy to them
- Identifying their needs and expectations related to the product
- Designing and developing products according to their needs
- Periodically reviewing their needs and expectations and changes

Monitoring and Reviewing Information Related to the Interested Parties

The organization's business environment is ever-changing and dynamic and the needs and expectations of the related interested parties change with it. Thus, the organization is required to monitor and review changes and trends in the business environment and related changes in the needs and expectations of the interested parties. This can be done through a constant monitoring and analysis of the organization's business environment and the context of the organization:

- The collection and management of data and information related to the context of the organization (periodical update to PEST and SWOT analyses).
- Evaluation of the new results of the analysis.
- Identifying and understanding the different changes in the different aspects: social, economic, ecological trends, or local cultural.
- Determining whether new needs or expectations arise from those changes.
- Updating the relevant QMS elements: new quality objectives, changing products, launching new products, employing new workers, and so on.
- Informing its interested parties of changes and updates relevant to the QMS through appropriate channels.

Updating the needs and expectations of the interested parties is related to the different areas in an organization: human resources are responsible for identifying changes related to human resources, marketing is responsible for identifying changes in the needs and expectations of customers, and top management shall receive inputs from the investors. This type of update must be performed periodically; the organization shall define the interval. In order to create a systematic review, I refer back to chapter 4.1 (Understanding the context of the organization), where I recommended a method for documentation. This type of documentation will serve as the input for management review (as required in clause 9.3.2, c-1).

4.3 Determining the Scope of the Quality Management System

Boundaries and applicability of the QMS must be determined in order to enable the definitions of the scope of the QMS. The scope of the QMS refers to the areas, locations, product or lines of products (or services), and processes or activities of the organization to which the QMS is relevant and will be influenced by them. In other words, the scope of the QMS defines its domain. The ISO 9001 Standard requirements are that

- The organization shall determine the boundaries and applicability of the QMS to establish its scope.
- The scope shall refer to products and services of the organization as well as the processes and activities required to realize them and the various locations which the organization uses for the realization of the product or services.
- When determining the scope of the QMS, the organization shall relate to external as well as internal issues as discussed in clause 4.1.
- When determining the scope, the organization shall relate to the requirements of interested parties as mentioned in clause 4.2.
- The organization shall meet all the ISO 9001 Standard requirements as applicable to the QMS.
- While determining the scope, the organization shall justify any decision for not including any requirements of the ISO 9001 Standard in its QMS (i.e., exclusions).
- A justification for exclusion must prove that the exclusion of a standard requirement does not affect the ability or responsibility of the organization to deliver products or services with conformity to customer requirements and the ability to enhance customer satisfaction.
- Claiming conformity to the ISO 9001 Standard is only possible when the exclusions are proven not to affect the organization's ability or responsibility to ensure the conformity of its products and services and the enhancement of customer satisfaction.
- The scope will be documented and maintained as documented information.

Determining the Scope of the Quality Management System

The scope of the QMS defines in which areas and boundaries the organization is active and describes the services, resources, processes, operations, and products the quality system applies to. In practice, it is a text that will be maintained as documented information and also appear on the certification. An accurate and correct definition of the scope is crucial because it determines which products and services are included under the QMS and that determines the processes and activities that need to be planned, controlled, and documented according to the ISO 9001 Standard requirements. The scope will relate to the following issues:

- The products or services included under the QMS and that organization delivers to its customers.
- All the applicable realization activities that will be under the QMS: research and development, production, marketing, installation, service, and support.
- Locations of the organizational units that will be under the QMS (if applicable). Bear in mind that each location that is included in the scope must be audited separately.

Reference to External and Internal Issues

As said before, external and internal issues are significant factors, conditions, influences, situations, opportunities, or events that have an effect on the QMS and will affect the ability of the organization in achieving its objectives and the intended results. Such issues and their influences were identified when we discussed the context of the organization. These issues shall serve as inputs when determining the scope of the QMS.

For example, when the organization produces a product that is submitted to several regulations, this will be mentioned in the scope: *Developing and manufacturing a product while applying the relevant local, regional, and international regulations.*

Reference to Relevant Needs and Expectations of Interested Parties

Interested parties are persons or organizations that affect, or may be affected, or perceive themselves to be affected by the organization or its activities. The interested parties and their relevant needs and expectations were identified and discussed in clause 4.2 (Understanding the needs and expectations of interested parties).

For example, if one need or expectation of your customer is distribution of your product with a certain technology, for example, an online shop, the scope will indicate it: *Marketing of products on an online or an Internet platform.*

Application of All the ISO 9001 Standard Requirements in the Scope

The Standard requires complete adherence to all the requirements mentioned in the standard when they are applicable. In other words, you may not exclude or neglect any requirement of the standard unless this requirement is not applicable to your QMS and you have an acceptable justification. For example, if you are not purchasing any services or goods from an external provider, there is no sense in developing quality management tools and controls for those areas.

List of Exclusions and Nonapplications

In the case one of the standard requirements does not apply to your QMS, you are entitled to exclude it. Excluding means not implementing those requirements in the QMS. The scope of the QMS shall indicate a list of exclusions of the standard requirements—requirements from the ISO 9001 Standard that the organization for some reason decided are not applicable to the QMS and has chosen to exclude them. The reason for the exclusions and nonapplications is that those specific standard requirements do not apply to the operations, products, or services of the organization. For example, if your organization does not use any monitoring and measuring devices on its products or processes, it may exclude the requirements of clause 7.1.5—Monitoring and measuring resources.

Each exclusion requires a justification on behalf of the organization—a statement with the explanation of an acceptable reason for excluding the requirement.

For example, I would manage a table with the relevant clause from the standard along with the justification, which would look like this:

Exclusion	Reason
7.1.5—Monitoring and measuring resources	The organization does not use any resources, personnel, tools, or equipment for ensuring results of monitoring and measuring.
8.3—Design and development of products and services	The organization does not develop any products or services.
8.5.3—Property belonging to customers or external providers	The organization does not hold, store, maintain stocks, or inventory of property belonging to customers or external providers.

When you decide to exclude a requirement, make sure any activity, operation, or process is not included in the QMS and that the does not refer to those requirements. For example, if you exclude the requirements of clause 8.5.3 but do receive packages from a supplier, make sure that activities for handling packages or containers of the supplier are not included in the processes.

Including New Process Operations and Product under the QMS

When the scope of the QMS is already defined and documented, and the organization decides to develop a brand new product with new processes, operations, and activities, as long as the product is not included in the scope of the QMS, it is not officially included in the QMS, although the development and realization were done under the appropriate controls and all the records are maintained, and so on.

Documenting the Scope of the QMS

The scope of the QMS and the list of exclusions must be maintained as documented information; must be created, updated, and controlled under the requirements of clause 7.5 (Documented information). One option is to create a designated document for the scope. I (along this book) recommend to document and maintain the scope of the QMS and the exclusions in the quality manual, if you still want to maintain this type of documentation (the 2015 revision of the Standard does not require maintaining a quality manual anymore).

Example Scopes for the QMS

Here are some example wordings for QMS scope from certificates I have seen:

- The design and development, manufacturing, and service for motor units and control equipment for pumps, fire pumps, fans, blowers, and automatic water supply systems.
- Design, manufacture, supply, service, and customer support for diesel engines from type XXX to type YYY.

- Developing and manufacturing semiconductors components for the XXX industry.
- Developing and provision of construction engineering services, consulting services including project management services. Included sites: New York branch, London branch, Zurich branch.
- Provision of cleaning services at the customer premises and provision of cleaning products and materials.

4.4 Quality Management System and Its Processes

4.4.1

Clause 4.4.1 is a declaration of the standard's intention regarding the QMS. In this clause, the requirements and main principles of a QMS are presented. This clause can be regarded as a foundation for self-evaluation of whether the organization's QMS follows the general requirements. First let me review the basic requirements:

- The organization shall establish, implement, and maintain a QMS within the organization with conformity to the requirements of the ISO 9001 Standard.
- The organization is required to continually improve the QMS and its processes in accordance with the requirements of this International standard.
- The organization shall determine the processes needed for the QMS.
- The organization shall determine how these processes will be applied in the QMS.
- The processes included in the QMS shall be identified and planned, implemented, controlled, and improved.
- The organization shall determine the required inputs and expected outputs of every process.
- The organization shall determine the sequence and interaction of processes included in the QMS.
- The organization shall determine how these processes and their controls will be implemented in the organization.
- Methods, criteria, performance indicators, and measurements for ensuring effective monitoring and control of the processes will be defined and implemented.
- Resources needed to support these processes shall be allocated and available.
- Responsibilities and authorities for these processes shall be assigned.
- Risks and opportunities identified and determined in accordance with the requirements of 6.1 shall be addressed.
- Changes necessary to ensure processes achieve their intended results shall be evaluated, planned, and implemented.
- The organization will implement specific measures for the improvement of these processes (achievement of objectives and maintenance of effectiveness).

4.4.2

- According to its needs and requirements, the organization shall maintain documented information to support the operation of its processes.
- According to the needs and requirements, the organization shall retain documented information to ensure that the processes are being carried out as planned.

Establishing a QMS according to Clear Principles

The ISO 9001:2015 declares in clause 4.4 quite clearly with which principles the QMS shall be established:

- Establishment of a QMS according to the ISO 9001:2015 Standard Requirements
- Definition of processes and their interactions needed for the operation of this QMS
- Continually maintaining and the effectiveness of the QMS through improvement

The important message here is that in order to deliver a conformed product or service and to meet customer requirements as well as other needs and expectations of other relevant interested parties in an effective way, a QMS must be established and maintained—a QMS

- That is based on the quality principles suggested in the ISO 9001 Standard.
- That is defined, planned, implemented, and controlled.
- That is customer focused—the QMS should use methods to understand present and future customer needs, and shall develop processes to meet customer requirements and strive to exceed customer expectations.
- Whose activities and processes address the needs and expectations of interested parties.
- Whose resources are planned, allocated, and controlled.
- Whose processes and activities are managed and whose interrelations are clear.
- That is constantly analyzed and controlled—analysis of data and information is implemented and decisions are based on facts.
- Supports improvement through collection of evidences and their analysis.
- That is fueled by the top management leadership—through leadership the purpose and strategic direction of the organization are established. Leadership shall create the environment for establishing the appropriate quality policy, in which employees can become fully involved and quality objectives can be achieved.
- That persons in the organization are aware of.

Employing the Process Approach

The ISO 9001:2015 Standard employs the process approach in order to enable the organization to effectively plan its processes and their interactions (see clause 0.1 of the Standard—General). How effective? The effectiveness of an organization depends much on its ability to perform several interconnected activities simultaneously in order to achieve intended results—the expectations of the interested parties.

These relations should be managed, prioritized, and controlled. The ISO 9001 Standard requires adopting a system of processes within the organization. This system of processes requires the application and implementation of a method for the identification of processes in the organization, the definition of their sequences and interrelations, and the application of their controls. The goal here is to develop and plan processes and methods for the realization of products or services. The ISO 9001 Standard regards it as a critical element of product realization.

Terms and Definitions

Before we start to unveil the requirements of clause 4.4 and their implementation, it is important to know some terms and definitions:

- Process—A set of interrelated or interacting activities that convert inputs into outputs and accomplish a specific organizational goal. These activities require allocation of resources such as people and materials.
- Scope of a process—Scope of the process defines precisely where a process starts and ends, what its related inputs and outputs are, and which activities are included and excluded.
- Supplier of a process—The deliverer of inputs to a process (data, information, goods, or services). Supplier may be an external supplier that delivers, for instance, goods or material or an internal supplier—an organizational unit that delivers inputs to a process.
- Customer of a process—The receiver of the outputs of a process (data, information, goods, or services). The customer defines what outputs are expected according to its needs. Customers may be external customers, end customers, or internal customers.
- Inputs—Specified requirements needed to be put into a process in order to start the process. The inputs will be processed by a process or activity.
- Output—Specified expected or intended result of a process.
- Risk—Combination of the probability of occurrence of not fulfilling process specifications or customer requirements.
- Monitoring of processes—A continuous, sequential, and periodic examination of processes and its outputs.
- Measurement of processes—Determining a physical measurement of processes and their outputs based on data.
- Process owner—An organizational function responsible for a process or subprocesses.

Applying the Process Approach

Process approach or system approach refers to the act of implementing a method or rules that analyze, identify, manage, and measure the processes of the organization. These processes are necessary for the operation of the QMS and the realization of the product. The fundamental goal is to create standardization of processes in the organization and to ensure that persons or different organizational units in the organization work in a unified way. The objectives of the process approach are as follows:

- Creating awareness and understanding in the organization regarding responsibility for managing activities
- Implementing a method for the identification and planning of activities needed for the operation of the QMS
- Defining the sequences between processes
- Promoting a smooth and transparent flow of operations in the workflow

- Identifying and ensuring the interactions between processes, that is, activities in the organization
- Ensuring accurate delivery of inputs to processes
- Monitoring and controlling activities of the QMS
- Ensuring delivery of the right process outputs
- Ensuring achievement of intended results or process objectives
- Enhancing satisfaction of process customers
- Creating basis and environment for addressing risks and preventing errors
- Creating basis and environment for the planning, implementation, and analysis of improvements (PDCA cycle)

Determining the Processes in the Organization

Which processes are to be included in the QMS? Applying the process approach requires identification and determination of all processes needed to realize the product or service. In other words, you are required to determine all key stages and substages (processes or subprocesses if you may) necessary for the delivery and realization of the product or service. Identifying and determining the processes included in the QMS is the first practical step in applying the process approach. Until now we

- Understood the context of the organization
- Identified needs and expectations of interested parties
- Determined the scope of the QMS

Now, we must declare which operations are required in order to fulfill the scope (realize the product). The level of details and complexity of the processes depends solely on your organization and the nature of its products. But the thumb rule indicates that only processes and activities that affect the product, its intended use, and quality must be included.

There are many methods and ways to identify and determine which are the processes included under the QMS. For the ISO 9001 Standard it is important to have a clear definition of these processes because these are the activities that will be planned, monitored, analyzed, and controlled. It is important that the list of processes you come up with answers these questions:

- Do these processes reflect your ability to deliver your product or service?
- Are all processes, key stages, subprocesses, operations, or activities critical for the realization of the product identified?
- Are all areas of the realization of the product covered?
- Is the scope of each process clear?

The end result of this determination of the processes and activities included in the QMS may be displayed with a list that specifies all processes, or a diagram (or a set of diagrams) that illustrates the processes and the interactions between them. Again, the book is too short to suggest a certain method. You must identify the method most suitable to your organization and its processes.

Relation between Processes and the ISO 9001 Requirements

It is important that the defined processes will refer to the relevant ISO 9001 Standard requirements. What do I mean by that? The ISO 9001 Standard presents us with many quality management requirements such as management review, management of resources, but many quality requirements for the operation of the QMS. For example, when you design the process of offering or selling products to the customer, you must take into account the specification in clause 8.2—Determination of market needs and interactions with customers:

- Plan and implement activities to interact with customers.
- Plan and implement a method to meet requirements specified by the customer.
- Answering the requirements for the intended use, and so on.

While defining and designing the processes included in the QMS, one must include such operational quality requirements.

Outsourced Processes

Outsourcing a process is a situation where the organization has chosen to execute a certain process or activity by an external organization (external provider), that is, the organization delegates the responsibilities for this activity to another organization. Outsourced processes, that are part of the realization of the product, must be identified and included as part of the QMS because the organization is obliged to control them (see clause 8.1). In other words, handing over the task to the external provider does not dismiss the organization from the responsibility of conformity to all customer, statutory, and regulatory requirements. For example, when a manufacturer is realizing the product, but the packaging activities are carried out by a supplier, the following are required:

- To define the scope of those packaging activities that must be planned and controlled by the organization—This means to define what is being done by the organization and what will be taken care of by the external provider.
- To include these activities under the QMS—describing them in the diagram of the processes, for example.
- To apply the method for process approach to these processes—demand at least from the supplier to prove that these processes are being planned with the process approach.
- To monitor, measure, analyze, improve, and to control the effectiveness of the process—define how you control and measure the process outputs of the supplier.

Determining the Sequence of Processes

Determining the sequence of processes means determining the sequence of different activities of different elements involved in the process and constructing the workflow in the organization. The goal is to make sure that the processes achieve quality objectives, deliver intended outputs, and ensure conformity of products or services. In practice, you define how your processes flow in your workflow (Figure 4.2).

The sequence should allow an overview of your workflow and reflect the way you are doing business and operating in your organization. A correct sequence of processes will allow the information to flow effectively in the workflow, deliver the inputs to the processes as required, and provide the right outputs. Normally, these processes have subprocesses. Some processes or activities may be in sequence and some may work in parallel. But the end result should be a process map that indicates or describes the workflow in the organization.

You must determine the sequence of activities within a process—what has to be done, in which order, when, by whom, and which resources are required. One way or method to describe the sequence of activities in a process is through a documented procedure—documented information that includes

- Reference to a process
- Goal or objective of the process
- Reference to relevant documented information
- Target group—to whom is this document designated

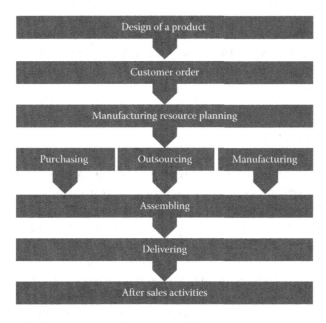

Figure 4.2   Example for sequence of processes of the organization.

- Description of activities
- The expected outputs
- Relevant documented information
- The required records

Types of such documentations:

- Management-oriented process—This method is designated to support the management of different areas of responsibility and enables core processes needed to achieve strategic objectives, such as defining quality policy and objectives, strategic planning, and management review.
- Process diagram/flowchart—A graphic demonstration of the separated steps of a process in sequential order indicating the entities involved in the process, the required inputs, and expected outputs.
- Documented procedure—A structured, documented, and formatted set of activities needed to achieve an objective.
- Work instruction—A list of documented actions that specifies what an employee is required to perform and what the expected inputs and outputs are. This type of documentation is usually used to define specific activities.
- Standard operating procedure—A detailed written instruction to achieve the objective of the performance of a specific activity.

Another tool that reflects and demonstrates the sequence of process is an ERP system, where processes are managed according to a defined workflow: customer offer, customer order, planning/scheduling (MRP), retrieving recommended purchase propositions, retrieving recommended production propositions, purchasing, outsourcing, manufacturing, delivering, invoicing, and managing after sales activities. Such systems dictate the sequence of activities for the user. In some cases, documentation of such systems may serve as process diagrams or work instructions. If you decide to use this type of documentation, ensure to document the gaps and loopholes—those activities and operations of the realization that are not covered in this documentation.

Interaction between Processes

Processes in the QMS must interact with each other. A process, by definition, is a set of interrelated or interacted activities that transform inputs into outputs. Interaction between processes refers to the delivery of inputs to processes, the acceptance of outputs from processes, and the transferring of these outputs as inputs to the next process. The interaction defines how inputs, outputs, or resources are transferred between processes and activities. Processes of a system exchange many types of information, data, material, goods, or services through activities. In order to make the system effective, the interactions between the processes in the system must be planned and known to the operators of the system. This interaction is influenced by many factors. I prepared here a table of the factors and their influence on the interaction between processes:

Factor	Influence of the Interaction between the Process
Supplier of inputs	The supplier initiates the interrelation by delivering inputs to the process.
The required inputs	The required inputs specify what is expected from the supplier. Here it is important to know that the inputs are handed or delivered to the process. This definition of inputs can be documented.
Required methods or activities for operating the process	Here is the reference to the methods that are used to operate the process. It includes the required activities, tools, technologies, facilities, or infrastructures for the operation of the process and the documentation that is needed for the support of the operation of the process.
Resources that are needed to operate the process	Roles or organizational units in the organization that operate the process should be clear—The unit the processes inputs into outputs. It need not necessarily be human resources, but, for example, software too. The knowledge and competence needed for the operation of the process must also be defined.
Customer of outputs	It must be understood who is the customer of the process and how the outputs will be delivered.
The outputs the process generates	The expectation of the process and the verifications, validations, and criteria that are required for the process must be defined. The evidence that a process delivered its intended results should be recorded.

If you define and plan all of these, you will be in a position to plan effectively the interactions between the processes. In practice, it must be clearly defined with the method that you are using to document describing how the interactions are taking place.

Defining Inputs to the Processes

After defining the processes and their activities included in the QMS, you must define for each activity which inputs are needed. As far as the ISO 9001 Standard demands, you must prove that each identified process has identified inputs and that the supplier of the process knows exactly what they should deliver.

Inputs of a process are defined as specified requirements needed for the operation of a process. Inputs are the fuel that drives the process: people, resources, materials, data or information, technology, or knowledge. Inputs may be tangible (raw material for production process) or intangible (information or data, e.g., results of a customer satisfaction survey or knowledge). In order to effectively analyze and identify the inputs, one must first know which activities a process includes. Let me review the important aspects of inputs related to the ISO 9001 Standard:

- Defined—For each process the inputs are defined.
- Deliverable—There is an effective way to deliver the inputs to the process.

- Measureable—Inputs of a process must be measurable in order to verify their availability.
- Planned—It will be clear when during the workflow inputs must be delivered.
- Known—The supplier and the operator of the process must know which inputs they must deliver (supplier) and receive (operator) for the process.
- Assigned—Responsibilities and authorities for the inputs are assigned.
- Located—Persons who operate the process must know which inputs are required for their operations and where they may find them or how they should request them.
- Verifiable—Persons who are responsible for the inputs have the means, criteria, and knowledge to verify or validate that inputs are as expected.

Defining Outputs to the Processes

An output is a deliverable result of an operational process aimed to address the expectations of the customer of the process. An output may be tangible (finished products) or intangible (services provided to a customer or information such as results of a calculation). As far as the ISO 9001 Standard demands, you must ensure that the expectations of the customer for each identified process are identified and understood by the people that operate the process. In practice, regardless of whatever methods you are using for analyzing your processes, make sure that the outputs for each process are as follows:

- Identified—The intended outputs of a process are identified and clear.
- Measureable—The outputs of a process must be measurable in order to verify their conformity.
- Assigned—Responsibilities and authorities for the outputs are known.
- Known—The persons who operate the process should know which output is expected from them.
- Verifiable—The persons who are responsible for the outputs have the means, criteria, and knowledge to verify or validate that outputs are as expected.

Availability of Resources

Resources that are needed to support the processes and to perform activities must be planned and be available. Resources may affect or determine the process capabilities. By providing personnel with the required resources, the organization ensures that processes are effective and efficient and will achieve their objectives. The specific requirements regarding resource management are well covered in chapter 7.1—Resources. Here in clause 4.4, we are required to define which resources are needed for each process, operation, or activity and plan their availability. Consideration of resources applies to the

- Identification of resources and technology needed for the operation of activities (tools, equipment, facilities, materials, energy, knowledge, and competence)
- Provision of these resources

- The evaluation of their adequacy to the processes
- The efficient use of these resources

Examples of resources include

- Human resources
- Infrastructure
- Work environment
- Information
- Natural resources
- Materials and components
- Financial resources

Responsibilities and Authorities for Processes

While defining the requirements related to the QMS for each process, it is required to determine the authorities and responsibilities for specific duties and obligations for performing the process activities for ensuring the implementation, maintenance, and improvement of each process and its interactions. I recommend the assignment of an organizational role, functional units, or authority to a process. This organizational role will relate to the organizational structure. By assigning a responsible person for a process, we create the relation between the organizational structure and the workflow. The objectives of this person are as follows:

- To decide and navigate a process (or several processes)
- To manage the resources for the process
- To solve problems related to the process
- To ensure the effectiveness of processes, that processes achieve their intended outputs

One very effective and recommended approach is defining a process owner. This approach supports the assignment of an organizational function regarding the responsibility and authority for a process or subprocess. The extent and complexity of the process dictate the degree of responsibility and authority of a process owner. The process owner can be a person, a group, or a team, depending on the nature of the process and the organization. Some of the process owner responsibilities may be

- Definition of the scope of the process, the interactions and interrelations with other processes, and process owners
- Leadership, motivation, and encouragement of employees involved in the process
- Navigating and operating the process
- Designing the process and its flow
- Measuring, monitoring, and analyzing the process
- Documenting the process
- Managing training and knowledge related to the process
- Improving the process and its effectiveness

Knowledge competence and skills of the process owner may be

- Basic and technical knowledge and deeper understanding of the nature of the process
- Understanding of the process structure in the organization and the interactions with other processes
- Full acquaintance with process specifications, resources, and flow of information
- Understanding who is the customer of the process
- Acquaintance with process requirements such as applicable standards, regulations, customer requirements, inputs and outputs, and so on
- Acquaintance with situations and events related to the process

These responsibilities and authorities must be communicated in the organization. The definition or documentation of the process owner responsibilities and authorities may appear in the

- Job description
- Documented information of the process, work instruction, process diagram, and so on

Other suggested roles regarding a process that may share the responsibilities of the process owner are as follows:

- Process sponsor—This role is one level above the process owner and refers to the person who navigates or directs the upper-level processes and authorizes or approves resource for processes. The process sponsor is usually a member of the top management.
- Process manager—This role is one beneath the process owner and refers to the person who is responsible for the operation of the process and works directly with the process workers, supplier, and customers. He reports to the process owner. Usually, the process owner assumes the process manager role as well.
- Process worker—This is the lowest level of process responsibilities. This person operates the process and accomplishes activities according to specifications.

Achieving Intended Results

Specifications of outputs are the intended results of the processes. Achieving intended results indicates whether activities were performed as expected and determines whether a process is effective. It is done through

- Controlling which inputs were delivered to the process
- Monitoring and evaluating what is being done in the process
- Measuring and monitoring its outputs

As we know, a process produces outputs—some desirable and some undesirable. This affects the interactions between processes because the output of one process is the input to the next process. And this fact makes this control critical. In order to ensure that intended results will be achieved, you must design the control that will demonstrate its ability to achieve intended results. The control covers the following aspects:

- The appropriate inputs are delivered to the process.
- The necessary actions of the processes are carried out.
- The output is desirable and meets the specifications.
- Undesirable outputs are prevented.

In practice, for each process the following are decided:

- What are its specifications for inputs and what are the expected outputs?
- Which resources are required to control the process?
- Which methods are used to control the process?
- Which documented information is needed to provide the evidences?
- Which actions are required when undesirable outputs are accepted?

Analysis, Measurement, and Monitoring of Processes

The organization is required to define criteria and performance indicators to the processes. The objective is to monitor and measure the processes included in the QMS effectively, as follows:

- To control and ensure the effectiveness of activities and the efficiency of the process
- To verify that processes achieve their quality objectives (quality requirements, schedules, intended use, etc.) and their intended outputs
- To ensure that the outputs of these processes meet the expectations of the customer (the receiver—whether internal or external)
- Ensure improvement of these processes

Monitoring is always applicable—one may sample a process and determine its effectiveness. But with measurement it is more difficult and less practicable because processes might not always provide objective data. The tactic of measuring processes is to

- Define a method for the measurement and the control
- Define an appropriate criteria
- Collect data
- Analyze and report data that are relevant to the process and the realization of the product

The analysis shall strive to allow the operators of the process to systematically detect trends and patterns in the processes and respond to events on time. As a result, the effectiveness of the QMS is maintained. Specific requirements for this issue are well detailed in clause 9.1—Monitoring, measurement, analysis, and evaluation.

Regarding the process approach you must ensure that the following principles are applied for the planning and analyzing processes:

- Processes are evaluated for their need to be reviewed, analyzed, measured, controlled, and monitored. Where subprocesses may affect the main process, they must be identified and controlled as well.
- The appropriate method for measuring and controlling is defined and applied—a method that provides you with quantitative or qualitative information about the effectiveness of the process.
- The appropriate criteria are defined and established. The criteria must allow a comparison to accepted process results. The criteria must reflect the quality objectives.

Addressing Risks

During the analysis of its processes an organization is required to address risks related to products and services that may occur when processes deliver unintended outputs or when the interaction between the processes is ineffective. QMS in its essential is a preventive tool aimed to plan controls and achieve objectives. Risk refers to the probability of occurrence of not achieving an objective and may result in a decrease of customer satisfaction due to unintended outputs or ineffective interaction between processes. The objectives here are as follows:

- To create awareness to risks when planning processes
- To indicate the consequences of the harm of the risks, and how severe it might be
- To initiate controls of these risks while planning your processes
- To plan corrective actions when errors occur

The difference between the last revision of the standard (2008) and this one (2015) is the event of this analysis. The last revision promoted the discipline of the preventive action—a quality tool for protecting a product or a service from potential nonconformities—that is, evaluating risks and potential events that may affect the quality of the product. In practice, it is done after the planning and during the realization process. The new revision takes it one step backward—to the planning of the process. In other words, when planning the processes you must address associated risks (or opportunities). This subject is widely dealt in clause 6.1—Actions to address risks and opportunities where the organization is required to identify risks that may prevent it from achieving quality objectives, delivering conformed goods or services, or fulfilling customer satisfaction—and this is what the ISO 9001 Standard is trying to prevent.

Applying the process approach, determining the processes included in the QMS, and identifying their owners, inputs, and outputs should put you in a position to understand what the intended results are for each process. The next step will be to understand the relation between the end product or service and the intended results of processes. Or, more precisely—what is the relation between the conformed products or services and the intended outcomes of the processes that realize them.

In practice, you need to develop or use a method that will identify all product or service characteristics that have an effect on the intended use of the product or service for which verification is required. These risks may occur throughout the supply chain, the realization of the product, or the provision of a service. After understanding which risks are potential, you must define the following actions to handle and control these risks:

- Identify the relevant risks.
- Define the verifications or validations.
- Define the appropriate criteria.
- Plan monitoring and measuring devices and their validations when needed.
- Plan actions for the limitations of nonconforming products.
- Determine the competence of personnel.

Let the experts and process owners decide what the risks to their processes are. You just need to ensure that it is done when they do the planning for the process. Take a look at the following analysis:

Process Owner	Process	Delivery of Unintended Results	Result/Impact
Sales manager	Acceptance of product specifications from the customer	Inadequate communication with the customer Failure to deliver data and specification to the planning	Inadequate planning of product realization
Development team	Development of the product according to the specifications	Lack of appropriate information for development Failure to deliver specification of development to the production	Realization of product not according to the specification of the customer
Production team	Realization of the product according to the planning	Lack of appropriate competence for production Delivery schedules not transferred to the production	Realization of product not according to the specification Delivery to the customer not according to the contract

Identifying Necessary Changes

Processes must be assessed in order to ensure that they deliver the intended results continuously. So far, for each process we

- Defined the required inputs and expected outputs of each process
- Defined which responsibilities and resources are needed to accomplish the processes
- Defined which criteria and methods for monitoring are necessary for determining whether the outputs conform to the specifications
- Identified risks for unintended results
- Monitored the processes and decided whether the intended results were received or not

In case a process fails to deliver the intended results, actions must be initiated in order to promote changes in the processes. The objective is to ensure that these intended results are achieved. Process change relates to the development of a systematical improvement of processes. This system of changes and improvement will be integrated into the business process. Let us now move on to the improvement of processes.

Ensuring Improvement of Processes

Improvement of processes enables the organization to align its business processes to its business strategy, leading to effective performance through the improvements of specific activities. Submitting specific processes for improvement is an effective approach to achieve quality objectives. Until now, we defined the processes, the interrelations between them have been determined, objectives have been set, and activities for monitoring, measurement, and analysis have been implemented. But something is missing. The loop has not yet been closed. The organization must initiate actions for improvement and systematically identify opportunities for improvement—situations and processes where planned results, quality objectives, or criteria were not met. These situations will be analyzed, the root causes will be identified, and actions for improvement will be applied.

In practice, when designing and defining the processes for the operation of the QMS, the organization shall implement methods and tools that will allow the identification of opportunities for improvement. Here are some methods that can be applied:

- PDCA cycle—Applying the plan, do, check, act discipline is an efficient way of implementing the improvement of processes. This method is dealt in details in chapter 10.3—Continual improvement.
- Statistical methods—Statistical methods are used to understand and then reduce or eliminate variability in processes.
- Quality control—Quality control samples process performers and indicates needs for improvements.
- Modeling of processes—Modeling is considered measuring the current state process performance and identifying gaps in the current process preventing from achieving the objectives.
- Redesign and reengineering—Analysis of processes and relevant activities with the aim to find what is preventing achieving the objectives and suggesting ways to improve them.
- Process selection—Selecting a small significant group of processes that have a great influence on achieving the quality objectives in order to improve them.

Usually, but not necessarily, the process owner is responsible for identifying the opportunities for improvements and initiating them.

Documentation of the Process Approach

The ISO 9001 Standard requires documentation of the process, operations, and activities that comprise the QMS. I promote it as part of standardization in organizations; the documentation of the process approach creates a system for designing, analyzing, and implementing processes. The extent and level of details of the documentations shall be determined by the organization according to its needs. But bear in mind that you will have to justify it. If you decided to maintain a low detailed documentation during an audit, you must show how this documentation is sufficient. Documentation is divided into two levels:

- The organization shall define how processes, operations, and activities shall be documented. Here the standard refers to process diagrams, procedures, work instructions, and so on.
- The organization shall define which process outputs must serve as evidence and be maintained in the form of records of processes, operations, and activities.

The methods and basics of defining and maintaining documented information is discussed thoroughly in chapter 7.5—Documented information.

Visualization of the Process Approach

I prepared a very basic diagram that illustrates all the ideas and requirements that were discussed in this chapter. You may use it as a concept when developing and planning your processes.

The first diagram (Figure 4.3) illustrates how of the general processes are planned.

The second diagram (Figure 4.4) is a detailed demonstration of the process approach elements and aspects that must be considered when designing a process. The diagram includes references to all requirements mentioned in this chapter:

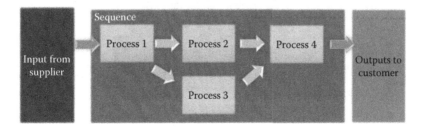

Figure 4.3   Definition of sequences between processes.

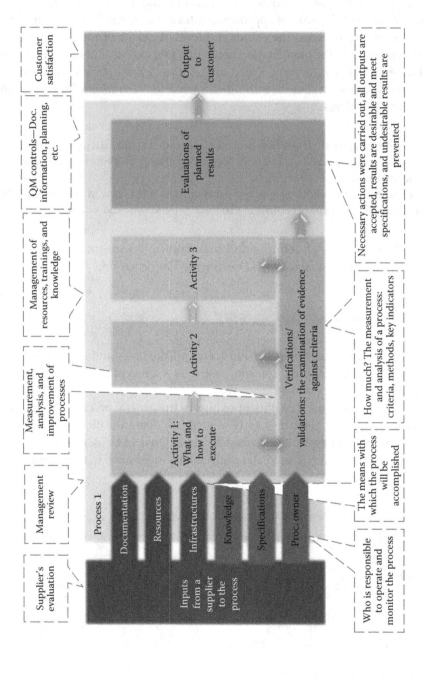

Figure 4.4   Quality management controls for internal/external processes.

5 Leadership

5.1 Leadership and Commitment

5.1.1 General

As far as I can say, the success of implementing a quality management system (QMS) depends on the commitment of the top management. This is no secret. The leadership in all organizational levels should create an environment that will initiate and promote conditions in which employees feel the commitment to achieving the objectives of the organization. But how can one evaluate whether the management is truly devoted to the QMS? According to ISO 9001, the top management is required to demonstrate how it assures that the QMS is effectively implemented. This clause (Leadership) lays out the principles and expected actions from the top management regarding their commitment to the QMS. This clause should be regarded as a basis for self-evaluation of your top management's status and its role within the QMS regarding its commitment. In order to prove its commitment, top management will supply evidences for its actions. Understanding the requirements first will allow you to understand the role of the top management in the QMS (the ISO 9001 requirements):

- The top management shall practice and demonstrate its ability to lead the organization in implementing the QMS according to the ISO 9001 Standard requirements.
- The top management shall assume the responsibility for establishing an effective QMS.
- The top management shall demonstrate its commitment by engaging in managerial activities while implementing the QMS.
- The top management shall ensure that quality policy and quality objectives of the QMS are established and are compatible with the context of the organization and the strategic direction of the organization.
- The top management shall ensure that requirements of this standard and the derived quality requirements are integrated into the business activities and operations of the organization.
- The top management shall promote awareness of the process approach while planning the QMS.
- The top management shall promote the approach of risk-based thinking while planning the QMS.
- The top management shall ensure availability of resources needed for the implementation of the QMS.

- The top management shall communicate among the organization the importance of an effective QMS.
- The top management shall communicate to the organization the significance of conforming to the QMS requirements and the requirements of their products and services.
- The top management shall ensure that the QMS, its processes, and related business activities achieve its intended results.
- The top management shall practice leadership, engagement, involvement, and support of persons in order to promote the effectiveness of the QMS.
- The top management shall promote improvement.
- The top management shall demonstrate leadership by supporting other relevant management roles in their areas of responsibility.
- Note: The term "business" used by the ISO 9001 Standard refers to those activities that operate the QMS and are central and foundational to the purposes of the organization's existence.

Practicing and Demonstrating Leadership

Why is leadership so important to the ISO 9001 Standard? According to the ISO 9001 Standard, leadership initiates unity of purpose, while establishing the direction and strategy of the organization. This unity of purpose creates environment and conditions that support the QMS and in which employees become completely involved in achieving the organization's goals and aims and assists employees in achieving quality objectives. The obvious conclusion is reaching an effective QMS. Let me put it another way: When an executive leads the implementation project and engages directly with employees at lower levels of the organization or shows high involvement, it has a strong impact because employees at those levels are getting the message directly from the top management, allowing them to know its perspectives and expectations. The ISO 9001 Standard promotes the proactive approach of the leadership—controlling the QMS by causing events to take place instead if reacting to quality or business events.

Employees that operate the QMS are a significant resource and their full involvement enhances the effectiveness and improvement of the QMS. In order to keep employees fully involved in achieving the organization's objectives, the top management should

- Promote activities for involvement in the QMS
- Promote the concept of an effective QMS
- Involve the entire organization in improving the QMS

Thus, the leadership shall support relevant management roles in their areas of responsibility and promote the empowerment of people at all levels of the organization to take relevant decisions regarding the QMS. How?

- Empowering people at all levels of the organization, authorizing them and delegating responsibilities of various areas of the QMS
- Inspiring and motivating employees to take an active part in improving the QMS

- Providing trainings needed for the development of human resources for the operation of the QMS
- Promoting the concept of an effective QMS and the awareness of improving the QMS
- Recognizing contribution of employees to the QMS

Here the standard lays out the principles for delegating authorities and responsibilities. The practice will be discussed in chapter 5.3—Organizational roles, responsibilities and authorities.

Practicing Managerial Activities

Managerial activities are those actions that the top management should consider for reviewing and controlling the QMS. These are the activities in which the top management is expected to be directly involved. The goal of these activities is to enable the top management to effectively review the status of the QMS. Low variability of managerial activities affects negatively the effectiveness of the QMS. Which managerial activities are expected of the top management?

- Review the strategy, status of objectives, the performance of the QMS, and the results of audits.
- Engage in activities that promote improvement.
- Analyze processes and operations of the QMS.
- Receive feedback from lower levels of the organization regarding the status of the QMS: indicators for performances, status of corrective actions, and so on.

For you, as an organization, it is important to demonstrate with evidence that the top management does managerial activities and that these activities

- Receive the appropriate inputs
- Produce the appropriate outputs

One way to institutionalize the matter is to include it in the quality policy or to put it down in a procedure although there is no requirement for any documentation.

Ensuring the Effectiveness of the QMS

The top management must ensure that QMS and its intended results are effective by ensuring that

- Quality objectives are defined and measured
- Processes and business activities are planned according to the ISO 9001 Standard requirements
- Processes and business activities are controlled and measured with the controls required by the ISO 9001 Standard
- Risks and opportunities are addressed
- Improvement is achieved

This requirement clearly emphasizes that the top management is not only responsible for developing the QMS but also for controlling its effectiveness. Effectiveness will be measured when carrying out the following fundamental activities:

- Defining a quality policy
- Establishing communication channels with the lower levels in the organization
- Implementing the quality policy throughout the organization using measurable quality objectives
- Fulfilling them by reviewing the QMS periodically and systematically
- Allocating the required resources

Quality Policy and Quality Objectives

The top management is required to define a quality policy with the goal of demonstrating the commitment of the top management and the organization to the implementation of a QMS and to the maintenance of its effectiveness. In practice, the top management shall initiate a written quality policy that will clearly demonstrate to the interested parties of the organization

- The mission of the organization
- What the top management expects
- Its values
- How the top management shall act in order to introduce and implement its policy

The content and nature of this policy will be discussed in chapter 5.2—Quality policy. After defining the quality policy, the top management will establish quality objectives and will ensure that these objectives are

- Appropriate to the strategy of the organization (referred by or presented in the quality policy)
- Continually achieved

This will be done by defining objectives that are relevant to the quality policy and the nature of the organization. After defining and determining the objectives, the top management must review their status, collect and analyze data regarding these objectives, and evaluate whether they have been achieved or not. Requirements regarding the quality objectives, their definition, and their planning are discussed in chapter 6.2—Quality objectives and planning to achieve them.

Communication Channels

The top management should consider different communication channels for different purposes. But all these channels have one goal—that the interested parties are exposed to the quality policy and objectives. Each different interested party of the organization should have its own channel with which it exchanges information with the top management (when needed). Each communication channel will be planned and implemented according to its nature and purpose.

Communication of the quality policy and its objective should reach the interested parties both vertically and horizontally.

- Vertically—The quality policy and the related quality objectives should be communicated in a downward direction from the top management to their subordinates with the purpose to distribute decisions and directives. Feedback, suggestions, and opinions regarding the status of the quality objectives will flow from the subordinates in an upward direction to the top management.
- Horizontality—Information needed for the achievement of the quality objective (the operation of the QMS and its business activities) flows between roles or persons at the same position. This communication is needed to coordinate activities of various departments in the organization.

Communication channels will be internal as well as external. The type of information that will flow in these channels is affected by interested parties' needs and expectations (the context of the organization) and the type of relations:

- Internal communication (with employees, stakeholders)
- External communication (with customers, suppliers)

There are many ways to publish this information:

- Publishing the policy and the objective on a bulletin board or in an organizational portal
- Maintaining a procedure that requires each employee to read the quality policy at least once, which they must sign indicating they have read and understood the policy
- Messages or notices to the public in media
- Letters for partners
- Website

I prepared a short review of which quality elements are expected by the standard to be communicated:

- The importance of effective QMS
- The quality policy
- Organizational roles, responsibilities, and authorities
- Quality objectives
- Customer communication
- Performance data related to processes, products, services, and customer satisfaction
- Communication of related data to the external providers

Awareness of the Process Approach

Adapting the process approach is highly recommended by the ISO 9001:2015 Standard. The top management from their side shall promote the awareness to the process approach.

- The top management shall refer to the process approach in the quality policy, define it as a quality goal, and if applicable, define relevant quality objectives.
- The top management can initiate benchmarking like reengineering projects or process mapping in order to evaluate the maturity and compatibility of the business processes.

- The top management shall allocate resources for the implementation of the process approach:
  - Responsibilities and roles
  - Plans and schedules for the implementation
  - Tools that allow process analysis
  - Required trainings
  - Outsourced consulting

How can one assess whether a process approach is integrated in the organization? I will rephrase the question: How can an auditor be convinced that the process approach is adapted in the organization? Would it be enough to present a pair of charts or diagrams? Apparently not. The organization must prove in action with evidence that

- A method for analyzing processes is implemented. There is no requirement for documenting the method but yet you will need to show evidence: tools and means with which you control your processes.
- All processes in the organization are mapped and are controlled according to this method.
- All processes have desired results that will allow assessment of the process.
- Responsibilities in the organization are authorized for these activities.
- Results of the process mapping exist and are reviewed.
- Process inputs and process outputs are available for review.
- Processes are regularly reviewed and improvement of processes is achieved.

In practice, I would include a capital in the quality policy that relates to the implementation of the process approach in the QMS and the business activities of the organization.

Awareness of Risk-Based Thinking

The risk-based thinking is the new concept introduced by the ISO 9001:2015 Standard. It promotes a new organizational culture that leads people to look for opportunities and to address risks. This concept advises that alterations produce opportunities as well as unanticipated consequences. Thus, the organization shall analyze and act in advance to address changes that may impact its objectives, expectations of interested parties, or product requirements.

- The top management shall refer to risk-based thinking in its quality policy, define it as a quality goal, and if applicable, define relevant quality objectives.
- While implementing the process approach, risk-based thinking will be integrated in the planning of the processes.
- The top management can initiate actions like small-scale risks analysis projects in order to identify risks that pose threats to the QMS or to detect opportunities that will improve the QMS and enhance desirable effects.
- The top management shall allocate resources for the promotion of risk-based thinking.

Ensuring Resources

No QMS can succeed without sufficient resources. The top management tends to take this matter not seriously because it involves costs. But still the top management must ensure allocation of all required resources in order to ensure the effective operation of the QMS. Which resources are needed and how to ensure their availability will be discussed in chapter 7.1—Resources. But the matter should be structured in the strategy of the organization and be expressed by its quality policy, for example, the policy shall indicate that

- The importance of allocating the right resources to the various activities of the organization is understood and considered
- The need for resources in order for processes and activities to be effective and efficient is acknowledged
- The top management is committed for providing the resources required to carry out the activities of the QMS
- Current process capabilities and the their resources are assessed
- Future resources and technology needs are identified
- Activities for the identification and provision of needed resources shall be determined

Very important is the link between the strategy and vision of the organization and the commitment to provide the necessary resources for achieving this strategy. In practice, I would include in each section, part, chapter, or capital of the quality policy reference to the allocation of the necessary resources. For example

- Responsibilities and roles
- The appropriate times resources
- Procedures and methods
- Tools and instruments
- External resources
- Training

Improvements

With a direct relation to the process approach and the needed resources, the top management must promote improvement in the organization. The requirements for improvement and its application will be discussed in chapter 10—Improvement. Here we discuss the strategic aspect. The ISO 9001 Standard views the improvement as vital to the effectiveness of the QMS. The expectance of the standard is a structured approach—by systematically improving the QMS, it would achieve its objective over the long term. Practicing improvement must address inputs such as data for analysis, changes in the context of the organization, and assessment of current risks and opportunities.

This topic must be forwarded to employees at all levels. It is important for them to understand the direct relation between their work and the effectiveness of the QMS. When individuals perform their activities as expected or according to specifications and as a result they produce the intended results, the goals of QMS are achieved. When they are acting for enhancing their outputs, then the QMS is improved.

Understanding this concept will motivate workers in striving for improvement. In practice, the top management must show evidences of actions for promoting improvements. Publishing it on the bulletin board is not enough. It is a cyclic process that includes

- Promoting improvement
- Initiating actions to collect inputs
- Analyzing the inputs
- Launching actions for improvement
- Measuring the effectiveness of those improvements

So if you maintain a suggestion box for improvements in an audit it will not be enough to show that suggestions were filled out on a controlled form but that

- The suggestions were reviewed
- A few of them were chosen for implementation
- They were submitted to a methodic process
- Their implementation was done under control
- The top management took the time to evaluate the effectiveness of the suggestions

Ensuring Compatibility of Processes and Business Activities to the ISO 9001 Standard Requirements

The ISO 9001 Standard requires from the top management to ensure the compatibility of QMS processes and the related business activities through integration of the standard requirements and the derived quality requirements, that is, how the top management can ensure that the ISO 9001 requirements are

- Understood
- Considered while planning the processes and business activities
- Implemented
- Controlled

The top management is required to prove with evidence its commitment to implementing a QMS. This requirement of the top management has two aspects: the establishment of the QMS and its continual control.

- The top management must ensure that the QMS is established according to the standard requirements.
- The top management must ensure that the QMS is maintained according to the standard requirements.

The first aspect could be covered with the quality policy mentioning that the QMS will be planned and designed with consideration of the ISO 9001 Standard requirements and will cover all the activities of the organization that are included under the QMS: marketing, sales, finance, administration, production, development, and logistics. The second aspect can be expressed with the commitment of the top management to evaluate the integrity and compatibility of the QMS through the application of the ISO 9001 Standard requirements. The standard requires various tools and techniques for that kind of control:

- Internal and external audits
- Risk-based thinking approach
- Quality objectives
- Monitoring, measurement, analysis, and evaluation
- Management review

5.1.2 Customer Focus

Customer focus is one of the quality principles that the ISO 9001 Standard is based upon. The Standard sets requirements to ensure top management's leadership and commitment to meeting customers' as well as regulatory requirements. Clause 5.1.2 sets the scope and the level of involvement of top management regarding its customers and their expectation of the organization. There are three main goals:

- Understanding and implementing customers' and regulatory requirements
- Development of the strategy related and linked to the needs and expectations of interested parties (customers)
- Realizing a product that will meet the customers' and regulatory requirements

What are the requirements? Top management shall demonstrate leadership and commitment with consideration to customer focus. This will be expressed by (the ISO 9001 Standard requirements)

- Ensuring and maintaining activities for the consistent provision of products and services that meet customer and applicable statutory or regulatory requirements
- Assessing and evaluating related risks that may affect conformity of products and services and the satisfaction of the customer or opportunities that may enhance customer satisfaction
- Ensuring activities for the determination of customer requirements
- Focusing on enhancing customer satisfaction

Clause 5.1.2 lays out the foundation for customer focus. Specific requirements and expected actions are motioned in the different standard requirements as follows:

- 4.2—Understanding the needs and expectations of interested parties
- 5.1.2—Customer focus
- 5.3—Organizational roles, responsibilities, and authorities
- 6.2—Quality objectives and planning to achieve them
- 8.2.1—Customer communication
- 8.2.2—Determining the requirements for products and services

In clause 5.1.2, the main requirement is to define the controls for the top management regarding the commitment for meeting customer requirements, regulatory requirements, and reviewing process' output (products or services). The nature of the organization, its business operations, will set the framework and structure—how these controls will be applied. For example, one instrument that the ISO 9001 Standard requires for implementation is the management review for evaluating the QMS (clause 9.3). Answering the requirements of clause 5.1.2 requires the understanding of three aspects:

- Defining the customers of the organization—Customers may be external as well as internal.
- Understanding the expectations of the customers—Each type of customer has a different set of expectations.
- Evaluating the perception of customers of the organization and how the organization complies with customers' expectations.

Meeting the Needs and Expectations of Customers

Meeting the needs and expectations of customers is crucial for the development of the strategy of the organization and for the implementation of an effective QMS. The organization may have different types of customers that have different needs and expectations. Very important here is to understand the expectations of the different interested parties and to examine conflicts between their expectations. The differences and the conflicts may occur because of various social, organizational, or cultural needs. This is why the organization must anticipate any potential conflicts arising from the different needs and expectations of its different customers.

Internal Customers

Internal customers include those members of the organization who operate the activities and rely on assistance from other internal entities to fulfill their duties. A good example is the sales cycle—the sales department (sales personnel) are the customer of logistics (the department that processes deliveries) and logistics must provide them with various services: deliver goods according to the agreement with sales. The requirements of the top management here are to initiate actions for understanding the relations between the different organizational entities, understanding who serves who (who is the customer and who is the provider), and whether their expectations are met. Examples of actions for the identification are process mapping or process engineering.

External Customers

External customers are the entities that receive the end products or services. One of the main goals of the ISO 9001 Standard is to enhance their satisfaction by meeting their requirements. Only through their purchase and the revenue they provide can the organization survive. By not meeting the expectations of the external customers, the organization risks losing them. Recollecting the sales cycle, the end customer that receives the goods is the external customer. The requirements of the top management here are to initiate actions for identifying the external customers and reviewing and understanding their expectations through market research or surveys.

Assessing and Evaluating Related Risks and Opportunities

The top management shall plan activities to assess risks and opportunities that may affect products and services (directly) and customer satisfaction (indirectly). The top management shall demonstrate its leadership and commitment by

- Initiating plans for actions for identifying and addressing such risks or opportunities
- Demanding and controlling the integration of those actions and their implementation in the processes of the organization
- Evaluating their effectiveness

Practical and precise requirements will be discussed in the designated chapter 6.1— Actions to address risks and opportunities. The top management is required here

- To promote the proactive approach by initiating involvement of organization's units at all levels in identifying and assessing the risks or in developing opportunities
- Reviewing the effectiveness of these actions with the instruments that the ISO 9001 requires, such as internal audits or management review

Ensuring Delivery of Products and Services That Conform to Customer Requirements

Top management shall ensure the delivery of products and services that conform to customer requirements through implementation and integration of actions and activities for identifying and meeting customers' requirements. The objective is to define actions that will

- Identify the requirements
- Review further necessary requirements that were not mentioned by the customer but are necessary for the realization of the product or service
- Publish these requirements at the appropriate stages during the realization of the product or service
- Review whether the requirements can be met
- Evaluate whether the realized goods or services meet their requirements

The method shall include which roles are involved, what their activities are, which information is to be maintained, frequencies and intervals, and data analysis actions. Precise requirements will be discussed in chapter 8.2—Determination of market needs and interactions with customers. The top management is expected to create the organizational conditions for the fulfillment of these customer requirements by

- Promoting and demanding the actions for meeting customer requirements
- Promoting and implementing actions for delivering conforming goods and services
- Creating involvement of employees by delegating the authority to take decisions
- Reviewing the effectiveness through the instruments that the ISO 9001 Standard suggests

How shall the top management create these organizational conditions?

- By communicating the strategy of the organization
- By publishing the values, symbols, and cultural behavior expected from the employees on how work should be carried out
- By creating awareness of the significance of meeting customer requirements
- By promoting comprehension of employees to the significance of delivering products or services according to the requirements (customer and regulatory)
- By promoting comprehension of employees to the consequences of not delivering products or services according to the requirements (customer and regulatory)
- By defining appropriate quality objectives that will support the goal of delivering goods or services according to the requirements
- By creating awareness of customer needs and customer feedback

Enhancing Customer Satisfaction

Top management shall promote the enhancement of customer satisfaction. This is achieved by following two main ISO 9001 Standard requirements:

- The implementation and integration of a method for monitoring customer satisfaction
- Reviewing the results of this monitoring

The first part refers to the collection of information; the top management shall promote the implementation of a method for collecting information regarding whether products or services meet customers' and regulatory requirements. The method shall indicate types of data that will be collected and analyzed, the responsibilities, the activities, and the interval. The second part refers to the conclusion of the information—reviewing the results and initiating actions to enhance customer requirements based on this information. Precise details are discussed in chapter 9.1.2—Customer satisfaction. By demonstrating leadership and commitment, the top management shall

- Define which method is needed (and most effective)
- Verify whether the method is integrated in the processes and is effectively implemented
- Validate that the information and data that were collected and analyzed are providing the appropriate situation report

In practice, the issue will be reviewed by the top management during the management review (clause 9.3).

5.2 Policy

A quality policy is the general guidelines, intentions, and goals of the organization referring to its quality intentions. The policy will be implementable, available, active, communicated, and documented in the organization. The policy will demonstrate what quality signifies to the organization with emphasis on the main goal of your QMS—it will express your intention to provide your customers with quality goods or services according to predefined requirements using a set of systems, processes, procedures, and controls.

5.2.1 Establishing the Quality Policy

The ISO 9001 Standard requirements regarding the quality policy are quite strict and clear:

- The organization will create, implement, and maintain a quality policy.
- The quality policy will be written with suitability to the purpose of the organization and its context and must suit the organization's nature and type of activities.
- The quality policy will provide the basic intention, vision, or plan for setting the quality objectives.
- The quality policy shall demonstrate the organization's commitment to meeting applicable requirements such as customer or regulatory requirements.
- The quality policy shall demonstrate the organization's commitment for continually improving the QMS.

5.2.2 Communicating the Quality Policy

- The quality policy will be available as documented information.
- The quality policy will be distributed, communicated, and presented throughout the organization.
- The quality policy will be distributed, communicated, and presented to the interested parties.

Suitability to the Purpose and the Context of the Organization

The quality policy is a document describing the organization's strategy and policies clearly and serves the different interested parties of the organization. And because the quality policy sets the foundations for the quality objectives, it must include references to the next issues:

- Mission and purpose of the organization—The mission of the organization describes the purpose of the organization: why the organization exists, who it serves, and so on. This was defined and discussed in chapter 4—Context of the organization.
- Vision of the organization—The vision of the organization describes the desired state that the organization wants to achieve: its general goals. This is important for the next section—relation to the quality objectives.
- Interested parties of the organization—The policy is based on the needs and expectations of interested parties.
- Business objectives of the organization—The policy must clearly relate to the business objectives of the organization (discussed in the next section).

One basic role of the policy is defining areas, subjects, and scopes where the organization would like to implement its quality efforts. Thus, the quality policy shall refer or be applicable to all activities of the organization that are included under the QMS: marketing, sales, finance, administration, production, development, and logistics. In practice, you need to be creative in formulating a policy that will convince the reader (or auditor) that your organization is willing to invest efforts in delivering a product that comply to customer requirements, to improve the QMS, and to enhance customer satisfaction (which is not an easy task).

Relation to Quality Objectives

The quality policy shall lay the foundations for the definition of the quality objectives. This relation between the quality policy and the quality objectives will turn the quality policy into a useful and practical document rather than just a documented statement. Usually the policy includes strategic objectives that refer to more specific and measurable objectives. One main goal of the revision in the ISO 9001 Standard is to create coherence between the business objectives and the quality objectives of the organization. Until today, they were usually separate and were always discussed by two different forums or panels of people. The new standard set the framework for setting business objectives and quality objectives as one.

The quality policy defines areas and scopes in which the quality will be active and valid. The quality objectives will be derived from these areas and scopes. For example, if the policy declares a desire to manufacture precision products to meet the most critical design requirements in a wide range of applications, the appropriate relevant objectives would be as follows:

- Promoting design and development in the organization
- Initiating communication channels with customer for a better understanding of their needs

Setting objectives correlated to the policy will assist in maintaining an effective QMS. But whatever the objectives are, there are two main goals that must be implemented regarding the quality objectives: meeting customer and regulatory requirements. The policy will include a clear reference to these issues.

Meeting Customer or Regulatory Requirements

One of the main goals of your QMS is consistently meeting customer requirements and striving to exceed customer expectations. And the quality policy should express that. Striving to meet customer requirements can be expressed by

- Initiating actions for understanding of needs and requirements for the satisfaction of customers or identifying new requirements, such as regulatory
- Initiating activities and controls that will generate products or services that will meet customer or regulatory requirements

In practice, the policy shall mention that such actions will be taken, resources shall be allocated, and the appropriate controls will be set. The requirements are mentioned in the designated clauses, for example, 8.1—Operational planning and control, 8.2.3—Review of the requirements for products and services, and 8.3—Design and development of products and services.

Improvement of the Quality Management System

The policy shall demonstrate the intentions of the organization regarding the continual improvement of the QMS and maintenance of its effectiveness, that is, the top management must initiate defined actions in order to improve the QMS and to identify opportunities for improvement. In practice, the quality policy shall include a reference to the intentions of how the top management shall promote the improvement of the QMS. For example, the quality policy may include references to the following issues:

- Evaluation of current resources
- Evaluation of capabilities of processes
- Evaluation of abilities of technologies

Communicating and Distributing the Quality Policy

Quality policy without the awareness of the interested parties is not much use. By awareness, I mean how the interested parties, in general, and employees, in particular, relate to the policy. Interested parties of the organization must be informed of the quality policy and the quality objectives in order to understand what is expected of them and what their mutual goal is. As you will shortly see, publishing the quality policy is not enough. The top management must involve the interested parties in the QMS through exchange of information of many kinds. This is why the top management must initiate an action that will ensure that the quality policy is distributed and communicated within the organization.

You might think that employees at lower levels do not have to understand the quality policy of the organization but only the quality aspects of their job. For example, an assembly worker does not need to know the quality policy of the organization but only the impacts of his job on the quality of the product: assembling parts according to the specifications of the product. That is not what the ISO 9001 Standard expects of you. The top management must create the conditions in which employees will feel committed to achieving the objectives of the organization. An employee must understand the implications of the statement on their work, how he or she helps the company in their job, or how his or her job relates to the company's policy. In other words, the employee must understand that their work and its outputs do not end up in their work place but has implications on later stages. This can be achieved if the employees

- Are aware of the quality goals of the firm: what is the vision of the organization, what are the values upon which the organization acts
- Know what is expected of the organization (by the customers)
- Understand the importance and of quality requirements and the significance of its practice
- Understand the meaning of an effective QMS
- Are familiar with their direct quality objectives
- Understand the significance of the quality of products and services and the implications of delivering nonconforming products or services

I would formulate a policy statement with an easy-to-remember idiom. In that way, everyone can identify with it and remember it. The employee then has the message in order to make the relation between his job and the quality goals of the organization. The ISO 9001 Standard takes it one step further and requires that the top management ensure that interested parties (employees) will understand the importance of an effective QMS. Effectiveness means that the QMS has achieved its quality objectives and the goods and the services were delivered according to the intended results. And this concept must be understood by the employees. Always remember—you must provide the employees with conditions in which they feel committed to achieving the objectives of the organization. Usually, employees tend to recite the quality policy of the organization. But do they strive for an effective system? Do they understand their position in the quality policy? Do they understand the significance of their quality objectives to the organization? Do they know all of the customer requirements? Furthermore, the top management must ensure that employees will understand the significance of delivering goods and services according to the expectation of the customers or contrariwise the implications of delivering nonconforming products or services. For example

- Communicating quality costs
- Publishing issues with customers
- Publishing and discussing quality problems or quality investigations
- Publishing results of audits or results of customer feedback
- Initiating quality incentives
- Initiating discussions between employees regarding quality

If we go back to the example of the assembly worker, he or she is expected to understand what will happen in other process stages when a nonconformity occurs on his or her activity: delivering a defected product forward to later processes, causing quality assurance problems, not meeting customer schedules, and affecting customer satisfaction. If you think about it, the ISO 9001 Standard expects here a bit of coaching for your employees—the top management must initiate training and development in which the employees are taught about the QMS:

- The employees must understand who are the organization's customers and what are their expectations.
- The employees must understand who are their direct customers and what are their expectations.
- The employees must always remember that they are expected to improve their performance and the results that they deliver.

Other interested parties must be informed on the policy and thus have access as well:

- Customers
- Suppliers
- Stakeholders
- Stockholders
- Legislators

Quality Policy as Documented Information

The quality policy must be maintained as documented information, that is, a documentation (of some kind) with the quality policy must be maintained and controlled on a defined medium. This documentation will be submitted to the controls suggested in clause 7.5—Documented information. Usually the quality policy was documented in the quality manual but the quality manual is not required anymore. You may though keep and maintain the quality manual and the quality policy inside it. I would do that. But if you decide not to keep the quality manual, just copy the part about the quality policy from the quality manual and generate a new controlled document.

Structure of the Quality Policy

There are no specific requirements regarding the structure of the quality policy. But the content must include all the topics that are mentioned in this chapter 5.2—Quality policy. If you still need guidance of how to formulate your quality policy, I suggest the following structure:

- Details of the organization
- Details of persons who wrote or approved the quality policy
- Corporate profile/purpose of the organization
- Scope of the QMS—to which departments, facilities, branches, or affiliates the QMS applies
- Organizational structure

Until now you provided details for readers to get acquainted with the organization and its nature. Now we begin to fill the ISO 9001 Standard requirements:

- Short description of the purpose of the organization, its mission, and vision
- General statement regarding the quality policy of the organization—this statement must be aligned with the mission, vision, business, and quality objectives of the organization
- Reference to the needs and expectations of interested parties
- Reference or list of the quality objectives
- Statement regarding the promotion of improvement in the organization
- Statement regarding meeting customer or regulatory requirements
- Statement of how this policy will be communicated or distributed

Do not forget to submit this document to the control of documented information.

Organizational Structure

Although not required, I recommend including the organizational structure in the document of the quality policy. The organizational structure is the definition of hierarchy in the organization and is the most basic definition of your QMS. The structure defines who is superior to whom and who reports to whom. The structure will relate to the working processes and scope of the QMS.

This will help in charting the QMS and is very important to the charting of the processes and their interactions and helps to position all the participants of the QMS in the organization. The organizational structure will include all the departments and organizational units or entities that the QMS controls. The emphasis will be on the organizational dependencies. I have prepared a very basic diagram as an example (see Figure 5.1).

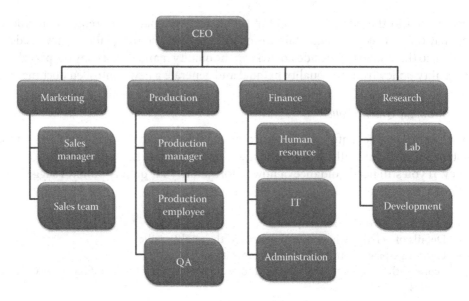

Figure 5.1   Example for organizational structure.

5.3 Organizational Roles, Responsibilities, and Authorities

Implementing the ISO 9001 Standard demands the definition of authorities and responsibilities that will bear the obligation for implementing and operating the QMS, conducting activities related to the QMS, and product realization. The ISO 9001 Standard requirements are as follows:

- The top management shall ensure the definition of authorities and responsibilities throughout the different organizational levels for operating the QMS.
- The definitions will be communicated to all the relevant levels of the organization.
- The top management shall appoint a representative or representatives on its behalf that has or have the authority and responsibility for the QMS in the organization and the following roles regarding the QMS:
  - The representative will ensure the establishment, implementation, and control of processes and activities required by the ISO 9001 Standard for the QMS.
  - The representative will ensure that the QMS continually conforms to the requirements of this international standard.
  - The representative will ensure that processes of the QMS interact.
  - The representative will ensure that the QMS delivers intended outputs.
  - The representative will report to the top management on the performances, competence, and implementation of the QMS, opportunities, and the need for improvements.
  - The representative will act to promote the awareness of customer focus throughout the organization.
  - The representative shall ensure the integrity of the QMS is maintained when changes to it are planned and implemented.
- The personnel (internal or external) that bear responsibility will understand their role in the QMS.

Defining the authorities and responsibilities has several important objectives concerning the QMS:

- Defining and communicating the hierarchy of decision-making in the organization
- Clearly allocating human resources to processes and activities
- Relating authorities with the relevant quality objectives
- Defining training and qualifications needed for each function
- Allowing the orientation within processes regarding the realization of the product, data, and information
- Enabling a structural sharing of information and knowledge between the different parties of the organization

Organizational Structure

The organizational structure is the definition of hierarchy in the organization and relates to the nature of the organization. The structure describes all the functions, roles, and relations in an organization and demonstrates who is subjected to whom and who reports to whom. Defining the structure is the first step in defining the authorities and responsibilities of the QMS: the authorities and responsibilities that will operate the processes and activities are to be mapped. An example of an organizational structure is given in chapter 5.2—Policy. Although there is no specific requirement regarding the documentation of the organizational structure, it is highly recommended to include an organizational chart in the quality policy.

Job Description and Understanding the Role

It is expected that every person in the organization that bears a role in the QMS will understand their responsibilities and authorities regarding the operation of the QMS. In order to fulfill this requirement, I suggest managing job descriptions. Basically for every role defined in the organizational structure, the responsibilities and authorities must be described. It is important to mention that all job descriptions must correspond to the list of processes included in the QMS. This will ensure effectiveness of process realization. The job description specifies the daily functions of a role and organizes the list of responsibilities and authorities of specific roles: development manager, secretary, production manager, or quality manager. The job description will cover the following:

- Identification of the function. The title of the role or function is important to identify every role: operational manager, lab technician, administrative secretary.
- Dependence. Which role is the person subject to or to whom must they report at the end of the day? This dependence will be determined based on the organizational structure.
- Responsibilities. The responsibilities are activities that combine the function's work and have effect on the products or services. The responsibilities relate directly to the realization processes. Try to be as accurate as possible. At the end of the definition, each realization activity must be allocated to a function.
- Authorities. The authorities are the points and events in a process where the function is authorized to make a decision that will determine the flow of the process or affect

the quality of the products or services. For example, a production manager decides which machine will be allocated to a process. A quality manager decides whether a product conforms or not and therefore has control over the release of the process.

- Competence
  - *External qualifications*: The organization will decide which external qualifications and competence are required for the role necessary for the operation of the processes: mechanical engineer, electrical technician, certified lab technician, certified account manager, licensed forklift driver.
  - *Internal qualification*: The organization will decide which internal training and certifications are required for each function: working procedure training, ERP system training, and machine maintenance.

These types of competence may serve as introductory conditions for an employee to perform a job.

All employees at all levels of the organization will be aware of their responsibilities and authorities. All employees will be aware of their related quality objectives as well. The matter will be communicated to each role:

- It can be integrated into the training and certification process of each function
- It can be published through some kind of internal portal

Documenting the definitions is not required by the ISO 9001 Standard but I see no other way how you may maintain these definitions. The documentation will be used for the verification of the job description in audits or reviews. The documentation of the job description may be carried out in several ways: work instructions, procedures, or designated documents stated as job descriptions (I personally recommend an independent document).

One organization I consulted for placed a laptop in a public area in the organization (e.g., the cafeteria) that was connected to just one internal website—the documentation of QMS. This way, every worker could see exactly how the organizational structure looks, where he or she is placed in the organizational structure, what is their job description, and download relevant documented information.

Reference to Regulatory Requirements

Where regulatory requirements demand the appointment of specific personnel or functions for specific activities or processes, it will be included in the definition of the responsibilities and authorities. The regulations may demand certain qualifications to be nominated and certain activities to be carried out. When such requirements exist, the top management shall ensure that

- These requirements are mentioned in the job description
- Appropriate personnel are integrated into the QMS

Involvement of the Top Management

ISO 9001:2015 is definitely setting a new level of involvement of the top management, and the relations between the quality management and the top management are seen from a new perspective. The last revision of the ISO 9001 Standard (2008) allowed a bound between the top management and the management representative regarding the QMS—he or she should take care of the QMS and report once a year that all is well. If things went badly, the management representative took the responsibility. Generally, the ISO 9001:2015 Standard is shifting duties related to the QMS from an appointed member of the management to the top management, that is, the responsibility of the effectiveness of the quality management lies directly on the top management.

And there is a logical explanation to this change: The purpose of the organization and its context defined by the top management is the base for the QMS. This is why the controls that the QMS is setting and their results must be of interest to the top management, and, therefore, more involvement is expected. As a logical result, the right information must flow to the top management and more than once a year. Which information should the top management expect?

- Status of implementation of the ISO 9001 Standard
- Status of processes
- Status of goods and services
- Needs for improvements
- Issues regarding customers and their satisfaction

The top management must create some kind of reporting system that will deliver this information on a regular basis.

The Representative(s)

The ISO 9001 Standard takes a different approach when discussing the management representative. The last revision of the Standard (2008) demanded that a member of the organization's management serve as a management representative. The ISO 9001:2015 takes a different approach. The top management on its behalf must appoint a person or a number of persons that their roles and authorities will be discussed. It is important to understand that the responsibility may be delegated to several functions in the organization whose responsibility, duties, and area of authority are defined. However, they shall have the authority to decide in certain situations. This authority must be delegated from the top management. Please review the following considerations when selecting the representative:

- The management representative must not be a member of the top management.
- Although it is common, the management representative must not be a quality manager. In cases when the management representative is not the quality manager, it is necessary to define the interfaces between the representative and the quality manager—how will they interact?—because they will have to exchange a lot of information. A periodic meeting is one idea.

- The representative may be an external responsible body in which the top management has confidence, for example, a consulting firm.
- Delegation of assignments and tasks of the representative to other roles and parties in the organization is possible.
- It is recommended that the representative will have the appropriate background and knowledge relevant to the area, technologies, and nature of the organization and the field of the QMS for which he is responsible.
- When appointing a role that performs or is involved in more activities in the organization related to the quality of the product, it is important to ensure that no conflicts arise due to its obligation to the QMS.

With this approach, the standard promotes the formation of a quality circle that will steer the QMS, take responsibility, and report directly to the top management. Before this approach, quality in the organization and the QMS was the responsibility of the quality manager; he or she used to take care of everything that is related to quality management and there was clear separation between quality and other managerial areas of the organization such as sales, purchase, manufacturing, finance, and so on. By delegating quality authorities to other members of the organization, the quality becomes their obligation as well; they participate directly in defining the terms and they are responsible for implementation.

It is important to document these nominations. If it's a quality manager or another role in the organization, then document the fact that they are the representative on their job description. If it's an external party, you must produce some sort of document that specifies its responsibilities to the top management and the organization. An agreement is one option.

Reference to the Process Approach

The ISO 9001:2015 Standard requirements employ the process approach, which enables the organization to plan its processes and their interactions. The specific requirements regarding planning the processes and their interactions are discussed in chapter 4.4—Quality management system and its processes. Here I would like to refer to the relation between the process approach and the roles and authorities of the QMS. One of the responsibilities of the management representative is to ensure the principles of the process approach:

- Ensuring planning of all process and activities
- Ensuring that processes interact
- Ensuring that processes are delivering their intended results

This requirement refers to skills and qualifications of nominated persons in the QMS—they should have some expertise in business process management or at least understanding of the processes that they are managing. This is why I definitely recommend you to include in the job description for each job or role a list of its related processes. One common practice is defining a process owner for each process—a person who is given the responsibility and authority for managing a particular process. This includes the implementation, maintenance, and improvement of the specific process and its interactions.

Another aspect to be considered is the interfaces between these process owners with the intention to ensure that processes interact. The interactions between the roles and the authorities define how a process detail flows between the different areas in the organization:

- The required information is flowing between processes.
- Each process receives the necessary inputs and releases the expected outputs.

The standard demands that these points, events, and requirements, where an interaction between two functions is initiated are defined and controlled.

Examining the QMS

The management representative has two very challenging tasks that no one in the organization (usually) volunteers to do—to examine and evaluate the status of the QMS and to report it to the top management:

- The representative will ensure the establishment, implementation, and control of processes and activities required by the ISO 9001 Standard for the QMS, that is, the implementation project.
- The representative will ensure that the QMS conforms to the requirements of this international Standard, that is, the maintenance of the QMS.

The ISO 9001 Standard presents us with several tools and instruments for evaluating the status of the QMS and its implementation, for example, performance evaluation, internal audits, and management reviews. These tools evaluate the performance of the QMS with different applications and generate different types of results. The management representative should verify that these controls are deployed and shall have access to their results. The main tasks of the representative are to

- Ensure whether the main objectives of the QMS are achieved
- Control the status of the implementation of the ISO 9001 Standard requirements
- Monitor the status of improvements

The examination will be done through the assessment of the processes identified as relevant to the QMS and that are defined in the quality manual.

Ensuring Intended Outputs

Intended outputs are required to meet the needs and expectations of the interested parties of the organization, to support the achievement of quality objectives, and in the long term fulfill the strategy and the policies of the organization. The intended outputs and the processes needed for achieving them were identified while employing the process approach (see chapter 4.4—Quality management system and its processes). The responsibility of the representative is to determine which additional processes and controls are necessary in order to ensure that intended outputs are achieved.

Processes may be assigned to the relevant process owners. The management representative can then receive the information from them. In practice, I recommend a reporting system that will deliver this information to the management representative on a regular basis. Once the delivered results are not satisfying, that is, the intended results are not being achieved, the representative will have the ability to identify such events and initiate actions in time (the proactive approach). Which information might be of interest for the management representative?

- Delays in deliveries and status of production orders
- Quality status of realized products or services
- Status of customer complaints

It is quite obvious that the management representative cannot cover all the topics (particularly in big organizations). This is why we must learn to delegate these activities to the appropriate parties in the organization but maintain a communication channel with them. A good example is an ERP system that has functional reporting tools for different levels of management.

Identifying Needs for Improvements

Another important role of the management representative is identification of opportunities for improvements in the QMS. There is a direct relation between achieving an effective QMS and the identification of the need for improvement. This is why the ISO 9001 Standard lays it at the feet of the top management and the management representative.

This duty is an outcome of the responsibility for ensuring intended outputs; through the control of the intended outputs, the management representative should know whether an improvement is required, to which extent, and in which area. Once an opportunity for improvement is detected, it must be submitted to the requirements of improvement as specified in clause 10—Improvement (submitting the improvement to methods using measuring, monitoring, and analyzing activities). The following inputs shall be considered:

- Issues of interested parties
- Data of processes
- Data of products or services
- Nonconformities
- Audits
- Self-assessments

A good example is the initiation of process mapping or process reengineering where the results may indicate a need for improvement. Then the management representative must take this information and forward it to the top management.

Promotion of Awareness of Customer Focus

Another role for the management representative is to promote the awareness of the employees of customer focus. Customer focus is one of the quality principles that the ISO 9001 Standard is based upon. The specific requirements are listed and discussed in chapter 5.1.2—Customer focus. The issues of customer focus are supposed to be identified on each organizational level or unit which customer expectations are relevant. The management representative has the role of evaluating how the ISO Standard requirements regarding customer focus are perceived by the members of each organizational level or unit and how they are fulfilled in each of this organizational level or unit. The next step for the management representative will be to verify that measures for promoting the awareness are being taken. The ways to promote the issues will be planned according to the nature and activities of the organization: training, lectures, publications, certifications, distributing information, and so on.

Ensuring Integrity of the Quality Management System during Changes

Changes may occur in different QMS elements such as planned processes, documented information, operations of the QMS, requirements or specifications for products and services, changes in the design and the development, and so on. When such changes occur they will be submitted to the controls suggested in clause 6.3—Planning of changes. The role of the management representative is to verify with the quality controls that the standard offers that

- The changes were identified on time
- The changes were submitted to the controls
- The integrity of the QMS is still maintained after the change was introduced

These information the management representative may get through the interactions with other relevant parties of the organization, such as process owners, or during the meeting of the quality circle. Ensuring the integrity means ensuring that after the change

- Needs and expectations of interested parties are still maintained
- Quality objectives are still met
- The QMS is effective

Creating a Reporting System for the Management Representative and the Top Management

As you probably already noticed, the responsibility of the top management and the management representative includes a lot of gathering of information on various subjects and topics of the QMS; the management representative will need to collect information and data regarding the QMS and its activities, assess it, and forward it to the top management. This information should flow constantly in order to seize the status in real time and to promote the proactive approach for managing the QMS. There is no exact requirement regarding the reporting methods or intervals. They may occur as management reviews, as complaints reports, or as internal audits reports. The reporting may vary according to different areas and activities. Another issue is the definition of what will act as evidence and findings to support the reporting. For example, as a management representative, aside from the management review and the internal audit I used to submit to my CEO weekly reports with different issues concerning the quality of the products or customer issues.

I support the establishment of a reporting system that will serve the management representative in completing his duties as required by the ISO 9001 and shall provide the top management with the necessary information. This kind of system can use the top management for the management review. The following are the data expected from such a system:

- Status of implementation of the ISO 9001 Standard
  - Accomplishment of tasks related to the implementation
  - Results of audits
  - Status of actions for improvements
  - Status of nonconformities
  - Status of corrective actions
  - Issues related to training
  - Status of processes related to the realization of the product or service
  - Delays in deliveries or schedules
  - Status of stocks
  - Status of production orders
  - Issues with suppliers
- Status of products and services
  - Quality status of purchased goods
  - Quality status of realized goods
  - Quality status of services provided to customers
- Information regarding customers and their satisfaction
  - Results of surveys
  - Information regarding complaints

Establishing such a system is not easy and each area and type of product or service demands other properties or characteristics. You need to locate the resources in the organization that can provide the management representative with this information. The next step will be to see how different types of information can flow to the management representative on a regular basis. The types of organizational sources of information are as follows:

- ERP/CRM systems can provide information related to orders, delays, production orders, service calls, or issues with suppliers.
- SPS inspection systems for the status of goods.
- Survey tools for customer satisfaction.

- Customer support centers for customer complaints.
- Quality tools (like form or systems) for the status of the QMS (audit results, corrective actions, and improvements).

The users of such a reporting system are the top management and the management representative. But there is a difference between the information that the management representative will use and that interests the top management; the management representative needs a very practical information regarding processes while the top management needs to know only the status and the trends.

6   Planning

6.1 Actions to Address Risks and Opportunities

In clause 6.1, the ISO 9001:2015 Standard presents a new concept: risk-based think-ing. Risk-based thinking directs people to look for opportunities, address risks, and come up with actions either to develop those opportunities or to mitigate or eliminate those risks. Risk-based thinking is actually a development and broadening of the well-known "preventive action" concept. Organizations should migrate from preventive action to risk-based thinking. Let us start by reviewing the ISO 9001 requirements:

6.1.1

- The organization shall determine which risks and opportunities may affect its ability to achieve the intended results.
- The objectives of this determination:
  - To ensure that the quality management system (QMS) can achieve its intended results
  - To enhance desirable effects
  - To prevent or reduce adverse effects
  - To achieve improvement
- The context of the organization and the needs and expectations of interested parties (as specified in clauses 4.1 and 4.2) shall be taken into account when considering the risks that may affect the ability of the organization to achieve the intended results.

6.1.2

- Actions to address risks and opportunities shall be planned and these actions shall be integrated into the processes of the QMS.
- The effectiveness of these actions shall be evaluated.
- The actions shall be planned in proportion to the potential effects on conformity of goods and services and customer satisfaction.
- Note—options to address risks may include actions that avoid risks, take risks in order to pursue an opportunity, eliminate risk sources, change the likelihood or consequences, share risks, or retain risks by informed decisions.

- Note—Opportunities can lead to planning and implementation of new practices, launching new products, opening new markets, addressing new customers, building partnerships, and using new technology and other desirable and viable possibilities to address the organization's or its customers' needs.

Throughout the standard, it is indicated where and how risk-based thinking will be applied in the QMS:

- Clause 4.4, QMS and its processes—You are required to address risks to conformity of products and customer satisfaction if process interaction is ineffective or develop opportunities that may enhance this conformity.
- Clause 5.5.1—The top management is requested to promote the concept of risk-based thinking when developing and implementing the QMS.
- Clause 5.1.2, Leadership and commitment—The top management is expected to demonstrate leadership and commitment with respect to customer focus by ensuring that the risks and opportunities that may affect the conformity of products and services, and customer satisfaction as a result, are determined and addressed.
- Clause 6.3, Planning of changes—Potential consequences must be addressed while changes are considered.
- Clause 8.1, Operational planning process—For planning changes, their consequences are considered.
- Clause 8.4.2, Type and extent of control of external provision—The organization considers the potential impact of externally provided processes, products, and services on the organization to provide products or services according to their specifications.
- Clause 8.5.5, Post-delivery activities—The extent of the activities is determined considering the potential undesired consequences associated with the delivered products and services.
- Clause 9.1.3, Analysis and evaluation—The organization analyzes and evaluates appropriate data and information arising from monitoring and measurement regarding the effectiveness of actions taken to address risks and opportunities.
- Clause 9.3.2, Management review inputs—During the management review, it is required to discuss the effectiveness of measures taken to address risks and opportunities.
- Clause 10.2, Nonconformity and corrective action—It is required to update risks and opportunities determined during planning when addressing nonconformities.

Terms and Definitions

Before we start to unveil the requirements of clause 6.1 and their implementation, I would like to present some terms and definitions that are used in this chapter:

- Risk—The probability of arriving at an unexpected state where requirements are not met. This state is represented by potential events.
- Opportunity—The possibility for improvement due to a favorable combination of circumstances or conditions in the QMS.
- Risk evaluation—The process of comparing identified risks against given risk criteria to determine whether the risks are acceptable or not.

Risk-Based Thinking

Risk-based thinking is a new concept introduced by the ISO 9001:2015 Standard regarding risks and opportunities in an organization. It promotes a new organizational culture that leads people to look for opportunities and address risks. This concept indicates that changes produce opportunities and unanticipated consequences; that is, an organization analyzes and acts in advance to address changes that may impact its objectives, expectations of interested parties, or product requirements. Therefore, before creating a change, one must identify and assess the associated risks and opportunities. An organization is required to incorporate this concept in the planning of the processes. These new requirements should not be considered as a burden but rather as a contribution to the QMS that helps the organization become more effective by eliminating adverse events.

A QMS in its basis acts as a preventive tool that initiates activities for identifying situations where intended outcomes, for example, objectives, would not be met or identifying opportunities that help an organization in achieving its objectives. This new concept is a development of the preventive action and requires a routine and more systematical analysis of risks and opportunities.

The difference between risk-based thinking and preventive action lies in the fact that in preventive action, a QMS relies on interested parties (most likely employees) to detect and report potential risks and initiate a preventive action. In other words, the approach is a reactive one—one reacts when a risk is already there. The ISO 9001:2015 Standard suggests risk-based thinking as a strategic instrument for the planning of a QMS and adopts a proactive approach toward errors and nonconformities by anticipating and identifying the root causes that may emerge during realization and initiating actions, which helps address the errors in advance. Risk-based thinking is more effective than preventive action because the identification and elimination of risks and opportunities are defined in planning stages and are incorporated in the process approach. The probability of achieving quality objectives improves because obstacles, interferences, and interruptions are identified and removed in advance or opportunities initiate improvements. Objectives of risk-based thinking are as follows:

- Establishing a proactive approach of identification and elimination of errors in advance
- Ensuring the QMS can achieve its intended outcomes
- Preventing or reducing undesired effects
- Ensuring long-term normality, security, and credibility of processes and business activities
- Achieving improvement
- Enhancing customer satisfaction

When we take a look again at the objective mentioned earlier, these are the classic quality objectives. Risk-based thinking enforces changes in issues such as

- Definition of specifications—According to the results of the analysis, specification may be changed.
- Definition of processes and activities—According to the results, activities and processes may be modified.

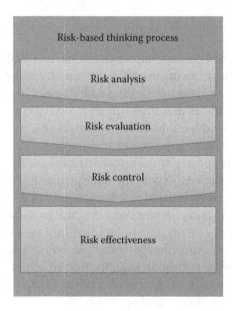

Figure 6.1 Suggestion for risk-based thinking method.

- Quality activities—Quality controls and activities may be redesigned, according to the results of the analysis.
- Documented information—Documented information might be modified in order to support the results.

The ISO 9001 Standard specifically implies that it does not formally require a certain methodology for applying risk management and an organization can decide for itself whether and how to develop a more extensive risk management methodology than required as long as the principles from clause 6.1 are maintained. I personally recommend the next process for implementing risk-based thinking (Figure 6.1).

When planning and applying risk-based thinking, you must refer to the entire lifecycle of processes or products:

- For existing processes or products, you should prove that risks are addressed in the frame of the ISO 9001 Standard requirements, the results (actions needed to address risks) are implemented, and processes, goods, and services are controlled.
- For the planning of changes in processes or new processes, you should demonstrate how you analyze risks and address them.

Responsibilities

An organization should identify and appoint an appropriate team of parties (may include external as well as internal persons) with appropriate competence, experience, and knowledge for planning and implementing risk management activities. Such team is required to

- Receive different views and inputs from different areas and expertise with respect to identifying risks and opportunities, setting related criteria, and performing the evaluation
- Ensure that necessary knowledge is available
- Ensure that all risks are identified
- Ensure the effectiveness and efficiency of action controls for addressing risks
- Appoint responsible parties for the implementation
- Delegate assignments regarding risk-based thinking

I would involve in this task the process owner, because he is supposed to have needed knowledge and experience.

Risk Analysis

All the constituents of a QMS that affect the quality must be analyzed for their probability to impact the ability of an organization to provide conformed products. The goal is to identify these aspects, their impact, and consequences on the QMS. The result will be the planning of their controls.

During the analysis, you identify and document the qualitative and quantitative characteristics of those system elements. For example, when allocating an employee for a critical activity, you may identify the level of its competence, qualifications, and experience. The type of identification is an essential step in recognizing all the aspects of the QMS that may affect the quality or the ability to meet the requirements. During the analysis, the next issues must be clearly defined:

- The description of the QMS element that is analyzed: a process, an activity, a resource.
- The identification of the party that performs the analysis.
- The scope of the analysis, the characteristics, properties, or activities of which are analyzed.
- The objective of the element—what is its intended outcome?

While analyzing the associated risks, ensure that you review the next aspects related to the realization of the product. At the end of this review, you will have a comprehensive list of potential events that hinder achieving the objectives, which must be addressed:

- Product specifications—Identification of product specifications that an organization might fail to meet. It may refer to semi-products, finished goods, services, or activities in a service.
- Resources—Identification of the resources that are related to the realization of the product and that might affect the quality of the product when not maintained or provided, for example, the deployment of a new technology or a new machine.

- Human resources—Identification of the qualifications of human resources that might affect the quality of the product when not provided, for example, the employment of new personnel.
- Process environment—Identification of the process environment conditions that are related to the realization of the product and that might affect the quality of a product when not maintained, for example, storing the goods in a new storage area.
- Handling activities—Identification of activities such as traceability, packaging, transport, and storage that might harm or affect the quality of the product. Handling materials or components that are used in the realization of the product are also considered here.
- Monitoring and measuring devices—Identification of monitoring and measuring device requirements needed to verify intended outcomes that might affect the quality of a product when not maintained. For example, are the results of the measurements done by this monitoring and measuring device interpretative? Could it be possible that due to a wrong interpretation of the results, product specifications would not be met?

Methods for risks analysis:

- FMEA—failure mode and effects analysis
- Fishbone
- FTA—fault tree analysis
- ETA—event tree analysis
- Design of experiments
- Reasons for variance
- Review for conflicting requirements
- Pareto-analysis

Risks Evaluation

After identifying risks, it is necessary to evaluate them using available information or data. The goal of the evaluation is to assess which objective will be impacted, what the significance of the risk is, and whether the risk is acceptable or not.

The first step in the evaluation is defining criteria. The criteria are used to determine the acceptability of risks: which risk is acceptable and not significant to the business activity and which risk is unacceptable and significant:

- Acceptable risks—Risks with which an organization can live, that is, the organization can still deliver products to customers without any nonconformity, without planning actions to control them.
- Unacceptable risks—Risks with which an organization cannot live, that is, when these risks are not addressed, there is a high probability that products delivered to customers do not meet the customers' expectations.

When risks are accepted, it is important that the risks be informed to all the relevant parties.

This kind of evaluation assists in focusing your efforts and resources on addressing risks. It is possible that two risks coexist; in that case, eliminating the first risk is a priority whereas the second risk can be managed by the organization. The criteria should enable you to balance your objectives against the associated risk. The criteria provide a range with which the organization decides whether a risk is needed to be controlled and reduced and to what degree. The criteria

- Are designed to ensure the achievement of intended results
- Provide the ability to decide whether an outcome is acceptable or not
- Ensure the conformity of goods and services
- Ensure customer satisfaction

While defining the criteria, one should consider the context of the organization, the organizational objectives, and the expectations of interested parties. The criteria may be based upon a defined set of prerequisites such laws, policies, regulations, relevant standards, qualifications, guidelines, or other specifications. When the probability of risks cannot be estimated, the probability of the risks cannot be quantitated; so special criteria shall be applied.

According to the determination of the criteria, the tools and techniques for the evaluation shall be developed. The tools enable the relevant parties to obtain relevant data and information needed to be compared with the criteria. The techniques and methods will translate the data and information into terms that could be measured against the criteria. Thus, must they be suitable to the nature of the process, activity, or product that they are measuring or controlling. The end result should enable you to decide whether the risk is acceptable or not. There are two types of methods:

- Estimated—This method enables the end user to choose the degree of the risk (usually predefined, for example, high, medium, or low).
- Calculated—Based on the method for calculation, the results indicate the degree of the risks, for example, with scale.

The evaluation examines

- The initiating event, circumstance, situation, and condition that causes the risk
- The sequence of events that could lead to the situation
- The probability of occurrence of such a situation
- The probability that such situation will cause not meeting requirements
- The characteristics, properties, or specifications (of the process or product) that might not be met—nonconformity

The output of the evaluation provides the following information:

- The probability of the occurrence of the event of not meeting requirements

Common Terms or Possible Description	
High	Likely to happen, often, frequent
Medium	Can happen, but not frequently
Low	Unlikely to happen, rare, remote

- The consequences of the risks, that is, how severe it might be

Examples of Qualitative Severity Level	
Significant	Failure to meet requirements
Moderate	Reversible or minor nonconformities
Negligible	Will not lead to nonconformities

Methods for risk evaluation:

- Risk evaluation matrix
- Diagram for decision process
- Formulas for calculating risk probability
- Comparing data or information against a known applicable standard
- Comparing data or information with disciplinary data

After this evaluation, all identified risks should have been assessed and their implications known. This is how you ensure that no risks are left out.

Opportunities

Though risks are considered negative, risk-based thinking suggests a positive aspect—opportunities. Opportunities are not always directly related to risks but they always refer to objectives, that is, developing an opportunity improves our activities and assists in achieving objectives. The analysis and assessment of certain situations or risks show that further opportunities may arise as options for improvement. An organization must prove that such situations are analyzed and opportunities are identified.

In note 2 of clause 6.1, the standard suggests business possibilities that may arise when analyzing opportunities:

- Adopting new practices
- Launching new products
- Opening new markets
- Addressing new customers
- Building partnerships
- Using new technology

Reference to the Context of the Organization

The context of the organization includes the purpose of the organization, that is, the values to which the firm is committed; quality policy, goals, and objectives with which the organization fulfills its mission; and strategies with which the organization would like to achieve its goals and objectives. These constraints and boundaries define the frame in which risk-based thinking is implemented. The context of the organization is used to determine relevant risks and opportunities. When planning the QMS, the organization determines which risks may prevent it from achieving these objectives or which opportunities may help fulfill the context of the organization better.

As mentioned earlier, one of the main goals of risk-based thinking is handling changes in the business and operational environment of the organization. The business environment is ever changing and organizations need to adapt to changes frequently. These changes affect the products and business activities of organizations. Changes are associated with opportunities and unanticipated consequences—risks. Thus, organizations should

- Review which business aspects may change over time
- Evaluate what the associated the consequences will be
- Adjust criteria, evaluation methods, and controls in accordance with the changes
- Review the effectiveness of these adjustments over time

Which context of the organization's aspects might hide risks and opportunities?

- Strengths—Strengths are the advantages of the organization when realizing the product or service. The organization derives opportunities from these strengths and tries to leverage them into improvements.
- Weaknesses—The weaknesses are attributes that may interrupt the organization in achieving its goals and objectives or may generate nonconformities. The organization is to identify them and suggest ways to mitigate, avoid, or eliminate them.
- Opportunities—Environmental, technological, economic, infrastructural, or social conditions or circumstance that the organization can draw advantage from in the future.
- Threats—Threats are environmental, technological, economic, or social conditions or circumstance that may prevent the organization from achieving its quality objectives. The organization is to identify them and find ways to cope with them.

Functions or responsibilities involved in the realization of the product must know the context of the organization and the relevant quality objectives. This is necessary to understand and effectively evaluate which risks or opportunities lay within realization processes. In practice, I would include this type of analysis in the management review and submit its results as inputs to the management review.

Reference to the Needs and Expectations of Interested Parties

The expectations of interested parties of the organization are taken into account when determining relevant risks and opportunities. Identifying the concerns of interested parties and addressing the associated risks or promoting opportunities improve their confidence and trust. The interested parties of the organization are in this stage known and defined:

- Owners and investors, who represent business requirements
- Employees, who represent qualifications and training requirements
- Customers, who represent contractual, specification, and quality requirements
- End users, who represent functionality and quality requirements
- Suppliers or other external providers, who take part in the supply chain, which represents contractual requirements
- Regulators, who represent legal or statutory requirements
- Auditors, who represent standard requirements
- Other interested parties, whom the organization finds relevant to the QMS

The organization must identify which risks are associated with interested parties, that is, when, where, and why their needs will not be met, and suggest ways and actions to avoid these situations or which expectations of interested parties might bring opportunities and whether developing these opportunities is feasible. Risks or opportunities associated with the expectation of interested parties are as follows:

- Owners and investors—risks associated with revenues and implementing strategies
- Employees—risks associated with errors or nonconformities related to the operation of a QMS
- End users—risks associated with the specifications or requirements from the product due to technological, sociological, ecological, economic, or environmental changes
- Suppliers—risks associated with the performance and ability to supply goods and services according to specifications
- Regulators—change of applicable regulations that might affect realization processes
- Auditors—change of standard requirements (e.g., the revision of the ISO 9001 Standard)
- Employees—risks associated with safety during the realization of the product

The analysis may affect the processes of the QMS and would require the implementation of certain controls on one hand or may lead to improvement and innovation on the other hand. A very good and known example (which will be discussed in detail in chapter 8.4—Control of externally provided processes, products, and services, section 8.4.1—General) is the evaluation or suppliers where activities are implemented to select, evaluate, reevaluate, and rank suppliers. If we analyze these actions,

- The expectations of a supplier are known
- His performances are monitored
- Related data are analyzed, trends are indicated, and risks are monitored
- Actions to mitigate those risks are taken

Risks Related to Intended Results

Intended results relate to the objectives of various business operations that constitute the QMS. One of the objectives of risk-based thinking is securing the delivery of indented results while addressing risks that might prevent the organization from achieving the objectives or identifying opportunities that may enhance the objectives. In other words, risks or opportunities that affect the reliability of processes, resources, process environment, and infrastructures.

Implementing risk-based thinking when planning processes increases the probability of delivering intended results. Furthermore according to clause 4.4, you are required to address risks to the conformity of products and customer satisfaction if unintended outputs are not delivered. In this stage, you are to determine and address those risks. Which objective might not be achieved?

- Strategic objectives
- Quality objectives
- Product specifications
- Process parameters

In order to eliminate situations where these objectives might not be met, the relevant outcomes of the related operations must be analyzed and the causes that may prevent the organization from achieving its intended results must be identified. The organization can do it through the application of different controls in different areas and aspects of the operations:

- During design and development phases, addressing risks related to the realization of products enables the planning, in advance, of necessary verifications and validations for realization stages. During the development stage, some critical process elements are determined and ensure intended results:
  - Required resources for the realization
  - Necessary training for the personnel
  - Definition of process outputs
  - Required verifications or validation of activities
- Statistical methods for the analysis of processes serve as tools for the evaluation of processes. Such methods indicate which process parameters or product characteristics must be controlled.
- Audit and review of resources related to realization processes may indicate which risks are associated with those resources (in terms of availability and usability).
- A control over purchased products or services is necessary and the performance of external providers must be supervised.

Implementing these methods generates data and information and indicates trends in the processes. The next step is to identify whether any risk is associated with those processes and activities and to decide whether those activities must be controlled. When risks are identified, the organization is to develop plans to address the risks with respect to the context of the organization and the expectations of interested parties:

- Would the quality objectives be achieved?
- Would the expectations of interested parties be fulfilled?

Enhancing Desirable Effects (or Risks Related to Product Characteristics)

The standard specifically requires addressing risks in order to enhance desirable effects. Desirable effects in this context mean the achievement of business or product characteristics resulting from a previous activity. Product characteristics represent the expectations of customers and relate to the materials or components that the product is made of, the usability and the intended use, and other specifications such as storage or transportation. The organization needs to identify products and special characteristics where the verification of their quality is required and determine for them their verification. In other words, if these characteristics are not verified, there is a danger of undesirable effects. Addressing the related risks enables the development of controls for preserving these characteristics, ensures desirable effects and achievement of process objectives, and ensures the delivery of products to customers according to their expectations. One might say that those controls will be implemented during the quality assurance phase of product realization. Which controls may we find?

- Loading tests
- Product life tests
- Crash tests
- Tests of materials
- Tests of usability

Actions to Address Risks and Opportunities

At this point, it should be clear which risks or opportunities are associated with the realization of the product and in which context they might emerge. The next stage is to plan actions to address and control them. Actions for addressing risks are needed to eliminate, avoid, or control the risks and reduce the probability of errors in the processes and their effect. Actions for addressing opportunities are needed to promote improvements in the QMS. These actions must be planned in advance (the proactive approach—identifying in advance rather than acting upon events after their occurrence).The organization is required to show which methods it develops and implements to avoid, mitigate, or eliminate risks. The objectives of those actions are as follows:

- Avoiding situation and events that may avert the organization from achieving its objectives
- Creating understanding among the relevant parties of risks and their implications
- Informing the relevant parties when such events occur
- Preventing the organization from delivering unintended outputs
- Developing opportunities into improvements

When planning the actions, the following issues will be referred

- The responsible parties for performing the action.
- Identification of a process, activity, output, good, or service that is controlled.
- What are risks, what are their implications, which objective might not be achieved, and how to identify the situation or event in which objectives will not be achieved?

- How to analyze such events and find the root cause.
- Which methods are applied to control and measure these events or product specifications?
- Which resources are needed?
- Which information or data are to be used?
- Relevant documentations.
- Expected outputs such as forms or records.

The activities may be supported by procedures, work instructions, guides, etc. I personally recommend documenting these activities; where a work instruction or procedure already exists, you might add a capital or paragraph for risks, and where such does not exist, create one.

The processes, activities, and tasks for addressing risks (how risks are controlled or which actions are to be implemented in order to control them) depend purely on the type and characteristics of the processes, the nature of the product, the complexity of the product, the risk associated with the realization of the product, and the objectives or expectations of interested parties. How to plan the activities? There are several types of actions that are applied at different stages of the realization:

- Identification of events or situations, during the realization, that may pose a risk
- Eliminating risks—identifying which activities are needed to drive activities away from those situations and events
- Avoiding risks—identifying which activities are needed to create conditions that avoid these events or situations
- Control risks—implementing solutions that reduce risks and minimize the impact or likelihood of the risks and ensure that those situations will not occur and when they occur, actions are applied

Let us see how these types of actions are applied in a real-life situation such as assembling a product:

Action	At Which Stage of the Process Life Cycle It Is Applied	Example
Identification of events or situations, during the realization, that may pose a risk	These situations should be identified during the planning phase.	A product can be wrongly assembled during the realization.
Eliminating risks	Planning should suggest actions that eliminate risks.	Planning a quality test that includes all the parts of the product.
Avoiding risks	Planning should suggest actions to avoid risks.	Automating the process of assembly. Training employees.
Mitigating risks	During the realization, actions to mitigate known risks are to be applied.	Setting all the components in a row: a worker cannot advance to the next product when not all components in one row are used.
Controlling risks	Appropriate controls are applied to ensure that risks are handled during the realization and requirements are met.	Applying a quality control that verifies that all parts were used during assembly. Counting the parts that were used, with the amount of products that were assembled and deducting the scrap.

Moreover, it is needed to implement actions for the tracing and retrieval of defective products according to the identified risks. Examples for such actions may include

- Identification and traceability of products or components of the products
- Batch of charge management
- Identification of products during transport and storage
- Management of validation date

Proportion to the Potential Effects

Conformity of goods and services as well as customer satisfaction play a main role when determining which actions are needed to address risks or opportunities and to what degree the actions should be applied. This is a stage where the organization must consider whether it should address risks or opportunities—controlling the risks combined with the impact of the risk or developing an opportunity combined with its benefits is worth handling. The organization may consider a number of options for actions or combine them. On one hand, the actions must be sufficient to enable the conformity of goods or services and customer satisfaction, but on the other hand, the efforts must be balanced against the benefits derived from reducing a risk or developing an opportunity. It means that the expected investment in terms of resources and efforts from the organization must be proportionate to the implications of those risks or opportunities in two levels:

- Effects on conformity of goods and services—it is necessary to measure the impact of a risk or an opportunity on the ability of the QMS to achieve or support conformity of goods and services
- Effects on customer satisfaction—it is necessary to measure the impact of a risk or an opportunity on customer satisfaction

The organization assesses the degree of the actions according to the nature and characteristics of the risks. This assessment of proportion assists the organization in determining which actions to choose (when there are more alternatives) and is important for the planning of those actions and for the allocation of the needed resources. Consideration of these implications determines the extent and degree of the criteria for accepting or rejecting processes. What might influence the decision?

- Regulatory requirements
- Effects on product requirements related to goods and services
- Effects on customer requirements and product specification as determined by the customer
- Effects on processes and activities needed to realize the product

Evaluating Effectiveness of Actions to Address Risks and Opportunities

After planning and implementing actions for addressing risks, it is necessary to evaluate their effectiveness and evaluate whether these actions do mitigate, eliminate, or reduce the risks to an acceptable level. The review of the effectiveness creates an iterative process of improving the controls. If you decide to document risk-based thinking, include this stage as well. Two distinct verifications are expected:

- A verification that the required controls have been implemented as planned
- A verification that the risks are controlled as planned

If the review shows that the action is not effective and the risk is not removed to the expected degree, the organization must determine

- Whether further controls are required
- Whether to accept the risk or to plan the QMS element again

For a practical review of the effectiveness, all the required data must be collected and examined. Therefore, the appropriate methods must be defined:

- Internal auditing is an effective tool for measuring the effectiveness of actions because it is a system that evaluates the improvement of processes over time and can deliver results regarding the ability of the organization to fulfill its objectives for the long term.
- Analysis and evaluation of data is an effective method for identifying the trends of processes. Evaluating the effectiveness of actions for addressing risks involves the integration of statistical methods.
- Many standards may include specifications or requirements regarding products or processes. By applying a standard, the organization can analyze the effectiveness of the actions. This is a good option where objective data or evidence for evaluating the effectiveness is absent.

6.2 Quality Objectives and Planning to Achieve Them

Quality objectives have a strategic role of carrying out the quality policy and its implementation through a quality management system (QMS) and provide a means to assess whether the QMS achieves its goals. The effectiveness of a QMS depends on the extent to which quality objectives have been achieved. In other words, quality objectives report the status of the effectiveness of the QMS. Without quality objectives, the organization cannot promote some of the ISO 9001 Standard basic goals: meeting customer requirements, achieving the improvement of the QMS and its products, and enhancing customer satisfaction. This statement is valid to all areas in the organization. Let us review the ISO 9001 Standard requirements:

6.2.1

Clause 6.2.1 prescribes how quality objectives are planned:

- Quality objectives shall be planned and allocated to the appropriate and relevant functions and processes on different levels of the QMS throughout the organization.
- Quality objectives shall be consistent with the quality policy.

- Quality objectives shall be measurable.
- Quality objectives shall be planned with reference to, and in accordance with, applicable requirements.
- Quality objectives shall be relevant to the conformity of goods and services.
- Quality objectives shall be planned in order to ensure customer satisfaction.
- Quality objectives shall be monitored at planned schedules and time intervals.
- A system for communicating the quality objectives shall be planned and maintained.
- Quality objectives shall be updated as appropriate.
- Quality objectives shall be maintained on as documented information.

6.2.2

Clause 6.2.2 dictates how quality objectives are implemented in the organization:

- The organization shall determine which actions are needed to achieve its quality objectives.
- The organization shall determine which resources are needed for achieving the quality objectives.
- Responsible parties shall be appointed for the quality objectives.
- A system for monitoring and reporting the status of the quality objectives shall be planned and maintained.
- A method for evaluating the results of quality objectives shall be defined for each objective.

Consistence to the Strategy of the Organization

Before we start with the ISO 9001 Standard requirements, I would like to point out the strong relation between quality objectives and the strategies of the organization. Quality objectives are strategical as well as operational steps that the organization defines. On achieving these objectives, quality goals are reached and the strategies of the organization are executed. The type and hierarchy of the objectives are to be planned in order to accomplish the strategies of the organization, and the quality objectives are determined according to the organizational goals and should support the achievement of the goals. It begins at setting strategic objectives and continues with defining operational objectives. The quality objectives are derived from

- The context of the organization—quality objectives adjust the organization to its relevant business and environmental conditions:
  - Promoting strengths
  - Handling weaknesses
  - Promoting opportunities
  - Reacting to the different economic or social environments
- Needs and expectations of interested parties—through the achievement of the quality objectives, aspects related to the needs and expectations of interested parties are ensured:
  - Customers' requirements
  - Customers' satisfaction
  - Competition
  - Regulatory requirements

- • Performance of suppliers
- • Expectations of investors
- QMS requirements—quality objectives support the promotion of quality management requirements:
  - • Relations and performance of processes
  - • Trainings, competence, and encouragement of human resources
  - • Management of knowledge
  - • Allocation of resources
  - • Addressing risks and preventing side effects

I add here some important principles to consider when planning quality objectives:

- • Quality objectives should support decisions making—the data and information generated and gathered through the implementation of quality objectives allows different managerial levels in the organization to make decisions regarding the QMS.
- • Quality objectives should be relevant to the organization—quality objectives express the business expectations of the organization concerning its business and operational performances.
- • Quality objectives must be achievable—it does not do any good to define quality objectives that look good on the paper and may make a good impression in the audit but in fact no one can actually measure or reach.
- • Quality objectives should not be exceeded—it is not quite recommended to define too many objectives as this may cause frustration and dissatisfaction among employees rather than positive results.
- • Quality objectives should be practical—they must relate to the processes, activities, and operations of the QMS.

Consistence to Quality Policy

The determined quality policy represents the general guidelines, intentions, and goals of the organization referring to its quality intentions. The policy demonstrates what quality signifies to the organization and defines the areas and scopes in which the quality is active and valid. Quality objectives are to be developed for those areas and scopes. Quality objectives are a result of the quality policy; quality goals are to be analyzed and operational measurable steps that assists in achieving them are to be planned as objectives. Those objectives must relate to operational and practical activities. Practical quality objectives should lead the organization to achieve more high-level objectives until strategic destinations are achieved. For example, if the quality objective is to enhance customer satisfaction and you know that customers are complaining about the package of the product, the operational quality objective should relate to those activities and quality elements that handle the package of the product (directly or indirectly). In other words, the strategic goals and the quality policy are to be broken down into specific steps. By doing that, the goals and the quality policy will be fulfilled. One should take a look at the quality policy, review the quality objectives, and would be able to say, "This objective will achieve this aspect of the policy." Examples:

- • If your quality policy is, Our company is obligated to provide the customer with products and services at the agreed schedules, an appropriate quality objective may be 95% delivery reliability.

- If your quality policy is, Establish, maintain and improve compliance to environmental regulatory requirements, an appropriate quality objective will be Certification to ISO 14001 in the next 2 business years.

Reference to Applicable Requirements

When requirements are part of the QMS and demand the achievement of certain goals, quality objectives refer to those goals. Applicable requirements:

- Regulatory or statutory requirements
- Quality requirements
- Sectorial requirements
- Market requirements
- Social requirements

Working in certain industries requires meeting certain requirements such as maintaining designated management systems. For example, in the construction industry, there are many safety and health regulations that must be applied. The quality policy must certainly refer to those requirements and the quality objectives must as well.

If your policy is, Establish, maintain and improve compliance to health and safety regulatory requirements, an appropriate objective will be, Educating all employees about health and safety measures until the end of this business year or certifying the organization to the OHSAS 18001 Standard in 2 business years.

Measurable Quality Objectives

Quality objectives must be capable of being measured and must enable the effective assessment of processes, activities, and their outcomes. Measurable objectives refer to the ability to compare the results of actions or activities to the objectives—did the activity achieve its objective? For example, it is hard to measure a quality objective such as "Promotion of human resources." Such an objective must be developed into a measurable plan, for example, planning three training programs in certain areas for the next business year. At the end of the year, one can determine, during an internal audit, for example, whether this objective has been met.

Quality objectives are determined with absolute figures (quantitative) or with scopes or ranges that measure the extent of achieving the objectives (qualitative). Quantitative objectives use a specific scale that may suggest whether the objectives have been met or not. Qualitative objectives use the weight of evidence, are based on professional judgment, and usually do not require a high analytical process or a method. Examples of quantitative objectives are as follows:

- Objectives related to customer satisfaction
  - Reducing reaction time to customer call to 48 hours
  - Preventing the loss of five customers this year
  - 95% on-time shipments by the end of this quarter
- Objectives related to business development
  - An increase of 50% in sales revenue in this business year

- Objectives related to quality
  - Reducing 50% of customer returns
  - Reducing 80% of quality costs until the end of this year

Examples of qualitative objectives:

- Objectives based on yes/no criteria
  - Updating all work instructions until the end of the current business year
  - The installment of a new quality portal in the premises of the organization
- Defining boundaries or thresholds of the results of activities
  - Standing within the threshold of certain process limits
- Initiating actions or activities
  - Using only REACH (Registration, Evaluation, Authorization and Restriction of Chemicals) authorized materials until the end of the next business year.
  - Developing a line of web-based products in the next 3 years
  - Implementing a service treaty until the end of the year

Relevancy to Roles and Responsibilities

Quality objectives relate to persons and responsibilities in the organization because the personnel are expected to achieve quality objectives. The relation between quality objective and a function ensures that each quality objective relates to an organizational role that through its work, activities, and actions the objective may be achieved. This assignment of responsibility has two levels:

- The managerial level
  - Defining quality objectives by translating the organization's strategic objectives into practical job objectives
  - Implementing actions for achieving these goals
  - Making sure quality objectives are measured and monitored
  - Analyzing the status of objectives.
- The operational level
  - Acting in order to achieve quality objectives

The same person or function may act in both roles. Reviewing the organigram may assist in locating necessary quality objectives by relating them to specific roles, stages of realization processes, or specific activities. Considering the organigram also facilitates the interfaces and relations between the objectives. Let us review the next examples:

Objective	Responsibility and Authority
Reducing reaction time to customer call to 48 hours	Customer service manager Manger of help desk
Reducing 80% of quality costs until the end of this year	Quality assurance Production manager
Scheduling an interaction with each customer at least once a year	Sale manager/key account manager
Implementing a service treaty until the end of the year	Vice president of services
Reducing 80% of production scraps	Production manager Production employee

Although it is not explicitly required, I would include in the job description the relevant quality objective.

Ensuring the Conformity of Goods and Services

Another very important goal of quality objectives is the relevancy to the conformity of goods or services. Quality objectives promote the requirements of products, services, and processes. Processes and activities related to the realization of the product must guarantee the conformity of goods and services. This is why the performance of those processes must be monitored. And in order to determine their effectiveness, they must have objectives that cover all levels of the realization of the product.

One aspect refers to the elimination or the reduction of the occurrence of nonconformities. If you are aware of the root causes of the past quality problems where nonconformity occurred in your organization, it will be much easier to set appropriate quality objectives to eliminate or reduce them. This is a reactive approach where you use quality objectives as a corrective measure, for example,

- Implementing a new visual quality control system in the next business year
- Improving design issues that affect product performance

The other aspect is a bit more complex; quality objectives lead to the achievement of conformity of goods and services. This is a proactive approach to handling and improving quality issues related to the performance of processes, products, or customer satisfaction. In this case, you must set objectives in such a way that by achieving the objective, errors will be avoided in advance and thus conformity will be attained. Setting those quality objectives leads the organization to achieve conformity of goods and services:

- A more comprehensive study of customer requirements
- Implementing the process approach for improving and optimizing processes

This aspect has a direct relation to the prior clause 6.1—Actions to address risks and opportunities; setting appropriate quality objectives may address issues needed to secure the delivery of indented results.

Enhancing Customer Satisfaction

One of the main goals of a QMS is to enhance customer satisfaction. We are aware that the organization depends on its customers and must strive to exceed their expectations. Thus, the ambition to fulfill the expectations of interested parties and enhance customer satisfaction can be effective only when relevant objectives are set.

The first step in determining appropriate quality objectives is to understand and identify what the expectations and needs of customers are. Meeting customer requirements is achieved through the initialization of actions for understanding the needs and requirements of customers, identifying new requirements of customers, or handling situations where potential conflicts with customers may arise. This is defined, discussed, and planned in the next standard clauses:

- Clause 4.2—Understanding the needs and expectations of interested parties
- Clause 6.1—Actions to address risks and opportunities

In practice, those expectations are usually documented; a market research, on a contract, an order, etc., and the handling of those situations are to be identified and controlled through specific actions.

The second step is to identify those QMS elements that may affect the needs and expectations of customers. Factors that affect customer satisfaction vary from one organization to the other and are influenced by the nature of goods or services. After identifying those QMS elements, you may declare appropriate quality objectives. The appropriate quality objectives enable the organization to respond to the needs and expectations of customers.

Quality objectives also serve the top management in the matter of leadership and commitment with respect to customer focus where trends of data related to the quality objectives indicate the efficiency of resolving issues and problems related to customers.

Let us look at some examples of the objectives that promote customer satisfaction:

Aspect of Customer Satisfaction	Example for an Objective
Quality objectives should enhance the ability to solve the problems of customers as soon as possible.	Reducing reaction time to customer complaints. Establishing plans to reduce customer complaints based on statistical analysis.
Understand and anticipate the needs and expectations of customers.	Scheduling an interaction with each customer at least once a year.
Quality objectives should increase the ability of the organization to meet customer satisfaction in the long run.	Informing customers of the progress of their contract or order. Improving the interface between the development and customers concerning their expectation from the product.
Quality objectives should motivate actions to eliminate situations where customer satisfaction is not achieved. The ability to ensure conformity of goods and services.	Increasing quality tests before delivering goods to customers. Handling nonconformities and corrective actions with the goal of achieving customer satisfaction.
Quality objectives should create the conditions for applying sufficient resources needed to achieve customer satisfaction.	Implementing a new system for a better communication of customer requirements throughout the supply chain.
Developing products that achieve customer satisfaction.	Improving the review processes of product characteristics. Performing a survey among customers for obtaining feedback and ideas for improvement in the next business year.

Setting Actions Needed to Achieve Quality Objectives

Activities needed to define, implement, and measure quality objectives must be determined. This requirement refers to the operational side of developing quality objectives in the QMS:

- Definition of quality objectives—who is responsible for defining the quality objectives and for which organizational units
- Implementation of quality objectives—which activities are needed to implement quality objectives, for example, training
- Measuring quality objectives—how quality objectives should be measured: which method, how data should be collected and gathered, in which time frame, and which report

The ISO 9001 Standard expects a kind of a plan to deploy and implement quality objectives. The plan facilitates the deployment and achievement of the quality objectives. I suggest the next hierarchy:

- Top management—providing the leadership and commitment needed to implement quality objectives:
  - Leadership—demonstrating the commitment to ensure the definition and control of quality objectives
  - Commitment—allocation of resources and periodical review
- First-level management—identification of issues that are related to the QMS and must be observed
- Second-level of management—setting practical objectives and gathering the data
- Operational level—performing the actions and activities needed to achieve quality objectives

Resources Necessary for Achieving Quality Objectives

When determining quality objectives and the actions needed to achieve them, the top management must allocate and provide sufficient resources: time, personnel, and budget (or purchasing software or computers, trainings, etc.). For each quality objective, it must be clear which resource it requires:

- Infrastructures—systems, facilities and equipment
- Process environment—work conditions
- Competence of employees or external providers—appropriate qualifications and skills
- Knowledge—concepts, work methods, technical solutions, and user applications

Let us allocate the resources to one quality objective:

Quality objective	Reducing reaction time to customer call to 4 hours
Infrastructure	Purchasing and implementing a new CRM system
	Improving communication channels and interactions between the customer service and other departments of the organization
Process environment	Allowing structural changes in offices in order to allow better communication between different departments
Competence of employees	Allocating and training specific employees to specific types of customer complaints
Knowledge	Developing processes and methods for communicating with customers

Furthermore, the allocation of resources needed for quality objectives should be reviewed periodically in order to ensure that the resources are still available and suitable. This review is part of the management review. When outsourced resources are needed, they are considered as well.

A System for Monitoring and Reporting the Status of Quality Objectives

In order to implement and maintain quality objectives, you must establish a system for monitoring and reporting the status of the quality objectives. By system, I mean processes, activities, and interactions for monitoring, controlling, and reporting the effective of the organization's objectives. There is no explicit requirement for such system, but during an audit, you are required to prove that all the data and information regarding the quality objectives are available, organized, and monitored. When planning or describing such a system, refer to the next aspects:

- Gathering data and information—The system must indicate clearly which type of data and information must be collected and processed. When the system refers to figures like turnovers, percentage of accepted offers, customer returns, and errors in the production, it must be clear where these data and information are available.
- Tracking and monitoring—The system should refer to the relevant QMS elements that must be monitored: processes, human resources, documentations, and records.
- Observing—The system should ensure that activities for maintaining the quality objectives are performed, the objectives are measured, results are communicated to the appropriate persons, and those persons invest the resources to review the results.
- Reviewing the results—The system must provide the view of trends indicating the business progress of the objectives and opportunities for improvement and include the comparison of periods.
- Reporting—The system should define the type of reports of the results of monitoring and measuring and should communicate them to the relevant persons in the organization.
- Resources—The system should refer to the allocated resources needed for the achievement of the quality objectives: trainings, software, tools, etc.

Examples of systems for monitoring and reporting the status of quality objectives:

- Objective card—a manual Excel sheet that summarizes information and data concerning an objective
- BI system—a tool that draws operational as well as financial data from a data bank and presents it in a way that allows the organization to make decisions
- Internal audit—the internal audit is a very effective quality tool for providing information regarding the progress of the quality objectives
- Balanced score card—a structured report that reports business and process indexes related to the strategy of the organization

Monitoring Quality Objectives in Planned Schedules and Time Intervals

To implement effectively a strategy, the organization establishes timelines and timeframes for each objective. The standard expects an arrangement of events used to manage and follow the achievement of the objectives. It refers to the scheduling of progress and review of the objectives.

Each quality objective must have a particular period of time for its completion—until the objective is achieved. This timeline is used to control the status of the achievement. The schedules should be stringent, and when an objective is not achieved in its defined period, an analysis of the cause must be performed and corrective measures must be applied:

- Allocation of more resources
- Update of the objective
  - Modification of the objective
  - Change of the variables for measure

Let us take a look at the next objective and its planned schedules—Certification to the ISO 9001 Standard: 2015.

Task	Time Objective
Adjustment of the QMS documentation according to the new ISO 9001:2015 Standard requirements	June 2016
Adjustment of all operational documentation according to the new ISO 9001:2015 Standard requirements	Jan 2017
Training to all the employeeson the new Standard and the transition from the ISO 9001:2008	March 2017
Starting preparation for the internal audit	March 2017
Internal audit	June 2017
Scheduling an external audit	June 2017
Internal audit—second round; corrective actions and improvements	September 2017
Starting preparation for the external audit	January 2018
Final external audit	April 2018
Certification	June 2018

As you can see, you will need to draw a plan with schedules for achieving the objectives. The schedule of each quality objective shall be determined according to the availability of resources and the nature of the objective.

Communication and Awareness

The organization must initiate the communication of quality objectives; the quality objectives and their status of achievement must be consistently communicated among employees and submitted to a review through the use of a designated tool that compares the objectives and the actual situation (target–performance comparison) and declares the results. This communication promotes the awareness of the QMS and the quality objectives of the organization.

Employees must be aware of the quality objectives hierarchy and understand the relevancy, relation, and effect of their actions and activities on the achievement of the quality objectives. Employees must understand the link between their activities and the objectives of the organization. They must understand that they bear the operative responsibility of achieving the objectives and experience and how their activities and the expected outputs promote the quality objectives. Such awareness enhances the identification of employees with the quality objectives and initiates the motivation to meet them.

Information regarding quality objectives that is to employees:

- What are their objectives (quantitative or qualitative)?
- What is the meaning of those objectives?
- How can these objectives be reached?
- How are these objectives measured?
- What happens when they achieve these objectives?
- What are the consequences when they do not achieve these objectives?
- What is the situation in comparison to these objectives?

A method for communicating quality objectives and relevant information, which I frequently encounter and is quite simple and effective (I might add), is publishing the quality objectives on the tabloid put up in the halls of the organization which employees daily pass by. Another method is to publish this information on the organizational portal.

Evaluating Quality Objectives

Quality objectives must be reviewed in order to evaluate their effectiveness—to what extent the quality objectives have been achieved. The evaluation of the quality objectives demonstrates trends related to processes, activities, and outcomes. The goal of the evaluation is to detect obstacles and issues that prevent the achievement of the quality objectives and therefore the improvement of the QMS. The result of the evaluation initiates decisions about further or corrective actions needed for achieving the quality objectives. I suggest the next practice for evaluating quality objectives on a continual basis:

- Operational objectives are to be frequently evaluated (once a week or once a month) on a lower managerial and operational level—a weekly team meeting or a quality circle meeting where regular business and operational issues are discussed.
- Strategical objectives are to be less frequently evaluated, but the evaluation should be conducted on a higher managerial level, such as top management, during the management review, for example, where the results of the QMS are presented to the top management.

Updating Quality Objectives

Quality objectives should be updated, on a regular basis, when changes in relevant QMS elements occur

- Changes in the context of the organization, for example, changes related to the strategy or the quality policy of the organization such as new priorities or changes in trends, which may have an impact on the quality objectives
- Changes in applicable requirements such as regulations

- Changes in the needs and expectations of interested parties, for example, changes related to the expectations of the product
- Changes in processes, resources, infrastructures, or activities, for example, technological changes that require new quality objectives or changes in the availability of resources that may make it impossible to achieve objectives

What shall be updated?

- The relevant documented information (quality objectives) that describes the quality objectives
- The method for monitoring and measuring the quality objectives
- The method for communicating the quality objectives
- The method for evaluating the quality objectives

Ensure that the relevant documented information of the change in the quality objectives is updated since this type of documented information must be controlled according to the requirements of clause 7.5.2, Creating and updating.

Documented Information for Quality Objectives

The organization must maintain documented information in order to prove that the quality objectives are defined, documented, allocated, measured, reviewed, and updated. There is no particular format of the documented information, and I recommend designing one that suits your organization. With the documentation, it is easier to manage and follow the quality objectives and it is much nicer to present during an audit. I recommend the next structure:

- First level—declaring that you have quality objectives that are part of the QMS, and committed to fulfilling the quality policy; for example, if you maintain a quality manual, you may mention the quality objectives.
- Second level—a primary list of the definition of the organizational quality objectives. Methods of defining quality objectives:
  - Quality manual (as an appendix)
  - Road map
  - Action plan
  - Business plan

  People are often used to include this list in the quality manual. It is not prohibited but takes into account that those quality objectives will probably often change with time and that means you will have to update the quality manual each time the objectives are changed what I find inefficient.
- Third level—a more detailed presentation of the quality objectives referring to roles, functions, divisions, departments, etc. For each quality objective, I will detail
  - The goal and logic
  - The method for the measurement
  - The intervals for the measurement (each week, month, quarter, or year)
  - Who is responsible the measurement

- Who performs the measurement
- The method and tools needed for the measurement
- The records that provide the data

This kind of documentation makes it easier to review and change the quality objectives periodically. This level can be documented on a designated controlled list such as an objective card.

- Fourth level—the method for capturing and analyzing, which is the practical part of measuring the quality objectives:
  - Excel sheet
  - Balance score card
  - BI system
  - Process metrics

  This level enables employees to be informed of their quality objectives and their status.
- Fifth level—report and review of the quality objectives and their trends and status. This, for example, is the input for the management review, enabling the top management to review the overall status of the QMS.

Of course, you are not required to adopt and implement all the levels. But this structure assists you in navigating through your system of quality objectives. Ensure that this structure is submitted to the control of documented information as specified in clause 7.5—Documented information.

Example of Quality Objectives

I add here some rough examples for quality objectives: their context, type (qualitative/quantitative), and how to measure them.

Quality objectives for the organization:

Quality Objectives	Type	How to Measure
Implementing and fulfilling regulatory requirements	Qualitative	Management review
Getting certified for certain standards (or their revision)	Qualitative	Management review
Securing or improving market position	Qualitative	Management review
Increasing revenues	Quantitative	Monthly turnover reports
Improving the QMS	Qualitative	Internal audit Management review
Benchmarking	Qualitative	Internal audit Management review
Reducing quality costs	Quantitative	Statistical measure Internal audit
Implementing a management system	Qualitative	Internal audit Management review
Expanding to other types of the QMS or other certifications	Qualitative	Internal audit Management review
Enhancing proposals for improvement	Qualitative	Internal audit Management review
Enhancing the effectiveness of the QMS	Qualitative	Internal audit Management review

Quality objectives for the product:

Quality Objectives	Type	How to Measure
Reducing scraps in processes	Quantitative	Percentage of the total output of a process
Enhancing the quality performance of external providers (of goods or services)	Quantitative	Supplier evaluation
Increasing product quality	Quantitative	Statistical measure of quality assurance
Increasing process capability	Quantitative	Statistical measure
Increasing customer confidence	Quantitative	Satisfaction surveys
Reducing recurrent errors	Quantitative	Statistical measure

Quality objectives for processes:

Quality Objectives	Type	How to Measure
Benchmarking	Qualitative	Management review
Objectives related to realizing new products or adapting changes in existing products	Qualitative	Internal audit Management review
Reducing recurrent errors	Quantitative	Statistical measure
Inventory turnovers	Quantitative	Statistical measure

Quality objectives for customers:

Quality Objectives	Type	How to Measure
Reducing reaction time to order	Quantitative	Statistical measure
Increasing customer satisfaction	Quantitative	Statistical measure
Reducing the number of customer complaints	Quantitative	Statistical measure
Reducing the reaction time of complaints	Quantitative	Statistical measure
Increasing delivery reliability	Quantitative	Statistical measure
Increasing satisfaction from service	Quantitative	Satisfaction surveys

6.3 Planning of Changes

Clause 6.3 deals with the management of changes and suggests measures that enable the organization to plan, in a methodical way, its modifications to the QMS. By managing and controlling changes, the organization develops the ability to react quickly to changes in circumstances: changes originating from customer demands, competitors, strategies, and revision of regulatory requirements. Managing changes refers to the actions of planning the activities for realizing the change and to the employment of controls of the result after the implementation. The ability in managing changes enhances the level of improvement in the organization, and failure to manage changes in the QMS is a major contributor to nonconformities. Following and implementing the requirements of this clause also ensures that changes to the QMS are consistent with the strategy of the organization and the needs and expectations of the interested parties. ISO 9001 Standard requirements are as follows:

- When changes to the QMS are required and determined by the organization, they are carried out in a planned manner (see clause 4.4)
- While planning changes to the QMS, the organization should consider the following issues:
  - The purpose of the changes
  - The potential consequences of the changes
  - The integrity of the QMS
  - The availability of resources
  - The allocation or reallocation of responsibilities and authorities

Identifying the Changes

A change is any modification to an element in the QMS or to a combination of several elements that may affect the quality of the product or the service or the ability of the organization to realize the product or the service according to the specifications: subsystems, regulations, processes, activities, components, documented information, resources, facility, software, infrastructures, training, personnel (in terms of organizational changes rather than dismissal). In order to evaluate the need for a change, one must provide supporting data and information. Process analysis may assist in determining where improvements are needed and identifying changes that may improve performance. Possible reasons for changes are as follows:

- Organizational changes may bring structural changes to the QMS.
- Processes may have to change in order to achieve the new desired performance.
- Regulatory or statutory requirements may mandate changes to the QMS.
- Changes requested by customers or dictated by the business environment.
- Change of process throughout the organization to bring about performance improvement (process engineering or redesign).
- Key process indicators may indicate the need for changes in the organization.

Type of changes:

- Changes in process input or output
- Changes in the activities of processes
- Changes in product or process specifications—changes in the design
- Technological changes
- Environmental changes
- Economical changes
- Organizational changes
- Changes in the context of the organization
- Changes of resources related to the realization of the product:
  - Personnel issues
  - Infrastructure
  - Monitoring and measuring devices

- Changes in documented information
- Changes as a response to nonconformity
- Changes in regulatory requirements
- Changes referring to specifications or expectations of external providers or relationship with external provider
- Changes originating from corrective actions:
  - Product specifications
  - Activities
  - Monitoring and measuring devices
  - Documented information such as instructions or procedures
- Changes originating form the modification of the business strategy
- Changes required in response to risk analysis
- Changes in the expectations of interested parties

Understanding the Purpose of Changes

Changes have a purpose—the reason why you initiate them in the first place. Changes may take place, or are expected, in economic policies, product demands, technologies, regulations, environmental conditions, or social and cultural conditions. The intensity of a change is dependent upon the nature and complexity of the modified element, for example, of a process or its output. The purpose of a change refers to the anticipated outcome of the change, and it guides your planned actions. Thus, understanding the purpose of the change is critical for the planning of the change. The lack of understanding of the purpose leads to planning and implementing wrong actions. The purpose of a change is derived from its objectives. When planning changes, it is important to answer the following questions:

- Why is the change necessary—which quality problem it will solve or which improvement will be achieved?
- What is the nature change—what will be done or changed?
- Where will the change be introduced—to which QMS element (or elements): a system, a process, a resource, etc.?
- How will the change be implemented—which actions will be undertaken?
- By whom will the change be executed—who are the responsible parties for implementing the change?
- Who will approve the change?
- When must the change be carried out—when must it be reviewed and when must it be approved?
- What are the impacts of the change on other QMS elements?

Planning and Implementing Changes in the QMS

The process of identifying the need for a change or planning a change in the organization is carried out in a planned manner and in accordance with the requirements of clause 4.4 (Quality management system and its processes). What are the principles of the process approach regarding changes?

- Objectives that are derived from the goal and purpose of the change are determined.
- Processes and activities, their sequences, and the interactions between them necessary to carry out the change must be planned using a defined method (including transfer of relevant data and information).
- For each planned change, all the required inputs must be identified. Inputs of a change refer to the need for the change, the QMS elements involved, and data and information that support the activities.
- It is necessary to plan controls for the change that analyze and evaluate the progress and the effectiveness of the change. Those controls may act as monitoring, measurement, analysis and evaluation activities, verifications, or validations of results.
- Criteria for monitoring and measurement necessary to ensure the effectiveness and the achievement of the changes must be planned.
- The resources required to implement the change and their availability must be determined: tools and equipment, applicable documentation, applicable knowledge, etc.
- The responsibilities and authorities for implementing the change must be allocated.
- Risks and opportunities relevant to the change must be identified and addressed.
- Where applicable, the changes should be submitted to the organizational arrangements for improvement.
- Changes should be communicated to the appropriate persons and relevant interested parties at all levels of the organization using appropriate methods and tools: training, publication, meeting, documentation, etc.
- Documented information supporting the execution of the change should be maintained: procedures, diagrams, and forms.

These requirements come to ensure that new processes, operations, and activities or process outputs originating from a change in the QMS must be planned and controlled like any other processes of the QMS. In practice, a kind of method for planning your processes and activities that operate the QMS should have already been determined and implemented. Now, it is time to introduce changes to this method. Other important issues are as follows:

- Changes in statutory and regulatory requirements should be tracked systematically. Such changes may have a great influence on the activities and operations of the organization.
- In some cases, it is necessary to get the approval of the customers of the process before the changes are applied.
- When the changes are technical, a consideration is given to gauge and measurement capabilities.

Understanding the Consequences of a Change

Due to changes, processes can become inconsistent to their original planning and no longer meet their objectives. An early analysis and understanding of the consequences of the changes ensure that those changes achieve their goals with a minimum impact on the QMS. An impact on the QMS refers to the negative effect that a change may have on the ability of the organization to provide conformed products. For example, when processes are modified, what would be the anticipated result in different areas in the QMS? Therefore, changes are reviewed, verified, approved, and reviewed again. In order to seize the real effect that a change might have, it is required to map the change in the overall process; the impact the change will cause upstream and downstream of the processes. Understanding the consequences refers to

- Understanding the business context in which the change takes place
- Identifying the relevant QMS elements that are affected
- Understanding how the change affects those QMS elements

Which QMS aspect that must be assessed?

- Must the quality objectives be modified after a change?
- Will the change affect realization processes?
- What is the impact on process outputs on one hand and process inputs on the other hand?
- Will the current controls, verifications, or validations still be valid or relevant?
- Will the product or service still conform to the agreed-to product requirements? Will the intended use be affected? Will the characteristics of the product be modified?
- Will different components of the product or system be affected by the change?
- Will customer requirements still be met? Will the product still meet its specifications?
- Will customer satisfaction be influenced?
- Will the change affect the regulatory status of the product?
- Must the interactions with external providers be updated?

Which controls may you implement?

- By providing as many examples or scenarios of processes where the change is implemented and may be impacted as possible, you may increase the control of the change.
- Reviewing and assessing the risks related to changes to the QMS.
- Periodic reviews should be done on all the changes to ensure that the changes achieve their objectives and that no adverse impacts (that were not considered) occur. The periodical reviews should be set according to the complexity and the risks that the changes pose.
- Introducing the change as an improvement to the PDCA cycle (see clause 10.3—Continual improvement).
- Comparison of situations, conditions, or results before and after the change allows a gap analysis, which shows the level and scope of the change. It allows the responsible team to revisit the existing state against the initial state and to ensure that the rearrangement in fact meets its expected objectives and resolves the issues discovered in the analysis stage.

Two practical suggestions are to introduce changes to the controls for addressing risks and opportunities (see chapter 6.1—Actions to address risks and opportunities, paragraph Actions to address Risks and Opportunities) or to submit the changes to the PDCA cycle (continual improvement—see clause 10.3).

Ensuring the Integrity of the Quality Management System during Changes

As mentioned earlier, changes may occur in different QMS elements. When implementing a change, there is always a risk that the integrity of the QMS is not be preserved or is affected. Controlling the integrity of the QMS is critical for the continuity of the QMS. Ensuring the integrity means ensuring that after the change,

- Needs and expectations of interested parties are still maintained
- Quality objective are still met
- Interaction between processes and QMS elements and the exchange of inputs and outputs are not disrupted or broken
- Resources are not negatively affected
- Availability of the information is maintained
- Performance of processes, operations, and activities is not negatively affected
- The QMS is still effective

Ensuring the Availability of Resources

Changes may affect resources in the organization and their allocation. When planning a change, it is necessary to review its effect: would resources need to be replaced? Added? Protected? Upgraded? If the review shows that resources necessary for carrying out the change are not available, they must be provided. Which resources may be affected?

- Process environment
- Human resources and competence
- Knowledge
- Infrastructure
- Monitoring and measuring devices

Allocation or Reallocation of Responsibilities and Authorities

The organization should consider the allocation or reallocation of relevant responsibilities and authorities when planning changes. Involving employees in the change is necessary to support and ensure the objectives of the change. The authorities must have an acquaintance with the processes, process output, or QMS elements related to the change that takes place and the area it affects. They must have the skills, qualifications, and knowledge to monitor and evaluate the progress of the change and its effectiveness. When training is required, it should be provided (with reference to clause 7.2—Competence).

These authorities could be the process owner of the relevant process, a manager, or an operator. The level of responsibility and involvement of each of the authorities depends on the scope and degree of the change. One person may serve as responsible and authorized at the same time when the organization finds it appropriate. The roles of these persons are to plan, implement, control, and improve the changes and to control their impact on other QMS elements. These authorities could be a person or a team, depending on the nature of the process or the change to be implemented.

Approving the Change

With the approval of a change at every stage before its implementation, you can ensure that the change has been identified, reviewed, understood, and planned. This surely makes a good impression during an audit. The approvals needed are as follows:

- Approval that the needed arrangements have been made
- Approval of the implementation of the change after the review of its implications
- Approval of the resources needed for the change

Assessing the Effectiveness of the Change

Although it is not explicitly required by the standard, I would consider actions for assessing the effectiveness of changes after they have been submitted for implementation. This is derived from the fact that a change is considered as an improvement and thus its effectiveness must be assessed. The effectiveness of a change refers to which extent its objectives have been achieved and indicates whether its purpose has been met. The effectiveness can be assessed through several aspects:

- Effectiveness of the planning of the change—assessing the effectiveness of planning meaning verifying that processes, operations, and activities have been implemented as planned.
- Effectiveness of processes and activities—assessing the effectiveness of processes and activities refers to the validation of performance of those processes and activities.

Implementing a Change

After the identification, understanding, planning, and approval of the changes, it is time to implement them. One option is to plan a designated process or an activity for implementing them in the QMS with all the necessary requirements. Another option is to decide that the changes should be treated as an improvement or handling of nonconformity and to submit to the controls suggested in clause 10—Improvement.

7 Support

7.1 Resources

Resources are one of the foundation stones of the quality management system (QMS) and, therefore, must be defined, managed, and controlled. The ISO 9001 Standard refers to four kinds of resources: human resources (people), knowledge, processes environment, and infrastructures. Resources are responsible for critical areas and scopes of the realization processes. Therefore, their appropriate definition and control are expected.

Resources are regarded as support tools for the QMS to meet the organization's goals. Emphasis is given to the implications of not conforming to the management's requirements. This means that the determination and definition of the resources must be aligned with the objectives of the QMS. The resources must be competent. Competence must be evaluated. In order to evaluate something, you need to have a scale of measurement. In other words, the standard expects you to be able to measure the compatibility of your resources to the expectations of the QMS. Furthermore, outsourced activities are regarded as external resources that must be defined and, subsequently, controlled.

7.1.1 General

Let us review the ISO 9001 requirements:

- The organization shall identify, determine, and define the required resources needed for establishing, designing, implementing, maintaining, and improving a QMS.
- The organization shall provide these resources.
- While doing so, the organization shall assess
  - Its current process capabilities
  - Its professional and organizational abilities and constraints
  - The conditions that limit the organization
  This assessment will indicate whether the organization is able to provide these resources or not
- The organization shall decide which resources will be provided internally and which will be provided by external providers.

Determining the Resources for the QMS

Clause 7.1.1 (General) is regarded as clarification of the requirements regarding the provision of resources required for the realization of goods or services.

The manufacturer of a product or service is expected to analyze its needs for resources while considering certain organizational aspects. In other words, to sit, think, and examine which resources they need in order to produce the product according to the expectations, which resources are available, how can they be deployed, and what are the limitations of its resources. The standard presents principles to help us determine effectively the resources that will support the organization or service provider:

- Requirements: The resources are to support the organization in meeting customer and regulatory requirements.
- Quality policy: A correlation between the resources and the quality policy is essential. In the quality policy, the nature of the organization and its processes is determined, there is a commitment to meet customers' as well as regulatory requirements, quality objectives (or principles for their determination) are set, and scope of activity is determined. Now, it is time to identify the necessary resources.
- Quality objectives: The defined resources must support the achievement of quality objectives.
- Realization processes: Processes are a crucial constituent for realizing goods or services. The processes determine the amount, extent, and complexity of the resources.
- Improvement: The resources are to support activities necessary to achieve improvement of the QMS.

These are strategic aspects of the QMS. The standard expects specification of what resources are needed to assist and support in achieving these strategic aspects. The following types of resources are considered:

- Human resources
- Infrastructures
- Process (work) environment
- Tools and equipment
- Information systems
- Suppliers and partners
- Natural resources
- Financial resources

Review of Resources with Reference to Capabilities and Constraints

The resources of an organization enable its capabilities for maintaining the QMS; how the resources affect those professional capabilities and abilities of the organization. Constraint (internal or external) refers to those conditions under which the organization is active, that limits the organization, and that may prevent the organization from achieving its goals. An example of an internal constraint is the number of machines the organization deploys. An example of an external constraint is a regulatory requirement that impels the organization to execute certain actions. The capabilities and constraints to be considered are

- Professional
- Financial
- Technological
- Regulatory

Reviewing those capabilities and constraints will indicate which resources are needed, which are available, and which activities are to be performed by external providers. The results of the review may serve as inputs for the management review and for the planning of the quality objectives (clause 9.3—Management review).

In practice (although documentation is not required), I would integrate this review in the method for documenting the processes of the QMS (part of the process approach) as required in clause 4.4—Quality management system and its processes.

Providing Resources for Improvement and Innovation

When planning activities for improvement (clause 10), the organization shall review and determine which resources are required for achieving the goals of the improvement. Activities such as identifying the need for improvement, analysis of data, planning changes, and risk identification will determine the required resources such as knowledge, competence, infrastructures, and tools.

Assessing Future Resources

When reviewing the required resources, the organization must assess the need for future resources, including the needed competence of persons who will take part in the future plans. The assessment shall be based on an analysis of trends and future planning of the organization (remember SWOT and PEST analysis?). In other words, when you are planning something, whether it is an improvement, removal of a risk, or the development of an opportunity or a new product, refer to the requirement of the resources. This reference shall consider all five aspects: requirements for the QMS, quality policy, quality objectives, realization processes, and improvement.

Documenting the Resources

There is no explicit requirement for documentation of resources—you are not required to deliver a controlled list of your resources. The requirements for determination, definition, and documentation hide beneath the various standard requirements and are needed in order to answer other standard requirements. When thinking about it, the QMS includes many definitions and determinations of resources in various places and forms: quality plan, standard operating procedures (SOPs), work instructions, test instructions, user manuals, forms, and maintenance plans, where you mention and refer to relevant resources; just add a paragraph named "required resources." The quality plan is a good example; the plan must include references to required conditions necessary for the realization of the product—among them, necessary resources. Thus, the documentation of the needed resources. We will deal with the details while discussing chapter 8.1—Operational planning process.

7.1.2 People

The ISO 9001 Standard requirement is as follows:

- The organization is required to determine and provide personnel necessary for the effective implementation, operation, and control of its quality management system and its processes.

Human resource has a significant weight and effect on the realization processes. In order to verify and validate the determination and deployment of suitable human resources when planning a QMS, the ISO 9001 Standard requires identifying specific requirements for certain personnel or functions related to its QMS and the realization of the product. Which kind of requirements can we expect?

- Regulatory requirements
- Business requirements

In order to verify this, it is required first to define the following:

- Which are the functions or roles necessary for the operation of the QMS?
- Which qualifications do those functions or roles require?
- What are the necessary evidences?

Specific requirements and how to answer them will be discussed in detail in chapter 7.2—Competence. In order to cover this requirement (7.1.2), I recommend the following:

- On the quality policy, mention that all requirements for human resources are reviewed, identified, and allocated, and add a reference to the organizational diagram or any other documented information where all the functions of the organization are described.
- If you do have specific requirements for human resources, add reference on the documented information of your processes.
- If you need another type of documentation, you may use the job description.

In this way, you can prove that all the requirements were reviewed and people required for the realization of the product were identified and allocated.

7.1.3 Infrastructure

Infrastructures are the stock of the basic facilities and equipment needed for realizing a product or providing a service. Infrastructures should provide the suitable conditions and accessories to perform the appropriate business tasks and activities and assist in achieving the desired conformity of product and service requirements. Thus, it is strongly related to the product or service and has a direct effect on their quality. The basic goal of the organization is to ensure the provision, availability, and sustainability of infrastructures. Infrastructures include all the means, applications, interfaces, and facilities necessary for the realization of products or services from the design stages through its delivery and post-delivery activities. Let us take a look at the ISO 9001 requirements regarding infrastructures:

- The organization shall identify, determine, and define infrastructures necessary for
  - Effective operation of the QMS
  - The realization of products or services
  - Meeting customer requirements
  - Enhancing customer satisfaction
- Infrastructures include all applicable facilities that participate in the realization of the product: buildings, work spaces, tools, equipment, machinery, and associated utilities necessary for the realization of the processes.
- The provision of infrastructures shall include the resources and the necessary services needed to support and operate them: transport, maintenance, communication, and information systems.

In general, the requirements of the ISO 9001 Standard are to ensure the availability of appropriate infrastructures throughout the realization processes. However, besides the provision of the infrastructures, the manufacturer is required to maintain and take care of them in order to ensure appropriate operation of processes and to avoid the probability of nonconformities. Maintenance of the infrastructures is a necessary precondition for the preservation of processes' long-term capability, to ensure reproducibility of processes, and to guarantee the achievement of the product requirements. And what better way to reach such quality management goals than to identify, plan, and control. The control over the infrastructures shall reach all levels of process support. According to this basic rule, even equipment whose failure might not harm or affect the product directly, but will affect, for example, the organization's ability to supply the product on schedule, a customer requirement, must be controlled and maintained.

Identification of Infrastructures as Process Equipment

What is infrastructure? Infrastructure is a structure that provides a framework that supports the operation of a system. In our case, structures that enable the realization of products or services. For example, a production hall enables the manufacturing of a product, a communication system at a call center enables the provision of services, a warehouse enables the storage of goods, and a CRM system enables data management of customers. According to the ISO 9001 Standard, infrastructures include software (the collection of functions and programs that provide instructions for a unit for the operation of activities) as well as hardware (the physical layout of components or parts of a system). Internet is a good example; the Internet infrastructure includes the telephone lines, the television cables, the routers, aggregators, and repeaters that channel the information that goes through the net, but also all software used to operate and manage the different Internet subsystems. In other words, the standard makes it clear that both areas are included under the definition of infrastructures, the virtual as well as the material. The definition includes services that support software and hardware and assist them in meeting the specifications.

The first stage in initiating the control will be the identification of infrastructures relevant to the realization of the product or service. The objective is focusing only on infrastructures that support directly or indirectly the realization processes. I suggest here an effective way for identifying the relevant infrastructures:

- In the first step, you are to review the main process and its subprocesses that are related to the realization and are included under the QMS: capture of customer requirements, planning of products and operations, purchase of goods or services, acceptance of goods and services, storage, transportation through various phases of the material flow, and manufacturing and delivery activities to the end customer. The review is part of the application of the process approach and should provide you with a long list of processes and activities that use infrastructures.
- The next step will be to map and list all the infrastructures that are being used for these operations and activities. Customers, as well as national or regional regulations, may set requirements for infrastructures. This listing is very important to the later planning of the maintenance activities and prevention measures.
- Next you will analyze the relations between the processes and the infrastructures and indicate which parameters may affect the processes and quality of the product. Significance will be given to the effect of an infrastructure on a given process, and this is measured on various levels: process parameters, elements of infrastructure, operators and responsibilities, and stage of process. This is the time to ask yourself the next question: How can infrastructures adversely affect process parameters, goods or service characteristics, and expected outcomes?

The review may come up with various types of infrastructures:

- Basic infrastructures like water supply, air systems, and electricity
- Buildings, structures, work spaces, and halls
- Design and development work spaces and tools
- Storage facilities
- Working tools, equipment, accessories, and monitoring and measuring devices
- Production facilities and machinery
- Computers, information systems, record-keeping systems and servers, and process automation and management systems
- Means of transportation and distribution
- Communication channels
- Means of security and safety

IT, ERP systems, and other information systems are regarded as infrastructures. Consensus indicates that today's IT systems are integral elements of the realization process; these systems process data, information, process characters, or parameters related to the realization of goods and services, and thus should be included under infrastructures.

Examples for infrastructures are ovens, sterilizers, computer systems, ERP systems, CRM systems, labs, clean room controllers, CAD systems, operators and phone systems, HVAC, printers and copiers, air compressors, industrial vehicles, normal vehicles, forklifts, aisle pickers, lifts, motorized pallet jacks, trucks, cranes, security and alarm systems, and filters.

In practice, the following table may serve as a tool for the analysis and determination of infrastructures:

Process	Subprocess	Type	Related Infrastructure
Customer requirements	Determination of customer requirements	Communications Management of data	Operators and phone systems ERP system (customer master data module and order module)
Production	Packaging	Halls and buildings	Main storage
Preservation of product	Storage	Halls and buildings	Storage halls and environment controls
Delivery to the customer	Preparing goods for delivery	Equipment	Wrapping machine
	Delivering goods to the customers	Means of transportation and distribution	Truck

Evaluation of Infrastructures

After determining the required infrastructures, we need to evaluate their suitability to the QMS. The evaluation shall be conducted with respect to quality objectives and quality planning. The goals of the evaluation are

- To ensure that the infrastructures are intact, sustainable, and stable
- To ensure that the infrastructures will support the organization in achieving quality targets and plans
- To verify that the infrastructures will not disturb the achievement of objectives or reduce the capability of processes
- To identify areas and ranges for control
- To determine which controls are needed
- When needed, to implement improvements or to update the infrastructures

The evaluation shall be conducted with six parameters:

1. Suitability: The quality of having the properties that are right for a specific purpose. This parameter evaluates how appropriate an infrastructure is to its final purpose and how much it can support this purpose.
2. Security: This parameter evaluates the ability of an infrastructure to guarantee that an expected outcome will be met.
3. Reliability: The quality of being worthy of reliance or trust. This parameter evaluates whether an infrastructure is reliable while playing its role in the realization process.
4. Maintainability: The capability of being kept in good condition. This parameter evaluates how well can an infrastructure be maintained.

5. Efficiency: The capability of the output to the input of any system. This parameter evaluates the results of an infrastructure with comparison to its objective.
6. Safety: The state of being certain that adverse effects will not be caused by using the infrastructures under defined conditions and controls the impact of the infrastructure on the work environment.

There is no requirement to include the evaluation of infrastructures in the documented information but keeping records may assist you during audit in proving that you are conducting a methodic assessment of your infrastructures.

Effective evaluation may occur with the help of acceptance criteria: checklists that ensure specifications, allowable tolerances of equipment and machine performances, and user or supplier's instructions. This will be performed as a kind of gap analysis between the requirements and the state of the infrastructures. The output of this activity will be the controls that are to be implemented in your infrastructure referring to required activities, responsibilities, and required tools, equipment, and intervals. Another option is to order an external evaluation that assesses your infrastructures. For example, a company that assesses the electrical network in the production halls. This is preferable because it will be an objective evaluation. The auditor would definitely appreciate it. The following matrix is a suggestion for a documented evaluation:

Parameter	Details	Measures	Required Controls
Suitability	How suitable is the machine for the production of product X?	Suitability is ensured because the product was designed to be manufactured on this machine	Records of test during development of the product
Reliability	How can we ensure that no collapse will occur during the activity of the machine?	Reliability is ensured through a weekly inspection of the elements of the machine	Periodical inspection plan
Security	How can we ensure that the products produced by the machine conform to specifications?	Security is ensured through regular control of machine outputs	Regular quality tests of machine outputs (products or parameters)
Maintainability	How can we ensure that the machine can be maintained?	The design and construction of the machine allows a periodical maintenance	Periodical inspection plan
Efficiency	How can we ensure that the machine will produce according to planned results?	A periodical inspection of the machine ensures that the efficiency is controlled	Periodical inspection plan Regular quality tests of machine outputs (products or parameters)
Safety	How can we ensure that the machine will avoid disruptions or failures?	Safety measures applied to ensure the appropriate function of the machine	Periodical inspection of safety

If certain infrastructures require no maintenance activities, mention it on the evaluation: such infrastructures were identified and evaluated, and it was determined that no maintenance is necessary.

Integration of Risk Analysis and Safety Measures

The use of the infrastructures throughout the realization processes may pose risks to the users as well as to the products. Such risks will be identified and evaluated for their impact on the expected results. Risks may appear as

- Process failures
- Risks to human resources
- Harm to the product
- Deviation from expected results

When such risks are identified, proper controls and prevention measures will be applied. For example, machine start-up must be carried out according to manufacturers' instructions. Failure to follow these instructions may lead to production faults. Risks like this must be identified and appropriate controls are to be applied at the appropriate process stages. Risks to human resources are to be referred to as well. An out-of-the-ordinary example that cost me a note in an external audit was an ordinary ladder; during one of my external audits, the auditor demanded to see records indicating that a ladder was safe for use (?!). To tell the truth, it never crossed my mind to examine the ladder. But in fact this ladder was used daily to fill the machine with material and required a yearly maintenance according to the manufacturer's specifications (oiling and fastening of screws). Imagine that. But a ladder is a simple infrastructure. What about machines, electro systems, cranes, conveyor belts, forklifts, etc.?

Outputs of this risk analysis may be used as inputs for the required actions for addressing risks and opportunities as requested in clause 6.1 of the ISO 9001 Standard: Actions to address risks and opportunities. Each type of infrastructure can be reviewed according to the following discipline:

- The relevant risks must be identified.
- The impact of risks on the expected outcomes must be evaluated.
- The actions necessary for avoiding and reducing these risks must be planned and implemented.
- The effectiveness of these actions must be evaluated.

The necessary controls and preventions may be integrated into the maintenance activities. A plan for a solution of a problem related to the infrastructures is not required but may be implemented in the organization. The purpose of the plan is to define for the relevant parties or employees in different areas or departments what is to be done in case of a quality problem or failure concerning the infrastructures. The plan will refer to different aspects of the realization process—production, quality, storage, and distribution—and will provide details and instructions according to the case. Such a plan will be available at the workstations of the appropriate employees.

Here is a suggestion for such a plan (Figure 7.1):

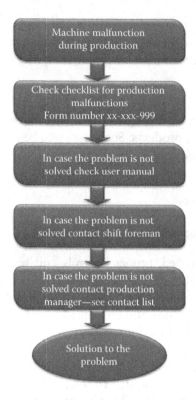

Figure 7.1   A suggestion for a problem-solution plan.

Implementing a Maintenance Plan

Before I start: there is no requirement for documented information regarding infra-structures. However, I propose here a plan for maintenance. Such a plan will assist you in fulfilling the standard requirements and will help in time of need—the need to investigate a failure. The concept is simple—any type of infrastructure that has a risk of failure, which may affect the integrity and quality of products or services as well as customer and regulatory requirements, will be submitted systematically for maintenance and prevention. The plan shall include definitions of responsibilities, maintenance activities, and prevention measures concerning the infrastructures. The plan will cover the following issues:

- The infrastructures required for control and maintenance will be identified and documented: machines, tools, systems, computers, servers, forklifts, and cranes. An identification method will be planned in order to identify infrastructures in a unique way: a number or a tag. The identification shall be visible on the infrastructure and will allow the responsible party to identify it during use and operation.
- Determination of maintenance activities and tasks including the definitions of intervals according to the specifications of the infrastructures will be done as a plan. The requirements and activities shall ensure stability of the infrastructure and the reproducibility of its processes over the long term. If water is supplied to the process,

periodical inspection of the water system is required to be planned (not the quality of water but the system that is delivering it).

- Assignment of responsibilities of appropriate functions and roles for actions and tasks will be determined.
- The plan shall refer to the allocation of necessary resources: services, spare parts, special tools and equipment, materials such as oils or lubricants, and external maintenance services.
- Necessary documents such as user manuals, technical instructions, technical schemes, or drawings shall be mentioned, including their location and availability.
- Inspections, controls, and examinations of the infrastructures will be planned and documented. These controls will verify that the activities are performed and validate (when needed) the results. When applicable, acceptance criteria will be defined. For example, specifications of machine or equipment provided by the manufacturer may serve as acceptance criteria for the comparison with the test results.
- The plan will initiate prevention measures through the identification of risks and the application of appropriate controls.
- When repair is required, it will be documented and revalidated.
- Required records and evidence indicating that maintenance activities and preventions that were performed and controlled will be defined.

As mentioned earlier, a suitable function or responsibility for maintenance and control will be assigned to each infrastructure. The documentation shall identify which department is responsible for which infrastructure elements (at either the corporate or site level):

- The IT department is responsible for all computers, IT networks, databases (including backup services and computer security), e-mail servers, program licenses and updates, antivirus software, user applications and passwords, interfaces between systems, etc.
- Maintenance is responsible for water infrastructures, forklifts, fire control equipment, electrical infrastructures, natural gas, transportation, HVAC, landscaping, grounds maintenance, pest control, disposal of waste, and recycling.
- Environmental health and safety is responsible for environmental issues and safety equipment related to the different areas in the organization.
- When the responsibility lies with a supplier, the activities will be defined, a responsible party will be appointed in order to evaluate the supplier's activities, and the supplier is obligated to provide you with the necessary recorded evidence.
- When distributors are required to maintain certain infrastructures or conditions, they will provide the organization with the appropriate evidence for these conditions.

The matter shall not be limited to the appointment of a maintenance person, electrician, or IT department but will spread down to the level of all employees using the infrastructures. For example, it will be defined for production employees how they shall safely use the infrastructures and what is required to be done at the end of each working day or job: turning off systems, cleaning instructions, daily maintenance, etc. The matter shall be integrated in the training program. Documentations may appear on a procedure (although this is not required), a form, a table, or a designated maintenance plan—whatever suits your organization and the nature of its infrastructures.

Please take a look at the next two tables. They depict the steps described earlier: Here's a suggested maintenance plan for a machine:

Internal Machine No.:	10	20	30
Manufacturer:	xxx	yyy	zzz
Type of machine:	aaa-bbb	aaa-bbb	aaa-ccc
Daily inspection and maintenance			
Hydraulic power unit	x	x	x
Safety devices	x	x	
Pneumatic facilities	x	x	x
Automatic central lubrication	x		x
Lamps and signals	x		x
Order and cleanliness	x	x	x
Weekly inspection and maintenance			
Programming unit	x	x	x
Fan filter	x	x	x

Here's a suggested list of maintenance activities:

Daily Inspection and Maintenance					
Internal Machine No.:		10			
Manufacturer:		xxx			
Type of machine:		aaa-bbb			
		Date	Name	Signature	Remarks
1	Hydraulic power unit				
2	Safety devices				
3	Pneumatic facilities				
4	Automatic central lubrication				
5	Lamps and signals				
6	Order and cleanliness				

However, today there are qualified information systems that provide you with the required planning of activities and follow-ups. Most of today's ERP systems provide such a maintenance module. It may be that your supplier has already planned a maintenance plan and provides it.

7.1.4 Environment for the Operation of Processes

The environment for the operation of processes (in the ISO 9001:2008 Standard identified as work environment) consists of the premises, sites, and locations where process-related activities (by employees or not) are being carried. I will refer to it from now on as "process environment." The process environment is supposed to provide the manufacturer or service provider with the optimal conditions for the realization processes and thus has a direct effect on products or services. The process environment includes not only the physical locations, but also the influence of equipment or materials used during the realization processes. The goal of controlling the process environment is to determine, define, direct, monitor, regulate, coordinate, and, when necessary, document the conditions of the process environment. It will be done by introducing the environment to predefined activities and controls. The activities cover specific areas, and their parameters are confined to specified limits or definitions.

The determination of process environment condition influences the characterization of tools, infrastructures, equipment, and human resources related to the realization processes. These elements are required to be synchronized in order to provide optimal qualifications to the process environment. Let us review the ISO 9001 Standard requirements for process environment:

- Process environment supports operations and activities related to the realization processes.
- Process environment shall ensure the provision of appropriate conditions required for the conformity of products, services, and customer satisfaction.
- Organizations will define, officially decide, make available, maintain, and preserve from failure or decline the process environment necessary for its operations.
- Parameters that influence process environment may include factors that affect the environment: temperature, recognition schemes, ergonomics, and atmospheric composition. Such factors may be social, psychological, or environmental.
- Besides the physical spaces or locations, social, psychological, and physical conditions are considered as process environment.

Environment for the operation of processes is the collection and integration of different conditions and influences (environmental, physical, ambient, and psychological) under which work and processes are carried out and that affect the process—a series of actions that realize or support the realization of goods and services. The environment is composed of

- The circumstances, objects, or conditions by which a process is carried out and process elements such as resources or outputs are surrounded
- The complex of physical, chemical, and biotic factors that surround a process and its elements such as resources or outputs
- The collection of social and cultural conditions that influence a process and its elements such as human resources

The ISO 9001 Standard requirements refer to all areas, spaces, and halls that serve the organization for operation and realization of products or services. These conditions are included under the following categories: physical elements of work spaces, factors that may influence the processes and activities, and resources that are necessary or operate activities in the work spaces. These three are measured using three parameters:

1. The ability of activities and operations to achieve expected results and product requirements according to customer satisfaction and regulatory requirements
2. The ability of the process environment to support the resources that operate business activities
3. The ability of the process environment to support the safety of products—the conditions of protection against consequences of failure, damage, error, accidents, harm, or any other event that could be considered as nonconformity

These will ensure conformity of products and services and customer satisfaction.

Processes Environment versus Infrastructure

Before we start to specifically decipher the requirements, I would like to start with distinguishing between infrastructure and process environment. When infrastructures include machines, the process environment is the area around the machine that is needed to realize products: a place for the worker to operate the machine, lighting above his head, the noise around the workstations that may affect the worker, and the area where the process outputs are being kept. Maintaining and controlling an infrastructure would require a periodic control plan—in our example a maintenance plan for the machine. Maintaining the process environment would require a control that certain conditions are being achieved—periodic inspection around the machine: cleanliness of the work space, sufficient lightning, appropriate storage conditions, and safe work environment. When infrastructures include storage halls, the process environment considers the conditions in which the goods are stored in these halls: temperature, humidity, safety, etc. Maintaining and controlling the storage halls as infrastructure will require building maintenance or electrical maintenance activities—ensuring that the building would not collapse. This would also require an environmental control system of temperature and humidity as well as an alarm system for the provision of safety. Examples of processes environments and spaces include laboratories, logistic centers, warehouses, clean rooms, production halls, stores, loading areas, meeting rooms, dressing rooms, and offices.

Definition and Determination of Process Environment

Defining the necessary conditions and applying the controls for process environment must cover all different areas and scopes of the realization processes. For example, storage of raw materials, parts, and components; aisles for transportation between departments and areas; manufacturing; and production halls for manufacturing and assembly activities. In order to reach and identify all the elements of process environment that may affect the goods, services, personnel, or processes, it is necessary to conduct a review, analysis, and assessment of all realization activities and processes and their outputs. The objective of the review is to understand the relationship between the product, personnel, and the environment in which they interact. This review of activities will be conducted as follows:

- Analysis of all employees' activities. The output is a list of all activities that are supported by work environment.
- In the next step, you will analyze behavior in and around workplaces or workstations. This is required in order to understand how the environment may affect activities and eventually the goods or the services.
- Now after understanding what influences are relevant to the process and may affect its outputs, you may define for each environment the conditions that must be controlled.
- In the next stage, you will define the necessary controls and the behavior around the work environment.

Another important aspect is the influence that environment conditions have directly on the products or services. The objective is to study the status of the goods and to examine which conditions may affect them and how they are integrated or influence the process environment. This aspect will be reviewed as follows:

- Sample of process output (a semifinished product, a subassembly, a product or a record of a service). The sampling will occur at various locations and stages of the realization.
- Examination of the product's characteristics and understanding the relation between them and the environment in which they are kept. The examination will recognize the environmental condition that may affect the products or the services.
- In the next step, you will define the controls necessary for the provision and maintenance of these conditions.

There is no requirement for the documentation of the reviews mentioned here. But their output will serve as inputs to the quality plan.

Review and Evaluation of Risks Related to the Process Environment

Process environment may cause a failure in the product when the required conditions are not provided and maintained. Actions necessary for addressing risks are referred in the ISO 9001 Standard in clause 6.1—Actions to address risks and opportunities. Let us discuss the principles in order to understand the effect of risk on the process environment.

The principle of risk management is evaluating situations that may set hazards to the realization of products and services at certain points, evaluating their degree, setting controls and measures for eliminating or reducing the hazards, and evaluating the effectiveness of these controls. Outputs of the risk review and evaluation shall be integrated in the planning actions for control of the process environment. I suggest here a sequence for identifying the related risks of the process environment:

- In the first step, try to identify the relation between the product and its surrounding environment. You must identify those elements of the environment that may affect the product and its quality.
- Then you must assess its extent and impact on the product; when the extent might pose a hazard or prevent you from delivering a conforming product, it is a risk.

- After identifying the risks for each element of the environment, you must find out which parameters in the environment must be controlled in order to alert you when risks emerge.
- When the parameters are known and clear, you may start to plan the appropriate controls and eventually implement them. The control must be proportionate to the extent and potential affect that the risks have on the intended outcomes.

Example: In the food industry, it is required to maintain a level of cleanliness in the production halls. Thus, the set of actions and controls are defined and implemented—from dress codes, where personnel must wear certain clothing items, to behavior procedures around production areas and automatic systems that control the environment.

While reviewing risks related to the process environment, you should cover the different aspects of the relation between the product and its surrounding environment:

- The identification of process environment conditions that may harm or endanger products, its components, parts, or materials. When referring to services—conditions in the work environment that may lead to the delivery of poor-quality service.
- Reference to factors and characteristics of the product that might influence or contaminate the process environment, such as emissions of toxic materials or the effects of products with an expiration date on the environment.
- Reference to the effects and results of disposal of the products on the environment.
- Reference to the influence of the process environment on the package of the products and its integrity.

Regulatory Requirements

Local, national, or international regulatory requirements may demand the establishment and maintenance of specific controls of your process environment. When such regulations are applicable and required, they are to be included in the planning of the controls. Try to review and identify under what relevant regulations you are realizing your product.

Conduct and Behavior of Personnel around the Work Environment

One of the most critical elements that affect the conditions of the work environment is the conduct and behavior of human resources. False actions may have adverse effects on goods or services. In order to promote employees' awareness of the issue, the organization must initiate actions and controls that will avoid such situations. In order to identify the activities and actions that may affect the conditions, the organization must identify all the roles and functions that have influence on the work environment on two levels: contact with the product and operation of realization processes and contact with the process environment elements that may affect its conditions. Functions and roles from all the realization processes shall be reviewed: production, transportation, storage, quality, service, maintenance, and cleaning. For each role or function identified, the following points will be addressed:

- When certain accessories are necessary for the maintenance of the process environment conditions, they will be provided. For example, dressing—clothing, gloves, hats, gowns.

- Work activities will be analyzed in order to verify that they are not creating a situation where the process environment is at risk.
- A set of activities and actions will be planned and defined to ensure that workers' behavior is appropriate around the workstations and their conditions are maintained. For example, during manufacturing processes, it is required to define how one shall handle the surroundings around the machines during and after any processes—material handling, cleaning requirements, and order requirements.
- If needed, the appropriate documentation and instructions necessary for the maintenance of conditions will be established: instructions for behavior in the work environment, dress code, cleaning instructions, and identification of nonconformities or potential situations that may create nonconformities. "If needed" means that you evaluated the environment and decided that for this particular environment documentation is needed in order to verify that conditions will be met.

After the identification of all relevant roles and functions, it is now time to define their training needs regarding the process environment. These needs will be integrated into the qualification and certification process of employees, and will be included in the training plan according to the specifications of clause 7.2—Competence.

Authorization and Access

In some cases, the safety of the work environment may be achieved by defining instructions for authorization and access to the work environment, sites, and locations: who is authorized and who has restricted access to or use of certain areas. The matter applies to external as well as internal users. External users may be customers, suppliers, auditors, or even employees from other departments.

For example, regarding manufacturing halls, the manufacturer may define who is authorized to go into the halls and who is not. Otherwise, you may find unwelcome guests. And when unwelcome guests are found, it should be defined what is to be done in order to verify that the defined conditions for the realization are still applied. The types of persons who could be moving around the facilities include the following:

- Internal: Manufacturing personnel, supervisors and productions managers, material handlers, storage workers, design engineers, quality control or assurance, management and administration, and service and maintenance personnel
- External: Suppliers, service providers, subcontractors, visitors, auditors, and customers

In order to manage authorization and access effectively, you may define and implement a method. A good example is an authorization system—a system that is normally implanted in the premises of the organization. Such a system manages the access of certain personnel to certain areas.

Setting Parameters for Control over the Work Environment

We reviewed and understood the different elements that may affect the process environment:

- We distinguished the process environment from the infrastructures.
- We have identified elements of process environment that are to be controlled.

- We have discussed the integration of the different inputs related to the process environment (risks, regulations, authorization).
- We have set the requirements for the human resources conduct and behavior in the process environment.

Now we need to discuss the crucial conditions that are needed to be controlled, which parameters are to be controlled, and how. These conditions are supposed to preserve the product, its functionality, its safety, and eliminate failures that may result in an adverse effect on the product. These conditions are derived directly from the characters and type of the product and will be controlled and verified through parameters that affect them.

Environmental conditions are the definitions of several parameters creating a situation or a status in a certain location to which the product and the workers are exposed. During the realization and through processes and activities, the product is subjected to various conditional stresses that may affect it. We need to identify these and control them. Determination of the conditions allows you to detect in advance which activities or lack of activities may create unwanted conditions that increase the probability of no conformity. For example, insufficient lighting in a production hall may affect the ability of employees to detect defects during production. As another example, for electronic components, levels of static electricity are harmful and it is very important to provide an environment free of electrostatic discharge. When such parameters are applicable to your organization, you will need to plan the suitable controls.

I would start by evaluating process environment conditions that may affect the product and its characteristics or employees that operate a process. The evaluation will cover the entire supply chain from the storage activities of raw materials and parts until the storage of finished goods. The evaluation shall detect extreme conditions that the products may be exposed to:

- Examples of environmental conditions that may affect product include air quality, aerosols, vacuum and pressure changes, temperature and humidity, magnetic fields, external electrical influences, electrostatic discharge, clinical storage requirements, dust and particles, and ESD (for electronic components).
- Examples for conditions that may affect employees include lighting, noise, temperature, safety, acoustic conditions, and work space layouts.

I suggest here a method for analysis:

- Review all the environmental inputs and requirements such as product, packaging requirements, human resource, and risks.
- List all the related processes and activities.
- Examine the flaws of the material throughout the realization processes and inspect all the outputs of the processes.
- Involve all the participants of the processes in order to cover all aspects and receive all the needed information.
- Determine at each phase what the required conditions for the process environment are.
- Plan and implement tools that will record the status of the conditions.
- Control the conditions through analysis.
- Maintain documented information of the process when you find it necessary.

The tool for the records of the condition should collect data regarding the process environment during the realization stages. It can be a form that an employee fills up or it can be a sensor that writes data every second of the day on a big screen with a beautiful graphic interface with lots of greens and reds. The function is exactly the same; data regarding the condition—good or bad—are collected and analyzed. The extent of the control will be set according to the potential effects that a condition has on the product or service. In other words, if failure to achieve the required conditions causes nonconformity or lacks in terms of customer requirements, the degree of control will need to be higher.

Defining and Planning the Controls

After understanding the requirements regarding process environment, identifying, and determining what the relevant process environments are and understanding the relations between the realization of goods or services and the process environment, it is time to set the appropriate controls. The objective of the controls is to ensure that all optimal conditions are being achieved while realizing goods or services. When setting the controls, you should refer to the following issues:

- When measurable conditions are controlled, like temperature or humidity, you may need to define acceptance criteria. The criteria will relate to the fluctuations of these parameters and to limits and tolerances.
- In some cases, it will be necessary to integrate the controls of the parameters in the validations of the product's provision (as required in clause 8.5.1) in order to prove that the process environment's conditions were as required during processes. For example, in sterilization and cleaning processes, it is necessary to ensure that environmental conditions at the manufacturing site achieve predefined criteria. This is one of the process validations. Through the control of these parameters it can be guaranteed that products were produced and packaged in a qualified, controlled environment. Other examples of environmental parameters, indicators, and controls are cleanliness of work surfaces, air filtration, pressure, and airflow.
- Psychological effects play an important role in the process environment. Agronomic and psychological factors affect the human resources and may prevent them from operating the process as expected. For example, long-term exposure to noise may result in distractions and disturbances, and certain shades of light may result in false identification of nonconformities. Such distractions may lead to inattention to the process. If preventions or accessories for these matters and issues are necessary, the controls will verify their availability.
- Environment conditions that may affect the packaging are to be taken into account. Here, I refer to conditions that may affect the integrity and quality of the package. For example, high temperatures may affect the material the package is made of and reduce its withstanding quality or even harm it.
- The use of tools and equipment necessary for the control shall be specified. The specification shall relate to safety measures of the environment as well as the personnel that operate it.
- When controls of the process environment are required by distributors of the goods, the requirements shall be communicated to them and they will implement them. The supplier is required to prove to you that these controls were implemented.

The next stage is determining the means and tools with which you will implement these controls and measure these parameters:

- Control devices—monitoring tools, sensors, particle counters, devices to measure temperature and humidity, and gas filters
- Instructions—work instructions, procedures, checklists, forms, and environmental review instructions
- Tests—the verifications and validations that the parameters were measured according to the definitions and that the required conditions were achieved through demonstration of results

The instruments for control will be planned to support and provide appropriate data regarding the conditions and in proportion to their potential effect on the product. For example, it is not enough to set a work instruction at the workstations; it is also required to verify that the relevant parties are aware of it and practicing it.

7.1.5 Monitoring and Measuring Resources

Monitoring and measuring resources are considered by the ISO 9001 Standard as key process elements of the realization process; measurements and monitoring activities are required to ensure that products meet specifications. The goals of controlling the monitoring and measuring devices are

- To control the monitoring and measuring devices
- To ensure that measurement and monitoring activities are planned according to requirements and are performed respectively

As always, we will begin with reviewing the ISO 9001 Standard requirements:

7.1.5.1 General

- The organization will define, officially decide, make available, maintain, and pre-serve from failure or decline monitoring and measuring devices that are necessary for its operations and needed to verify conformity of process outputs to product requirements.
- The organization shall ensure that the monitoring and measuring devices are suit-able and acceptable for their use activities and purpose.
- The organization will maintain the monitoring and measuring devices to ensure their continuing fitness for their purpose.
- The organization shall retain documented information as evidence, which shows that monitoring and measuring devices are fit for their purpose.

7.1.5.2 Measurement Traceability

- When traceability of measurements is a requirement or is considered by the orga-nization as significant for proving confidence of measurements results, monitoring and measuring devices shall be calibrated or verified (or both) prior to their use.
- Calibration and verification shall take place at specific predefined intervals.

- The calibration will be traceable to national or international standard measurements. In cases where such standards do not exist or are not applicable, the calibration and the basis for the calibration process (physical or calculating) shall be defined, justified, and recorded by the organization.
- Each of the monitoring and measuring devices shall be identified in order to determine the calibration status, validity, and intactness.
- The monitoring and measuring devices shall be safeguarded and protected in order to eliminate any situations or conditions that would invalidate them.
- When a monitoring and measuring device is found inappropriate for use, its calibration proves to be inaccurate, and measurement results has been adversely affected, the organization shall evaluate the validity of the accepted results, along with the ability of any products that were released with this device to meet their requirements.

Determine the Appropriate Monitoring and Measuring Devices as Resources

Clause 7.1.5 lays out the requirements and defines the controls for measurements and monitoring tools and instruments. These resources measure and monitor characteristics of process outputs that have an effect on the quality of the final product and are necessary to ensure that they meet their own specifications. The controls mentioned in clause 7.1.5 will ensure their accuracy and precision. The first requirement refers to the definition and determination of the needed monitoring and measuring devices. The goal is to ensure their fitness for purpose; providing valid and reliable results during monitoring and measurement activities necessary for the conformity of products and services to requirements.

How do we know if a device is considered as monitoring and measuring device and must be controlled? When the device measures a product characteristic that is relevant to its intended use and is needed to verify conformity to product requirements, it means that this device has some responsibility regarding the quality of the product. Each of such devices must be controlled—identified, maintained, preserved, and calibrated. The goal is to ensure that the device is able to perform a series of tasks.

Types of Resources for Monitoring and Measurement Devices

The different types of monitoring and measuring devices require different types of controls. If we take the classic example of a scale, the required controls are pretty obvious; a yearly maintenance and calibration usually done by a certified lab. The record of this control is a documented certification received with the device when the control is done. But there are normally other devices that are laying around in the organization or other methods for monitoring and measuring that are used to validate and verify processes and process outputs that do not have clear and defined controls. But still they must be controlled and verified before their initial use. For example, in the injection molding industry, each batch is normally required to be compared to a master product that was approved by the customer; as production of a batch starts, the quality assurance takes some samples and compares them to the approve master. This master is a kind of measuring device. If the customer delivers you a device to control the function of a product that you are manufacturing, it is also a kind of measuring device.

Guidelines for Controlling Monitoring and Measurement Resources

I would like to start with some guidelines and principles for determining monitoring and measuring devices. These guidelines shall assist us in planning the relevant activities and controls:

- Measurement and monitoring needs: First of all, we must identify how a process or a product will be reviewed, and which of the measurable characteristics will be measured. By identification, I mean the inputs to be considered when reviewing a process or a product, that is, product characteristics, customer requirements, regulatory requirements, intended use, risk management, safety, and performance specifications.
- Definition of monitoring and measurement activities: After identifying what is to be measured, we will define how the measurement activities will be carried out. We may refer to issues such as instructions, frequencies, areas, measured materials, products, and processes. Design and development outputs may serve as inputs to the definition.
- Determination of the appropriate criteria: The requirements are defined and so are the activities. The next step will be to define how the manufacturer set the criteria for measurement results; for example, tolerances, limits, expected test results, customer requirements, and regulatory requirements. When reviewing the criteria, one shall refer to relevant standards for comparison or assessment of the results.
- Allocation of tools and equipment: The following step will be to define and allocate devices and accessories needed for the measurements. Importance will be given to the appointment of a device to the predefined criteria; the device shall be able to provide the expected results.
- Identification of devices: After identifying which devices and equipment are necessary, we must plan a method for their identification, for example, names, types, serial numbers, catalogue numbers, and relevant processes and locations.
- Required qualification for personnel: Employees using the devices must know how to operate the devices and/or equipment correctly. It is critical for the control of processes and expected results. Thus, it is required to define the suitable responsibilities for monitoring and measuring activities. This matter will be referred when applying controls for the requirements of clauses 7.1.6—Organizational knowledge and 7.2—Competence.
- Controls: It is necessary to define how the required controls and validations of such tools and equipment are determined; for example, through calibration activities (external and internal), maintenance, storage, and handling.
- Nonconforming equipment: Actions shall be clearly defined to handle tools and equipment that were disqualified. The issue shall also cover the disposal of such devices.
- Records: The organization shall retain records as evidence, which shows that monitoring and measuring devices are fit for their purpose.

Maintenance of Monitoring and Measuring Devices

The ISO 9001 Standard specifically demands that you maintain the monitoring and measuring devices; preserve their fitness—the quality of being suitable or of being qualified to perform measurements and deliver reliable results. Bear in mind that the objective of maintenance is to protect a device's

- Intactness and functionality, including deterioration over time
- Status of calibration—the device shall be protected from any cause that may alter or damage its calibration status

Maintenance of monitoring and measuring devices refers to the following issue:

- Storage and handling of the monitoring and measuring devices
  - Status: The maintenance will initiate actions to preserve a certain state of the monitoring and measuring devices.
  - Safety: The maintenance shall keep safe and protect the monitoring and measuring devices from damage, harm, decay, loss, or deterioration.
  - Adjustments: The maintenance shall ensure that no adjustments that would invalidate the measurement results will happen to the monitoring and measuring devices.
- Preservation
  - Calibration: The maintenance shall ensure that the calibration of the monitoring and measuring devices shall be preserved.
  - Service: The maintenance will initiate (where applicable) service activities needed to preserve the monitoring and measuring devices.
  - Precautions: The maintenance will indicate which precautions are required to preserve the monitoring and measuring devices.

How to practice the maintenance?

- Risk-based thinking: You may conduct a small-scale risk analysis and identify the risks regarding monitoring and measuring devices. What should the risks analysis consider?
  - Adverse effect on calibrations
  - Damage during use, transportation of storage
- Protections: When determining the necessary protection, you shall refer to parameters that may affect the device and its calibration status, for example, temperature, humidity, transportation, operation, or storage. The extent of the protection depends on the device itself.
- Documentation: Where needed, special instructions will be produced in order to ensure maintenance.
  - Instructions of how to maintain the monitoring and measuring devices
  - Instructions of how to protect the monitoring and measuring devices
- Training: Where needed, employees and personnel will be trained in how to use and handle the monitoring and measuring devices.
- Maintenance plan: When the monitoring and measuring devices need a periodical maintenance, it will be documented and records of the maintenance activities will be retained as evidence that
  - The activities were performed
  - The results prove that the monitoring and measuring devices are fit for their purpose

Usually, the manufacturer of the tool provides instructions regarding the protection of the device. You may use these or refer to them. If such instructions are not provided, you may contact the manufacturer. In any event, you are required to prove the ability to protect the monitoring and measuring devices, their intactness, and their calibration status. The easiest example is the caliper gauge; it must be kept in its pouch, protected from mechanical damage like bumps or shakings, and usually must be calibrated once a year. There is a requirement to maintain documented information of actions taken to maintain the device. The documented information shall prove that

- Employees actually handle it with care (trainings, instructions)
- It is calibrated once a year (certificate of calibration)
- The results of the calibration allow the use of the gauge (certificate of calibration)

Identification and Traceability of the Devices (Resources)

Traceability of monitoring and measuring devices refers to the ability to link process outputs with the devices that verified or validated them. If a manufacturer produces a part with a serial number and this part was measured with an identified monitoring and measuring device, somewhere it must documented (with documented information) that this device measured this part with this serial number. This type of traceability is practiced with components, materials, or process elements as they may also affect the quality of the product.

To begin with, each device shall be identified, and the identification shall be clear and available for people to see. For example, the identification may appear on the device itself—an internal number, manufacturer's serial number, location, or a responsible party—according to the nature and type of device. The identification is important for the traceability of products that were tested and approved with the device.

Calibration

Calibration is comparing an instrument's accuracy to a known standard—for example, the length of a ruler A can be calibrated by comparing it to a standard ruler B, which has a known length and was tested and approved according to a predefined standard. Once the reference of ruler A to the standard is known and determined, ruler A is calibrated and can be used to measure the length of other things. Calibrated tools and equipment are necessary to achieve product quality objectives and ensure that processes will meet their requirements. Products or services (or realization processes) have measurable characteristics that must be controlled, verified, or validated. Through a set of operations that may affect the ability of the device and its measuring results, the calibration ensures that measuring devices conform to the requirements of their intended use. Each monitoring and measuring device has several characteristics that determine its performance and the influence of the results that it generates. The calibration and the set of operations will ensure that a device will conform to the requirements for its intended use. These activities usually include

- Calibration activities—a set activities needed to prove the ability of the monitoring and measuring a device to achieve its intended results
- Verification—comparing the results of the test to the applicable criteria
- Necessary adjustments or repairs in case the monitoring and measuring device does not answer its requirements
- Subsequent recalibration—for reverification
- Labeling—indication that the device is approved for use

The calibration of tools, monitoring, and measuring devices shall be traceable to predefined criteria. The criteria may be either international or national measurement standards, internal standards, or a predefined and documented basis established by the manufacturer, customer, or a regulatory requirement. This is an ISO 9001 Standard requirement. Where required, the results of the calibration shall be traceable to international standards. This is normally (and effectively) achieved by sending the devices to certified calibration laboratories that have the knowledge and the tools to perform, correct, standardize, and calibrate devices according to international standards. When calibration is performed externally, the organization will receive a document indicating the standard to which the tools were calibrated. This is the evidence for traceability to such a standard.

When such international or national standards are not applicable, and the organization is calibrating according to an independent or internal requirement for calibration, you are required to document these calibration requirements as reference. If you are using a gauge delivered by the customer in order to verify a part's characteristics, you need to establish a master or a reference part for the calibration of the gauge before each batch. In this case, it could be a product that the customer approved as a master product. By testing the gauge, you can ensure that it meets its intended use. I recommend you to document this issue (e.g., on the test instructions for the part). The supplier's or manufacturer's instructions may serve as calibration requirements. Bear in mind, however, that these instructions must be controlled.

Calibration and its verification will be performed prior to the use of the measuring device. That will ensure that only calibrated and verified devices were used to verify or validate goods, services, or processes. This means that if you introduced a gauge for the measurement of product characteristics, you may not use it as long as it is not submitted to calibration (internal or external). This way you prove by showing that the date of the calibration is prior to the date of the initial use. If it is a new gauge, make sure that it is already calibrated.

The records of the calibration are to be defined and maintained. The records shall cover the following details:

- A unique identification of the device tested
- A description of all the measurements and calibration activities that were performed for a single device
- Details regarding the measurements, that is, the date, location, conditions (if required), the person that performed the tests, and the tools and equipment used during the measurements
- A specification of the standards or criteria
- Reference to documentation, such as instructions (if applicable)
- The results of the measurements
- The status of the device
- Corrections, modifications, maintenance, or adjustments (if made)

Records of these actions must be maintained and shall be regarded as documented information and submitted to the control of documented information as required in clause 7.5—Documented information.

The following table suggests a method for documenting the calibration:

Internal Number	Description	Size/Type	Calibration Int./Ext.	Calibration Interval (Months)	Status	Last Test	Next Test	Location
C04A	Digital caliper	0–150 mm—digital	External	12	Calibrated	10/10/2015	10/10/2016	Quality Dep. Cabinet1
C05A	Caliper	0–600—mechanical	External	12	Not calibrated	23/05/2013	N/A	Quality Dep. Cabinet1
C06A	Depth gauge	300 mm	External	12	Calibrated	10/10/2015	10/10/2016	Production
C07A	Control gauge for product AX000234	Part no. 90012343	Internal	6	Calibrated	01/01/2015	01/07/2016	Quality Dep. Cabinet3

Controlling the Status of Monitoring and Measuring Resources

The objective of the control of monitoring and measuring devices is to provide information and determine the status of the monitoring and measuring devices—indicate whether the monitoring and measuring device is

- Calibrated and permitted for use
- Not calibrated and not permitted for use
- Permitted for use, but under restrictions

The control will manage the tools according to various parameters. While planning the controls of monitoring and measuring devices, you can refer to the following aspects (when applicable):

- Category: Try to sort the monitoring and measuring devices according to a family type; that is, according to use, range, department, process, or product.
- External or internal calibration: Define for each monitoring and measuring device its external and internal calibration requirements.
- Location: Document where monitoring and measuring devices are stored and under whose responsibility. When the monitoring and measuring devices are used by more than one party in the organization (e.g., by different departments), it is recommended that you manage their locations in order to maintain sufficient traceability.
- Calibration interval: Define for each monitoring and measuring device its calibration interval. This may be determined according to time periods or the number of uses; for example, for every 100 batches, the device must be calibrated. If you do not know what the interval is, try to look for them on the manufacturer instructions. This is an ISO 9001 Standard requirement.
- Calibration status: Indicates whether the device is calibrated and approved for use.
- Owner: You may receive monitoring and measuring devices or gauges from your customers in order to perform measurements on special products. In this case, you are required to document them under property belonging to customers.

The main idea is to manage a list of your monitoring and measuring devices. Choose the way that is most suitable to you (e.g., Excel chart, a form, or ERP system).

Advise: Some organizations may own many control and measurement devices that no one uses; these devices are either old (were purchased by previous employers) or are not relevant for the production anymore. However, the reason is that they should always be calibrated before the external audit because it is a standard requirement that all monitoring and measurement devices in the organization related to the realization must be calibrated. This results in high expenses. Here is my suggestion for handling devices of this kind:

- Separate the used tools from the unused.
- For the unused, define a status as unusable (not necessarily not calibrated).
- Wrap or pack them in a way that any tampering or use of them is immediately noticeable.

In time, you should be able to reduce the number of control and measurement devices without failing to meet the requirements.

ISO 10012-1 and ISO 10012-2

I include here a reference to the ISO 10012-1 and ISO 10012-2 Standards. These are standards for guidance on developing, planning, and implementing effective measuring equipment and measurement processes. The basic purposes of the ISO 10012-1 and ISO 10012-2 Standards are how to

- Establish a measurement management system for planning, implementing, and managing the monitoring and measuring processes and activities
- Determine the level of control needed for the effective execution of implementation

It is not required that you implement the ISO 10012-1 and ISO 10012-2 Standards, but it is suggested that you consult them when planning your monitoring and measuring activities.

Process Assessment

There is another type of monitoring and measurement that must be discussed: process assessment. It refers to the control of monitoring and measurement devices like software. During process assessment, evidence is collected with the objective to evaluate how well a test or a control has been carried out. Let us take, for example, software that is used to monitor and measure processes. You are required to ensure the fitness of software to reach its purpose in a methodic objective manner. A software is a set of actions that receive inputs, process these inputs, and deliver intended outputs. We cannot take the software and calibrate it. Therefore, when a software is used to monitor and measure processes or products, we must prove its effectiveness through assessment.

What is process assessment and how does it refer to monitoring and measuring devices? Process assessment approves the capability of a system or activity to deliver process attributes. Process assessment is needed for identifying critical measures of performance and applying controls associated with those measures related to a set of processes. A set of sequential activities are used to monitor or measure a process. The result of the assessment is receiving a detailed summary regarding the factors that enable the decision whether a process or process outputs has reached its objectives. When planning assessment of a process, keep in mind the following parameters:

- Clear description of the process and it purposes
- Input required for the assessment
- The action for the assessment
- The expected results (of your assessment)

I suggest surveys as a tool to evaluate the effectiveness of a set of processes. The survey performs basic steps with the objective of demonstrating the level of effectiveness. The basic steps to initiate a survey are

- Defining methods for the evaluation
- Defining the point of a process when the evaluation is needed
- Collecting data
- Analyzing the data
- Evaluating the data and deciding whether performance is sufficient or not

Validity of Previous Measurement Results

Another aspect of the traceability of the monitoring and measuring devices is the ability to undertake appropriate measures when a monitoring and measuring device was found not suitable for use in retrospect. Monitoring and measuring devices play an important role in the realization of products or services—according to the measurement results, it is decided whether the product is released or not. However, you must prepare yourself for the possibility that monitoring and measuring devices will be found to be defective, inaccurate, with malfunctions, or not calibrated. This means that there is a possibility that a quality problem, a fault, or a nonconformity has not been detected in time and that products that were released, based on measurements made with the defective devices, do not meet their requirements and specifications. When such a situation occurs, the organization must evaluate the effect of the inaccurate measurements on the quality of the product:

- It is necessary to trace back all the products that were tested with the defective device or all the products that are related to a process controlled with the defective device. The traceability and link between the device and the process outputs will serve you here.
- It is required that you assess the effect of the measurements on the quality of the product and whether a potential nonconformity could occur as a result of the improper measurements.
- In case it is found to be as a nonconformity, it is required to specify which measures must be applied, for example, recall of a product for remeasurement, issuing of an advisory notice, a sample remeasurement, maintenance or service actions, and so on.

The matter shall be recorded and submitted to the processes and controls as suggested in clause 8.7—Control of nonconforming outputs.

Records as Documented Information

Records and evidences for the fitness of the monitoring and measuring devices shall be retained as documented information. Generally, all of these activities must produce records that prove the fitness of the devices. Which types of records may we encounter?

- Lists of approved, disqualified, external, or internal monitoring and measuring devices used by the manufacturer
- Records of details of the monitoring and measuring devices
- Records of internal or external calibration activities, including results of measurements
- Certifications of monitoring and measuring devices according to specific standards
- Lists of approved laboratories for calibration of monitoring and measuring devices (can be eventually included in the list of approved suppliers)
- Records of service and maintenance activities applied to the monitoring and measuring devices, including repairs or adjustments
- Records required to maintain traceability of monitoring and measuring devices
- Records indicating the validity of previous measurement results

I would like to expand about records of internal or external calibration activities. The records of calibration and evidence that the monitoring and measuring devices meet their requirements shall be regarded as documented information and will be submitted to the control of documented information as required in clause 7.5—Documented information. The records of calibration shall carry the following details (if necessary or applicable):

- Description and unique identification of the equipment, including information for its identification such as manufacturer, type, model, serial number, etc.
- Date on which the activity was carried out and completed.
- The interval required for the calibration. This interval will be used to schedule the next test.
- In the case a procedure or protocol is required for the calibration, it will be mentioned.
- A list of any calibration certificates and reports, and other relevant documents.
- Maximum permissible errors where applicable. These may be determined by the person who performs the calibration or by reference to the measuring equipment manufacturer's specification.
- Relevant environmental conditions for the calibration.
- Identification of the person that conducted the test. Here you may include its competence.
- The results of the calibration obtained after the activities were carried out. If it is required, the results will be compared to the result before any adjustment, modification, or repair.
- Details of any maintenance, such as adjustment, repairs, or modifications, that were carried out.
- A statement about any corrections if necessary.
- Uncertainties if they occur during the calibration.
- Requirements for the intended use of the monitoring and measuring device in order to keep the device intact and fit.
- Evidence of the traceability of the calibration results.

7.1.6 Organizational Knowledge

Using knowledge as a resource and setting the knowledge of the organization in a knowledge base and allowing access to the relevant parties—this is the main message of clause 7.1.6—Knowledge. The ISO standard (finally) gives the much deserved importance to knowledge. In the eyes of quality management, knowledge promotes the main goal: to improve processes by predicting, preventing, and avoiding nonconformities and maintaining customer satisfaction through the use of knowledge. This is a new concept of the ISO 9001 Standard; until now, you were required "only" to determine which resources are necessary for maintaining the QMS effectively.

The knowledge that is required for realizing the product is considered as a resource. The ISO 9001:2015 refers to knowledge as a critical resource and requires a clear definition of processes or actions needed for identifying, obtaining, sharing, protecting, and maintaining knowledge that is necessary for the effective operation of the processes and making it accessible to the appropriate parties at the appropriate time and place. These actions should enable management of the knowledge. As always, before we plunge into the requirements of the ISO 9001 Standard and try and unfold them, let us review them:

- The organization shall decide which knowledge is necessary for the operation of its processes.
- The organization shall decide and determine which knowledge is necessary for obtaining and assuring conformity of products and services.

- The organization shall maintain that knowledge.
- The knowledge will be established in an existing state, available to relevant parties as necessary.
- The knowledge will be preserved and protected from decline.
- When changes to the QMS, its environment, or business activities are considered or addressed, the organization shall assess
  - The implication on the relevant knowledge base and the changes of the knowledge will be controlled in a systematic manner
  - The requirements for new knowledge necessary to support the changes or the new trends, its acquirement, and accessibility
- Note 1: Organizational knowledge
  - Includes the information that is used and shared to achieve the organization's objectives
  - Is considered as knowledge that is used internally by the organization and
  - Is generally acquired by experience
- Note 2: Organizational knowledge can be based
  - On internal sources (e.g., intellectual property; knowledge gained from experience; lessons learned from failures and successful projects; capturing and sharing undocumented knowledge and experience; the results of improvements in processes, products, and services)
  - External sources (e.g., standards, academia, conferences, gathering knowledge from customers or external providers)

Quality management has defined goals of meeting customer requirements and enhancing customer satisfaction. Knowledge as a resource must support these ultimate goals. The organization must identify which are the relevant practices, managerial activities, and technical considerations to be reviewed when managing knowledge. I claim that an organization will increase the effectiveness and efficiency of its QMS when it will manage its knowledge. Because it is a new discipline for the ISO 9001 Standard, I decided to promote in this chapter the discipline of knowledge management as I find the two disciplines (quality management and knowledge management) congruent. Today's business environment can very well support the knowledge management both ideologically and technically.

Identifying and Defining the Necessary Knowledge

But let us start from the beginning—in order to manage knowledge in the organization, you must first identify which knowledge or types of knowledge are required for the operation of the process and business activities. The standard is pretty much clear to which type of knowledge you shall refer to as necessary:

- Knowledge for the operation of its processes
- Knowledge for obtaining and assuring conformity of products

As a rule of thumb, the necessary knowledge should assist individuals in turning process inputs into expected outputs; whether it is a subprocess, an internal service that one department provides, or the end product delivered to the customer. This kind of knowledge can include work methods, instructions or trainings or experience—all the applications that a worker uses to operate an activity.

Practically identifying the necessary knowledge will be done through mapping, categorizing, indexing, and evaluating organizational knowledge assets. I suggest the following approach—while determining the processes needed for the QMS and their application (clause 4.4.1), you may include a reference to specific required knowledge per process or activity.

Identifying Future Knowledge Needs

An organization must identify future knowledge needs that will be necessary for the operation of its systems:

- If the organization is developing a new product or a new service, it must make sure that the relevant knowledge will reach the appropriate people.
- If the organization is aware of upcoming structural changes, it must ensure that future people in future roles will obtain the necessary knowledge. The classic example is a training program.

Leadership, Roles, and Responsibility

Success of knowledge management is heavily dependent on top management support. Knowledge management introduces a new way of sharing information. This new concept requires people to share knowledge that until today was considered almost private, and certain employees that "enjoy" a certain amount of prestige, due to their knowledge and experience, will find it hard to "share their prestige." In order to overcome this obstacle, you must present the employees the benefits to the organization from sharing knowledge (e.g., unity of purpose). This is where leadership comes into play (remember clause 5.1—Leadership and commitment?).

Capturing and distributing the knowledge should be included under certain managerial activities. The conclusion of this is that activities for managing knowledge are considered as management of resources and have a relation to the realization processes and an impact on the quality of products or services. This is why the ISO 9001 Standard expects that such roles, responsibilities, and authorities be assigned properly.

Maintaining Knowledge

The organization must demonstrate its ability to maintain knowledge and apply the necessary activities (normally with technology but not necessarily). We maintain the knowledge because we want to create an environment that shares knowledge with the relevant parties. For this, we must preserve the knowledge and be able to offer it in an appropriate manner to interested parties. Each area in the organization uses different kinds of knowledge that appears on different types of media and is distributed using different methods. The process and activities characterize the way knowledge is maintained. How do we maintain knowledge? Maintaining knowledge is establishing a set of activities that bring the knowledge to an existing state. In other words, maintaining knowledge is the ability to

- Capture the knowledge
- Describe the knowledge (metadata)
- Categorize the knowledge

- Store, retrieve, and archive the knowledge
- Process the knowledge
- Distribute the knowledge
- Update the knowledge

System for Maintaining Knowledge

The ISO 9001 refers to knowledge that is necessary for the operation of processes and elements that make up the QMS. A system consists of many elements that maintain interactions between each other and perform various activities. Each of the activities requires inputs and generates outputs. Each element has its objectives, but they all have some mutual goal, and each activity requires a certain degree of knowledge for its operation. In today's environment, business processes are becoming more complex and dynamic. Daily activities require a high level of skills and expertise. What was done manually 10 years ago is being replaced and automated by systems that require knowledge for their operation. In the (not-so-far) past operations, we were relying on data and information management in order to achieve goals. But today it is not enough and employees are required to obtain knowledge in order to operate a higher level of activities. And there is another challenge; the amount of knowledge available on any subject today is increasing to a level that requires a systematic management. Lack of such a system for maintaining knowledge will cause the loss of knowledge in the organization. This is why a new set of management is required in the organization—knowledge management.

Another aspect is the technology; the current technological and economic changes affect the factors that influence the QMS. Areas and functions in an organization are becoming more borderless. These changes bring a new factor into the equation—knowledge.

Making Knowledge Available with Knowledge Management

The explicit ISO 9001 requirement demands that knowledge be available to the interested parties of the QMS at the appropriate time and place. There is no specific requirement in the ISO 9001 Standard for a system for knowledge. But I chose to promote the discipline of knowledge management because I feel implementing such a system will assist in achieving the standard requirements.

Knowledge management is a set of concepts, work methods, technical solutions, and user applications that introduces a new way of information sharing and decision-making. Knowledge management allows the organization to manage its knowledge in an effective way, to promote sharing of knowledge between interested parties, and to ensure continuing improvement of the organization (the organization becomes a learning organization). Knowledge management uses the tactic of relating knowledge in the organization between individuals that need it and its processes and activities. In other words, knowledge management implies getting the right knowledge to the right people at the right situation and to assist individuals in sharing information and striving to achieve improvement of the QMS.

Why do we need knowledge management? Relying on data and information management in order to ensure conformity of products may not be effective once the conditions for realization are a bit complex or unpredictable. One may find themselves providing poor-quality products or services, being busy in redoing things, wasting time and money, and, most importantly, losing customer satisfaction. Deploying knowledge management as a strategic tool will reduce all these negatives symptoms.

It is important to understand the following approach—knowledge management shall be integrated in all of the business activities and will not be designated for one area. Each area and scope of the organization may have other knowledge needs that will set up its management of knowledge. That is why there is no precise requirement for a knowledge system but a requirement for implementing a knowledge system. The goals of knowledge management are

- Developing an organizational infrastructure for the management of knowledge to serve relevant parties in the organization
- Becoming a learning organization—training human resources in order to have the right knowledge and skills
- Developing processes needed to collect, store, retrieve, share, and leverage knowledge assets
- Allowing access to knowledge infrastructures
- Providing the required knowledge and information as resources for the operation of processes
- Obtaining necessary external knowledge
- Maintaining and preserving necessary knowledge required for the operation of processes
- Integrating knowledge of employees with knowledge of systems
- Fostering innovation and collaboration

Applying knowledge management achieves the following objectives:

- Allowing codification and categorization of knowledge
- Relating business aspects to their knowledge issues
- Allowing the support of human resources in how to translate knowledge into process outputs, goods, or services
- Maintaining knowledge relations with external parties of the organization such as customers, suppliers, stockholders, or governmental issues

In practice, knowledge management is a collection of activities for

- Capturing the organization's collective expertise on any media (database, paper, or intellectual)
- Developing channels for sharing of knowledge
- Provisioning of knowledge to interested parties at the right situation

And those are the exact goals of the ISO 9001 Standard. Examples of systems for maintaining knowledge are

- The Internet
- Intranets or extranets—a system that allows the sharing of files within or outside the organization
- Organizational libraries—repositories of organizational knowledge
- Organizational processes and information systems—ERP or CRM systems
- Document/content management systems—systems that enable control over documentation or content in the organization

- Expert systems—systems that support business activities, for example, CAD system
- Systems that support decision-making—BI, data warehousing, and data mining

Experience of developing, planning, and operating processes should be turned into knowledge and made available to all relevant parties. This can be used later on for the planning and developing of new processes. Such activity will serve the main goals of managing organizational knowledge: reproducing successes, elimination of repeating errors, and turning tacit knowledge into explicit knowledge. Such experience includes:

- Feasibility studies
- Risk analyses
- Knowledge related to tests and experiments
- Maintenance of infrastructures or devices
- Knowledge and Benchmarking or process engineering

Evaluating Knowledge

Employees or users are confronted daily with situations, activities, and problems that are unpredictable. Situations that do not allow a worker to proceed and where the acquired or provided knowledge is of little value and cannot provide with the appropriate answer. Making knowledge effective will be to anticipate which knowledge is needed or might be of use at a certain time or in a certain situation. For example, effective knowledge is combining the tacit knowledge with the explicit knowledge (more on that later); it is not enough to publish factual data. It is required to accompany it with experiences, best practices, argumentations, examples, and approaches. In order to decide whether the knowledge is effective or not, you must evaluate it. Each type of knowledge shall be evaluated according to the nature of its purpose.

An easy example is the work instruction for operating a machine; maybe it will not be enough to describe which buttons you need to press in order to operate a machine—this is merely an administrative data. Everyone can operate a machine, but life is dynamic and situations occur out of the ordinary. The knowledge on that work instruction should allow each worker to deal with all scenarios and situations. Maybe it will be more effective to add some capital for troubleshooting that will allow the employee to deal with problems or situations like what to do when the machine jams or how to handle the machine when a part was produced did not met expectations—this is the added knowledge that is missing. This is the tacit knowledge that is missing.

Explicit Knowledge and Tacit Knowledge

There are two distinctive kinds of knowledge in an organization. There is knowledge that is held privately by individuals and that is difficult to transfer to other persons in the organization. This type of knowledge is called tacit knowledge and is very hard to capture and document. The second type is knowledge that is articulated, codified, stored, and can be transmitted and shared among the interested parties of the organization. It is called explicit knowledge. We can view it in the following way: there is knowledge that was given to me and allows me to operate my business activities (e.g., a work procedure) and there is knowledge that I hold (inside my head) and is not organizationally documented (e.g., based on my experience). Another expression for tacit knowledge is knowledge that is being transferred in brainstorming sessions and one-on-one conversations where each individual brings their own tacit knowledge from experiences.

Systemizing the tacit knowledge is one of the challenges the organization is facing—transferring and distributing the tacit knowledge from private hands to the public and initiating activities for sharing this private knowledge.

Codification of Knowledge

Codification of knowledge relates to the way we collect, codify, retrieve, and distribute the knowledge in the organization. Knowledge then is considered information that is saved and the organization has developed a way to manage it: capturing it, making it existent, preserving it, and allowing access to it. The goal is to create a system that manages organizational knowledge and turns it into a medium accessible to those who need it. Through codification, people or employees can recognize and retrieve it. The following are the categories of knowledge:

- Organizational/managerial: Internal procedures, forms, and instructions.
- Cultural: Social elements of knowledge dependent on organizational and community culture. Such elements may determine and set the conditions for knowledge sharing.
- Technical: Work instructions, information of product or customers, designated software (like CAD).
- People related: Personal files related to the realization of the product, for example, private files that are saved on an employee's computer.

Personalization of Knowledge

The personalization of knowledge is the development of networks for linking individuals so tacit knowledge can be shard and be accessible to parties who need it. This is a new initiative for many organizations that require some cultural changes. The organization must be aware that people are the main factor that influences the success of knowledge management (not the technology). The tactic here is to develop channels between individuals with the same area of (professional) interest or the same knowledge needs in order to create the sharing of knowledge. The goals are focusing on developing and providing employees with channels or sources of valuable knowledge, which they can then use in particular situations. This can be achieved in many traditional ways: using communication traditional technologies (telephone, e-mail, video conferences) or with a new technology such as a system that channels knowledge from one to another based on predefined conditions such as webinars, content management systems, ticketing systems, Semantic networks, or decision support systems.

Communities of practice is a good example—a group of individuals with a common working practice who do not constitute a formal work team with the goal of sharing knowledge and experience that otherwise is unavailable. Each member is an owner of knowledge in his or her particular area of expertise. By exchanging ideas, problems, and solutions, they can share their collective knowledge. Initiating such communities in the organization will increase gaining of knowledge (by the members of the community). Communities of practice could be voluntarily organized in areas where there is a need for knowledge transfer.

Adapting and Controlling Changes

The systems for maintaining the organizational knowledge; the activities that operate it and the technology that supports it must be by nature flexible and ready for changes. This is a new discipline that emerges in a business environment that is ever changing. What kind of changes can occur?

- Changes in the product or service: Such changes concern the type, content, or nature of knowledge used by the organization. The organization might renew or revise its products or services according to trends in the market. Such changes require adjustments in the knowledge that serves the organization.
- Organizational changes: New responsibilities and authorities may need to accomplish new managerial activities for knowledge management.
- Technological changes: New tools and instruments for channeling the knowledge in the organization.

And changes are likely to occur often. For example, if today you are carrying out periodical meetings face to face for exchanging ideas with one person taking notes, tomorrow you might need to do it over a video conference and input your inputs into a central information system. The goal is the same but the circumstances are changed.

The ISO 9001 Standard understands the reality of ever-changing business environment and trends and insists that these changes be controlled. The reason is that such changes may have a significant impact on the processes, products, or services. Clause 6.3 offers a very effective method for planning and controlling changes related to the QMS. When considering a change relevant to knowledge, the following aspects are to be considered:

- Changes will be submitted to a process for planning and controlling.
- The risks related to the change, provided opportunities, and the implications of the change on the relevant knowledge base must be reviewed.
- When the knowledge directly affects products or services, documented information related to those changes must be updated.

Let us review these requirements with an example. Say that your organization is installing a product in the customer's premise. There are three kinds of activities regarding installation of a product: installation activities, configuration activities, and user activities. Each has its own knowledge needs. I summarize these in the following:

Activity	Required Knowledge
Installation activities	Work instructions
	Checklists
	Technical support
Configuration activities	Training
	Technical support
User activities	Training
	Technical support

One fine day the organization decides to change the technology with which the product is working. It means that there are changes necessary in the knowledge involved. In this case, the changes must be submitted to a process that will control it. The following table describes the different knowledge elements and their response to changes:

	Work Instructions and Checklists	Trainings	Technical Support
Identification of the required changes—which knowledge elements are affected?	Must the work instructions be changed? Must the checklist be updated?	Must the training be redesigned?	Must the technical support acquire the new knowledge in order to be able to provide the appropriate support?
Planning the change—How the different types of knowledge will be changed?	How and when will it update the work instructions or the checklists? Who will approve it?	Who is responsible for updating the training and when will they be planned?	How will the technical support obtain the new knowledge?
Evaluating the change—which risks or opportunities are associated with the change?	Must the work instructions and checklists be tried out before publishing?	Which experts must evaluate the training?	Must the technical support be provided with a troubleshooting document in order to address unexpected situations?
Reviewing the change—how can you ensure that objectives are achieved?	Planning actions for evaluation before the first installation of a new product. Addressing changes according to the conclusions of this test. Supervising and evaluating the first few installations and reacting according to the needs.		

Knowledge as Documented Information

There is *no* specific requirement for documented information regarding knowledge, but there is a direct connection between the two and I will prove it. Maintaining knowledge is necessary. That means you have to bring knowledge into an existing state, protect it, and make it accessible to parties who need it. You may maintain some kind of technology that enables management of this information and deliver it to the employees. Poor quality of information has a direct impact on decision-making and realization processes. Therefore, guidelines and principles must be planned and established for ensuring quality of information necessary to ensure integrity and consistency of data related to knowledge:

- A technology or a system that will serve the purpose
- Definition of persons who will enter the data or information into the system
- Identifying the elements of the data information required to preserve the integrity

Users have a significant role in the capturing of data and are responsible for its integrity and quality. This is why users must be involved in determining which information is going into the system and how it will be turned into knowledge. And that has a direct relation to documented information. Knowledge is considered as a resource required for the processes, is related to the competence of a person, and its maintenance is crucial to the realization of the product. For conclusion, although it is *not* required by the ISO 9001 Standard, I *recommend* to submit knowledge about the requirements for documented information as presented in clause 7.5—Documented information.

External and Internal Sources for Knowledge

In every organization, there are events or opportunities for obtaining knowledge from various sources. There are internal and external sources of knowledge. Under internal sources, we may find intellectual property, knowledge gained from experience, lessons learned from successes and failures, capturing and sharing undocumented knowledge and experience, and the results of improvements in processes, products, and services. Under external sources, we may find standards, academia, conferences, and knowledge gathered from customers or suppliers. You must find a way to seize such events and opportunities and identify what knowledge you can obtain from them. Here's a short review:

Type of Source	Event/Opportunity		Type of Knowledge
Internal	Event	Learning from successes and failures	Try to take cases (successes or failures) and see which part of the "story" could use other workers in order to eliminate the failure or recreate the success.
Internal	Opportunity	Experience of other employees	Experience of employees is a treasure of knowledge for less experienced employees. The organization must initiate sharing of knowledge through planned trainings or support meetings.
Internal	Opportunity	Know-how of other employees	Employee specialized in some areas can share their knowledge with other workers through planned training.
Internal	Opportunity	Sharing of knowledge in between areas	Communities of practice is a good example—a group of individuals with a common working practice who do not constitute a formal work team with the goal of sharing knowledge and experience that is otherwise unavailable.
Internal	Event	Information about innovations and updates	New technologies, new products, new product characteristics—informing employees through planned trainings.
External	Opportunity	External parties	Knowledge from external parties or partners such as customers, suppliers, or regulators—informing employees through planned trainings.
External	Opportunity	Academic and professional institutions, training courses	Sending your employees to external trainings or courses that are relevant to your organization.

7.2 Competence

Competence in the context of quality management is developing the ability to apply appropriate qualifications, skills, and knowledge to the right activities or operations with the goals of achieving intended results. Employees are considered to be a resource for the realization of products or services and have objectives that are derived from the context of the organization, its strategy, and the processes that operate the QMS. The needs for competences should be defined in a systematic way that can fulfill these goals and objectives. Training and qualifying should be planned in order to enhance the achievement of these goals and objectives. While planning the human resources for the realization processes, the organization shall refer to the following ISO 9001 Standard requirements:

- For each function or role that performs an activity that affects the quality of the product or its conformity to the requirements, the organization shall determine necessary qualifications, training, and certifications.
- The organization shall ensure that personnel and employees performing activities, tasks, and work that affect the quality and conformity of the product will acquire the adequate skills and be competent.
- Where needed, for the conformity of goods or services or where the organization finds it appropriate, it shall plan and implement training and certification activities in order to enhance the personnel to the required qualification level.
- Effectiveness of actions regarding training and certifications will be evaluated.
- The organization shall retain documented information and evidence of training and other actions related to the competence.
- Competence and adequate skills include hiring or contracting of competent persons, reassigning employees, training, education, qualification, mentoring, coaching, and experience.

Identifying Training Needs

One critical step in building competence is establishing an effective training and certification process. But before you set the certification process, you must identify and locate the relevant training needs. The trainings and the certifications of employees in a workplace are a process with the purpose of assisting the members of the organization in fulfilling their activities and tasks better and more effectively by introducing them to skills, knowledge, and professional and organizational approaches. Training and certifications may be needed in the following cases:

- Hiring and introducing new personnel
- Improving knowledge and skills
- Reassigning employees
- Introducing innovations in the organization
- Supporting organizational processes

By initiating actions, the organization promotes its personnel in achieving work objectives as well as quality objectives. The qualifications, skills, and knowledge required for the fulfillment of the product objectives need to be defined and determined according to the products characteristics and the nature of the activities involved:

- The customer requirements
- The characters and features of a product or a service
- The processes for the realization of a product or a service
- The raw materials or components it is made of
- The processes that operate the product and constitute its functionality or generate the service

In the organization, in each organizational unit (production, logistics, development, purchase, marketing), different employees or agents (workers that were qualified or certified by the organization to perform an operation) perform different tasks and activities that promote shared quality goals: realization of the product or service and meeting customer requirements. For each level in each organizational unit, it is necessary to define the level of qualifications and skills that those persons need to perform their tasks. As far as the ISO 9001 Standard is concerned, some elements and job characteristics must be defined and documented:

- Education: The level of knowledge and education required for a role will be determined: engineer, technician, programmer, biologist, and certified logistician. Some roles may not require an education or external certification at all. These definitions will be the preconditions for the hiring of personnel for specific roles.
- Regulatory requirements: When a regulatory requirement demands a certain certification according to a law or a standard, it will be mentioned as a precondition for hiring a person to the job. The matter will be verified with evidence prior to the hiring. That means that an auditor may ask to see different licenses or certifications of certain employees.
- Knowledge: The knowledge that is required for the operation of activities, realizing the product and meeting customer requirements.
- Experience: The extent of experience and background in parameters of time as well as areas and scopes of activities will be defined.
- Certification: The manufacturer will define a certification plan or process needed to introduce a person to a specific role: training about processes, procedures, work instructions, or activities related to the job. The plan shall cover both operational tasks related to the realization of the product or service as well as administrative and quality tasks.
- Training: The training (external as well as internal) for the role will be planned. It is necessary to identify the critical points or events in a process, which may affect the product and its integrity and to plan the training in accordance.

Certification Plan

Each employee that is recruited to the organization and will participate in the realization of goods or services is obliged to go through a planned and documented introduction of the organization and its relevant processes. The plan will cover specific areas and topics in the organization, such as

- Introduction to the organization.
- Administrative and human resources issues, for example, holidays, vacations, working hours, social conditions.
- Introduction to the different zones and areas in the organization and their use: cafeteria, parking places, toilets, and offices.
- Introduction to the organizational structure of the company and to different persons and functions, for example, work colleagues. This is very important for the employees in order to orient themselves in the organization.
- Their part in the realization of product and quality processes.

The plan will be documented and controlled. Records indicating the performance of the plan shall be maintained as documented information.

I present here the next recommended method for developing qualifications. The method is based on the principle of Plan, Do, Check, Act:

- Planning of the certification (Plan): Building a process for qualifying and certifying a person for a role.
- Instructing the employee (Do): Applying the plan to the subject of the certification, directing and leading the subject in the new role.
- Supervising the work (Check): Monitor the work of the subject and provide feedback.
- Further guidance (Act): Providing the feedback to the subject and seeing that he is implementing it.

Another aspect of certification is the provision of personalized knowledge and best practices through internal networking for linking the subject of the certification to tacit knowledge. The tactic here will be to develop in the certification plan channels between individuals with tacit knowledge and the subjects of the certification.

Training Plan

After defining requirements for qualifications and skills, training, and certification, the organization is required to plan the training or other activities in order to meet these requirements. The objective of the training plan is to fill in the gaps between the training needs and objectives and the current situation of employees and persons that realize the products or services (regarding qualifications and skills). The matter can be completed in four stages:

1. The determination of target groups in the organization: departments, workers, divisions, roles, or functions
2. The review and evaluation of the status with a defined method
3. The identification and mapping of the gaps in the organization
4. The development of a training plan according to results obtained

Training is a broad and complex concept. The subject is discussed many times in many forums. However, the ISO 9001 Standard lays down some basic principles on what constitutes a training plan:

- The plan will be based on the nature of the processes and of the organization and the training activities will be oriented to support the realization processes.
- Each function, role, area, department, or field in your organization that is related to the realization of goods or services or may affect their quality will have a designated training plan that will relate to their field of activities. In the plan, the subjects of the training—the audience or target groups—will be defined.
- The plan will be periodical: Monthly, quarterly, or yearly, and will include dates and schedules (you are not required to provide specific dates—periods of the year are sufficient). The period will be set according to the type of the training and the target group. For example, refreshing work procedures may occur once a year. But customer complaints and nonconformities must reach employees immediately.
- A minimal specification of the topics of the training will be included.
- The plan shall define the methods for the training. The matter will be discussed in detail in this chapter.
- The plan will define the tutor: Is it an internal (an employee training another employee) or external tutor?
- The plan will cover quality aspects such as quality procedures and quality controls implemented throughout the QMS. For example, each employee must know how to identify a nonconformity and respond to it. But the matter will be different for a production employee and a logistics employee. Therefore, for each target a designated plan is needed.
- The plan will relate to customers' and regulatory requirements. The plan may cover specific requirements and will indicate to employees how he or she may access these requirements during the performance of its tasks and activities, for example, customer requirements such as orders or signed offers and their location in the organization.
- The training shall be correlated with the quality policy and objectives. The matter will be discussed in detail in this chapter.
- The plan shall relate to training requirements derived from risk assessment based on risk-based thinking. The objective is to ensure workers whose activities or tasks may pose a risk to the products or services are identified and trained. It may include specific qualifications or certifications for particular processes or products.
- The plan will be evaluated for performance and results.
- The plan will define specific organizational events where training is necessary:
  - Refreshing work instructions and work procedures
  - Quality or work procedures
  - Training related to structural changes; transfer of employees between departments or roles (structural changes) within the organization
- Transfer of employees between processes or products (within a department):
  - Introduction of innovations within the organization; new products or projects
  - Introduction of new technologies or new developments in the industry
  - Introduction of new infrastructures, machines, or tools
  - Training required by regulations or applicable standards
  - Introduction of new regulations or refreshment of old ones

- Special, specific events that may affect the quality of the product (special projects or special visits at various sites)
- Coaching employees
- Trainings required by customers
- Trainings provided to suppliers
- Customers' complaints, nonconformities, or results of customer satisfaction surveys
- It is possible to define parameters within the training plan, such as
  - Location of the training: meeting rooms, auditorium
  - Required resources: computers, software, projector
  - Length of the training
  - Who the tutor is

One objective of training is to enhance personnel that are directly engaged with the products or services. The training should enable the development of skills and provision of knowledge in the area of product realization. The business environment (the external and internal issues) in which the company is active will be periodically reviewed in order to identify new training needs. The plan must be flexible enough to be updated at any time.

In order to create effective training that will assist your employees and provide them with tools and knowledge for their everyday work, I recommend that persons with the appropriate knowledge, background, and experience in a specific field supervise and navigate the training for each field or area of activity in the organization. It should be someone who possesses more than the explicit knowledge but also the tacit knowledge and best practices: someone who is aware of problems, special events, or situations, and particular cases that your employees will face during their daily tasks. The plan will then indicate to the employees how they should react. Here, just like by certification, I recommend developing channels between individuals with tacit knowledge and the subjects of the training.

Types of Training

There is no requirement to perform the training in the traditional way as frontal courses, lectures, or personal training. You may define training as self-learning or online tutoring, for example. The purpose remains the same however; only the method needs to be defined. Here are some examples of types of training:

- Frontal internal training performed by personnel of the organization
- Training provided on a digital format such as e-learning courses or webinars
- Purchasing literature
- External courses
- Exhibitions, conferences, or conventions
- Visit to suppliers' premises
- Trainings provided by supplier about his products

So when employees pay a visit to a supplier, or participate in a conference, the matter may be considered as training as long as it enhances their knowledge and skills regarding the realization of the goods or services.

Evaluating the Effectiveness of the Training

Training has objectives and goals. In terms of training, effectiveness is measured by whether the training has achieved its goal and to what extent. Specific training has its objectives, which support and promote the whole of the organization's objectives. The ISO 9001 Standard is aware of that and thus requires a systematical evaluation of training. The standard relates to certain aspects of training:

- The evaluation will determine whether the training was effective or not. It is necessary to define methods for the evaluation of the effectiveness of the training: reviewing processes, counting defective parts, measuring personal performances (qualitative as well as quantitative), and inquiry of employees.
- The evaluation will be done periodically.
- Each employee (on a personal scale rather than functional) will be evaluated for their competence for the training needs: required qualifications against current status of qualifications.

The evaluation of training will be done based on several parameters:

- Evaluation of the relevancy of the training to the products and services and its field of activity. For this reason, the training is to be planned by professionals who have background, knowledge, skills, and experience in the field of activity. This is exactly where their added value is required.
- Evaluation of the employees— Check whether the employees have assimilated the training, skills, and knowledge that were assigned to them and implemented them in the realization processes. You can check this by conducting an exam (written or oral) at the end of the training or analyzing processes in a defined period of time. It is always recommended to examine the results of the related processes rather than gathering information from the employees themselves. An employee can always claim that they have implemented what they have learned, but the question is what is happening during the realization processes. Inspect for yourself what is going on.
- Evaluate the tutor or trainer. Do they really understand what they are talking about? Do they have the required qualifications and skills? How are their training and communication skills? Can they really perform the training effectively? You may let the subjects of the training evaluate the tutor or trainer.
- Evaluate whether the training is updated. In our modern world, changes occur on a daily basis. What was relevant a year ago could be totally irrelevant today. Perhaps the organization has implemented new processes or technologies but the training refers to the old ones.
- Regulatory requirements. Regulatory requirements tend to change and be updated. In any case, it is required to update documentation regarding regulatory requirements regularly. This should also be done for training.

The most important evaluation is whether the training has achieved its goal. If the training is related to a certain customer complaint, it must be reviewed to ensure that that complaint will not be repeated. If the training is related to work instructions, the processes must be sampled and it must be verified that they are being done as specified. Such an examination will be planned in advance.

Evaluation of Employees

Another required evaluation of performance, effectiveness, and competence of the human resource is the periodical assessment of employees. This evaluation has several goals:

- Ensure that employees are competent to perform their job and assess their competence and quality of work according to predefined parameters.
- Identify needs and suggest measures for the promotion of the employee in the organization.

Parameters related to quality issues for evaluation:

- Professional competence of the employee: Does the employee have the necessary practical or theoretical knowledge and the technical experience and background to perform their tasks?
- Attitude for dealing with and solving problems: Does the employee recognize problems and suggest solutions? Do they constantly seek and suggest improvements and optimizations?
- Approach to the company: Does the employee display a high level of commitment to their workplace? Are they proactive and reliable?
- Performance and approach to the company's objectives: Are the performances of the employee satisfying? Is the employee committed to the goals of the company and achieving their personal goals?
- Approach to the work environment, tools, and equipment: Does the employee treat and behave in the work environment according to the firm's values and code of ethics or conduct? Do they utilize the working tools and equipment with responsibility and according to the specifications?

Such an evaluation will position the employee with regard to their qualifications and training objectives. The next stage is to assess the results and to determine whether further measures are required. Such evaluations can appear as a checklist or a form, with a qualitative or quantitative assessment. The results of the evaluation will be subjected to the requirements of documented information.

Expected Documentation and Records

As you have probably noticed, the issue of planning and managing human resources demands the establishment of certain documentations and the maintenance of records. For documentation, the following is required:

- Training needs: Definitions of qualifications and training requirements relevant to the role of function. I recommend including these definitions as part of the job descriptions. There is a direct relation between the list of activities and the required qualifications to perform it.

- Certification plan: A list of activities and training required to introduce a person to a specific job or function. It may appear as a form, a checklist, or a procedure.
- Training plan: Although there is no requirement to document the plan, I suggest you do so. It will assist in following the implementation of training activities and may avoid answering some tricky questions during audits.
- Evaluation of employees: A plan that specifies the parameters for assessing and evaluating employees and proposes topics for discussion between the employee and his supervisor. This is normally managed by the personnel management.

The documentation will be submitted to the controls required in clause 7.5—Documented information. The following records must be maintained:

- External evidence that proves the competence of an employee according to the training and qualification needs: diplomas, certification of education, regulatory certifications, or licenses. For each employee, a correlation is required between evidence of education and the definition of the job.
- Documented information and evidences of the internal certification activities for employees (certification plan). The records will identify employees who participated and specify with details the dates and the activities that were initiated in order for them to perform their job.
- Documented internal certification of employees that allows them to perform certain activities or relates to specific roles or functions. This may appear as an approval stamp on a copy of the certification plan or on a designated from. It is important to mention that the certification will be personal. Authority management in an ERP system is a good example.
- Records and evidences that prove that training activities have been carried out. The records will include details such as dates, tutor or trainer, participants, and topics of training.
- Evaluation of the effectiveness of the training actions taken. This may appear on the same records of the training or on a designated record. Records of employees' evaluations of performance, effectiveness, and competence. These records may include records of physical examinations necessary for the realization processes.
- Any other records required by regulatory requirements.

These records will serve as documented information and will be submitted to the appropriate control as required in 7.5—Documented information. A good way to manage it is to maintain a designated file for each employee with relevant quality records. My advice to you is to monitor how your human resources manage it—maybe they are carrying out most of the documentation already.

7.3 Awareness

Awareness is getting more importance from the ISO 9001:2015 Standard. It is so important that it even has its own clause. This I believe is because of the consensus that awareness of the QMS increases the motivation and devotion of employees. Awareness in the context of quality management refers to the understanding of the

- The context of the organization—conditions in which the organization is active and its values
- The quality policy and quality objectives of the organization
- The implications and consequences of actions of persons performing activities and operating in the QMS

However, because there are many subjects or topics that employees must be aware of, the standard specifies relatively clearly what the person who is operating the QMS must be aware of (the ISO 9001 Standard requirements):

- Persons doing work under the organization's control shall be aware, know, and understand the quality policy.
- The organization shall ensure the awareness of personnel of the relevance of their actions and activities
  - To the achievement of the quality objectives
  - To the effectiveness of the QMS
  - To the benefits of improved quality performance
- The organization shall ensure the awareness of personnel regarding the effects and importance of their actions and activities on the quality of the product and conformity to the requirements.

Awareness and Motivation of Employees

Awareness initiates identification, devotion, and commitment to the goals of the organization among personnel. By promoting the awareness, the ISO 9001 Standard tries to avoid the following side effects:

- Not understanding the context of the organization
- Not knowing exactly who the interested parties are
- Not understanding or knowing what the customer requirements really are
- Delivering poor-quality products or services to the customer
- Promising the customer what cannot be achieved
- Ignoring or missing quality problems
- Reacting to quality problems instead of avoiding them

Sounds familiar? Well, these are quality objectives in their essence. In practice, personnel should be aware of their

- Customers—identify who their customers are and to whom they are committed
- Duties—know which actions or activities are expected from them and what their duties are
- Requirements—know what the requirements are and how their actions affect meeting these requirements
- Quality of their job—know when a process is effective and how to perform the job correctly

- Relation to the quality objectives—understand how their actions contribute to the quality objectives
- Improvements—be aware of the need for improvement, know how to look and when to suggest improvements in the QMS

Awareness to an Effective Quality Management System

Employees must understand that they are part of the QMS and must be aware of their contribution to its objectives. Sometimes, the common consensus among workers is that the QMS revolves around the quality manager, the quality assurance, and involves some customer complaints, internal and external audits, and comes down to publications of the top management in the cafeteria regarding the ISO certification. These are their only touching points with the QMS. Even today, after 28 years of ISO 9001 Standard, many employees do not feel part of a QMS and ISO 9001 is an alienating term for them. Why should it be otherwise? The certification just hangs in the office of the CEO and most of them never see it; during an external audit, employees are asked to project all is well, but when there is a quality problem, the employees are to blame. The new revision of ISO 9001 Standard wants to change it and demands the creation of awareness among all participants in celebration of quality; employees must understand the whole story of the organization in order to assimilate the fact that their actions matter. This is not an easy task. In many organizations, ISO 9001 is perceived as a bureaucratic topic and some people are trying to avoid having anything to do with it.

So how can you market the ISO 9001 Standard and the QMS as an exciting thing that has a direct relation to the daily job of employees?

- First of all, I would explain very easily and in layman's terms and definitions that are used in QMS; quality policy, quality objective, product conformity, customer requirement, corrective action, etc. There is nothing more frustrating to an employee to be in a meeting and not understand what people are talking about—immediately you lose the attention of this certain worker.
- Second, I would present to the workers the structure of the QMS in the organization and explain to them where they stand and most importantly what is their contribution to the QMS.
- After covering the basics, I support the very simple method of demonstrating results, consequences, and implications of dos and don'ts.

The next one is a recommendation—I avoid using the words ISO 9001 Standard—I just feel that with some groups of people these terms create bad association or the wrong association.

Awareness to the Quality Policy and Quality Objectives

The QMS includes the quality policy and the quality objectives with the main goal of meeting customer requirements and improving it. Training is one means of implementing the policy and achieving these objectives. Quality objectives are divided into subobjectives that are obtained through operative objectives. The issue filters down to the level of the single process: each process has its objective—the expected output. The employees must be aware of this hierarchy and understand the relevancy, relation, and effect of their actions and activities on achieving the quality objectives, meeting customer requirements, and improving the QMS. Employees must be aware that their activities and the delivery of intended outputs promote the quality objectives of the organization. Such awareness will identify the role of employees regarding the quality objectives and motivate them as necessary. The employees must understand that they are bearing the operative responsibility of achieving the objectives. How? By demonstrating (e.g., during training) the relation between the results of their actions and the quality objectives.

The basic principle states that each employee must be familiar with the policy and objectives and therefore they are to be an integral part of the training plan:

- If one of the quality objectives is to meet specific personal customer requirements, employees must understand how their actions contribute to this objective; for example, during the delivery processes, an employee must identify these specific requirements and follow all instructions given to them by the customer. Plus, the employee must be educated on cases where customer requirements were not met and the consequences.
- If one of the quality objectives is to provide a product that conforms to a specific regulatory requirement, the employees must know these requirements and understand that a product that leaves the factory must conform to those requirements. Plus, the employee must be educated on cases where regulatory requirements were not met and the consequences.

In practice, I would consider developing a system that will inform persons in the organization about the status of relevant quality objectives:

- Publish customer complaints: Cases where customer requirements were allegedly not met.
- Publish customer gratitude: Cases where customer requirements were met and the customer is well satisfied.
- Introduce to the organization a system that presents the practical work objectives such as status of orders, delays in deliveries, status of production orders, status of service calls: A tool or some means by which an employee can view the situation. Such systems promote the proactive approach and allow employees to act in a way to avoid such situations rather than having to face unwanted situations.

Such publications achieve several goals in the context of awareness:

- They promote discussion about customer requirements and their achievement
- They evoke conversation among workers about the objectives
- They create identification, devotion, and commitment to organizational goals

Awareness of Product Conformity

Awareness of product conformity is necessary in order for personnel to understand their direct contribution to quality of the product. In practice, you must evaluate how well the employee is aware of the product or service requirements and how his or her work affects the quality of the product:

- Does the employee know what the product requirements are?
- Does the employee understand what a qualitative product is?
- Does the employee understand the consequences of a low-quality product?
- Does the employee know how they can increase the quality of the product?
- Does the employee know risks associated with the product?
- Is the employee familiar with known nonconformities?

Measuring the Awareness

Effectiveness is one of the most important principles of the ISO 9001 Standard and is relevant to awareness as well. Although it is not required by the standard, the organization may develop methods to measure the extent of awareness of personnel about their actions and how they are related to the quality objectives. Evaluating awareness, in my opinion, should be done on several levels: level of product, process, and QMS.

Level of the product: How well the employee is aware of the product or service requirements:

- What are the product requirements?
- What is a qualitative product?
- What are the consequences of a low-quality product?
- How can an employee increase the quality of the product?

Level of the process: How well the employee knows the processes they are involved in:

- Which are the activities (processes) of the employees?
- What are the required inputs?
- Which outputs or results are expected?
- When is the process considered effective?
- What are the results of an ineffective process?

Level of the QMS: How well is the employee aware of their place in the organization and their contribution to the QMS:

- Which quality objectives are related to the work of the employee?
- How can an employee contribute to achieving their quality objectives?
- Where can an employee identify opportunities for improvement?

In practice, you can conduct a survey among personnel to evaluate their knowledge and awareness regarding quality.

7.4 Communication

Communication channels (internal as well as external) play a very important role in an organization and in implementing an effective QMS and have a direct effect on the realization processes. Communication takes place everywhere and involves almost everybody in the organization. Communication encompasses all business activities and interaction between interested parties of the organization: employees, customers, suppliers, governmental offices, and other interested parties. These characteristics may cause disorder and confusion in the communication when not played appropriately. And here is what the ISO 9001 is trying to avoid through a definition of principles for communication. Let me put it in this way: If information does not reach the right people at the right time and place, you might not achieve quality objectives and you might deliver nonconforming products or services and decrease customer satisfaction.

The ISO 9001 Standard includes communication channels in clause 7—Support activities because communication channels are tools that promote your strategy and thus should support processes and encourage parties to share information and knowledge, and they do require resources. ISO 9001 Standard requirements:

- The organization shall determine the relevant communications for its QMS (internal as well as external).
- Communication means shall be defined for each QMS element.
- For each communication means, it will be decided
  - What will be communicated
  - When it will be communicated
  - Who will communicate
  - To whom it will be communicated

Communication in the eyes of quality management means a process or activity for exchanging information between entities for the operation of the QMS. This process of communication has a structure and direction and uses technology, tools, means, or instruments. Communication may be internal (that takes place within the organization) or external (between organizational units and external interested parties). Communication channels have important strategic goals:

- Ensuring that information and knowledge reaches the designated persons
- Boosting the processes in the organization and contributing to their effectiveness
- Promoting the sharing of knowledge and information between entities or units that operate the QMS
- Promoting and conveying to employees the importance of meeting customers' and regulatory requirements
- Helping identify problems and opportunities for improvement

Reference to the Process Approach

Which elements of the QMS must be communicated is a bit abstract. That is why I refer back to clause 4.4—Quality management system and its processes; when planning and implementing the processes of the QMS, you were required to

- Define the process of the QMS
- Determine for each process its necessary inputs and their sources

- Determine for the operation of each process its expected outputs inputs and their receivers
- Determine the sequence and interaction of processes included in the QMS

The requirement in clause 7.4—Communication is how these processes will communicate/interact and how the inputs and the outputs will be transferred between the different parties and in what sequence. In practice, I suggest that you include in the analysis of the process the aspect of communication; for each process covered by your analysis, include the aspects of communication mentioned in this chapter. The definitions of the required communications shall include communication with external partners such as customers, suppliers, and other interested parties.

Defining What Will Be Communicated

Defining what will be communicated must be planned according to the following concept: it is necessary to understand who is expecting what. Defining what will be communicated refers to the content that will be transferred: data and information relevant to an activity or a context. The content must be significant to the recipient and they should be able to make decisions based on this content or understand or be able to evaluate a certain situation. For example, a customer should be able to understand the terms of an agreement or a production manager should be able to have all the information in order to plan the resources for the next month. Referring back to the process approach, for each process you have defined what will be communicated:

- What the necessary inputs for each process are
- What the necessary outputs for each process are

Locating the means, tools, or instruments for the communication is essential to the definition, that is, if you are planning to deliver a report with certain information, you must make sure that

- You can summon the data related to the information
- You can work this data into the requested information
- You have the technology to present this information as a report to the person who awaits it

Referring to the process approach—for each process or subprocess, it will be clearly defined which information or data must be transferred and how it will be done.

Defining the Events for Communication

Delivering the message at the right moment is crucial for the effectiveness of its content. Process is built from many steps and sequences. Each step processes inputs and generates outputs. When planning the processes, one must consider at which step a communication must be initiated: incoming as well as outgoing. This aspect can have a significant effect on the objectives of the QMS; when a system fails to communicate customer requirements to the service technician, customer satisfaction may decrease; if a supplier does not receive the order on time or does not receive the schedules for delivery, he cannot deliver according to the plan and the organization may fail to deliver the product to the customer on time.

IT systems such as ERP or CRM can automate the transfer of such information:

- Each time QMS creates a nonconformity, a message is sent to the interested persons in the organization: CEO, product manager, key account manager, etc.
- MRP (material requirements planning) can automatically create and send purchase orders to suppliers with planned schedules.
- In many CRM systems, there is an option of automatically creating customer events based on business cases and designating them to users, for example, when customer sends a very low-ranked survey response, a message is sent to a representative that should contact the customer in order to understand why.

Referring to the process approach—the analysis of your processes will indicate in which business cases or events a communication will be initiated.

Defining with Whom to Communicate

Information and data delivered through the communication channel must find their target. In today's business environment, people tend to be flooded with irrelevant information. This is why the destination of the information for each type or channel of communication must be clear:

- Management: It is necessary to define which information will be communicated from the top management and back (vertically).
- Employees: It is necessary to define for each activity the communication between personnel (horizontally).
- Customers: It is necessary to define which communications will be established with customers. This requirement is dealt with in detail in clause 8.2.4—Customer communication.
- Suppliers: It is necessary to define which communications will be established for external providers. This requirement is dealt with in detail in clause 8.4—Control of external provision of goods and services.
- Other interested parties (such as governmental bodies or stakeholders): Communication for other interested parties must be clearly defined according to their expectations (which information they need and how they receive it—which technology).

Defining the Communication Channels

Each process or activity dictates the needed communication channel:

- Who initiates the communication? Who is the sender and who is the receiver?
- What information must be transferred and where is it available?
- Which technology will operate the communication?
- What knowledge, skills, or qualifications are required for the operation of the channel?

Examples of communication channels:

- ERP system: Business management software of integrated applications that use a common database that allows the transmission of data and information in real time between different departments in the organization.
- CRM system: Is a system for managing interactions with current and future customers for a company's that used technology to organize, automate, and synchronize internal as well as external communication.
- PLM system: A system that manages product life cycle by communicating its data with relevant interested parties: customer, design and development team, production team, etc.
- Pneumatic tube system: A system that propels cylindrical containers through a network of tubes by compressed air or by partial vacuum used to transport small, urgent packages (such as mail, paperwork, or money) over relatively short distances (within a building or at most within a city, mostly seen in hospitals).
- Brochures: An informative paper document used to introduce a company or organization and designated to a target group.
- E-mails: Well, I bet you are familiar with this communication channel....
- Internal news mails or organizational blog: Effective communication tool used to distribute information intended for employees in the company bearing relevant and useful information about important topics that employees should be aware of.
- Periodic meetings: This is the opportunity for mid-level shared employees to formally meet to discuss and improve visibility of issues, share ideas, and support each other better.
- Training and lectures: These might be used as a means of communication to transfer information and data to employees or receive inputs from them.

For each type of channel, it is necessary to determine the frequency of use and the types of data or information that it will handle. And it can be planned using different approaches: Review what the required information to be transferred is and then plan the channels appropriately. If it can be seen that the process of handling customer complaints in the organization is not effective and parties are neglecting their duties, a designated system may be implemented that will encourage the process.

Effectiveness of Communication Channels

Effective communication channels are important to an effective QMS because they create transparency in the organization and allow the efficient flow of data, information, and knowledge. The effectiveness of communication channels is measured with the following:

- Verification that all needed communications are identified
- Verification that appropriate or correct data and information and knowledge are transferred
- Verification that data and information and knowledge reach their designated destination at the right time

In order to achieve effectiveness, you should promote the following regarding the communication channels:

- Encouragement: The channels must be encouraged and be active at all organizational levels.
- Clear and understandable: Each role and responsibility must know its obligations; how it must communicate, when, and what.
- Bidirectional: If the information must travel in both ways (forward and backward—from sender to receiver and back), the channel must allow it.
- Adapted to language requirements: If employees of the organization speak more than one language, the communication channel must support it.

In practice, you may include all this in the qualification and training program.

Analyzing the Communication Channels

As mentioned throughout this chapter, the communication channels define the reporting methods, support the interaction of processes in an organization, and must be planned in accordance with its organizational structure and the workflow. I suggest here a simple method for analyzing the requirements for the communication channels. Basically, each employee, function, or role should know to whom they must report or who reports to them, with which tools they should be reporting, which inputs they should receive, and which outputs they should deliver. And these are the parameters that you should identify. I divide them in the following way:

- Sender: Who is responsible for initiating the communication—this definition will be on level of a role/process owner.
- Event: At which time point or which activity initiates the communication.
- Recipient: Who is bound to receive the information—this definition will be on the level of a role/process owner.
- Deliverable: Which information must be delivered—the expectation must be clear and known.
- Schedule: At which point in the process flow must the communication be initiated.
- Method: How the information will be communicated—telephone, e-mail, ERP system, automatically or manually.
- Acceptable: What information or data the sender expects in response.

Let us look at the following business cases: When a production manager needs raw material or components for the realization of goods, or when a service manager needs to plan work resources for the next week, communication channels with their suppliers (warehouse, team manager responsible for the technicians) will be determined and established in an effective way. The following table gives an overview of how information should flow between entities in the QMS for this example:

Parameter	Business Case Production	Business Case Service
The person or role that initiates the communication (sender)	Production manager	Service manager
Defined point of time (event)—when the communication must be initiated	According to the production plan (e.g., MRP), the production manager should know the schedules for production orders	According to the planned capacity (list of open service calls), the manager should know what tasks must be completed, what qualifications are needed, and their schedules
Target for the request (recipient)—to whom must the sender submit their request	The organizational unit that provides the raw material	Team manager—a person who is responsible for submitting people to work
Means of communication (method)—how shall the request be submitted	Supply request—through the ERP system, e-mail, or a form	Request through a CRM system, service system, e-mail, or a form
The required information (deliverable outputs)—which data or information must be submitted to the receiver	The person working in the warehouse will have all the necessary information: who sent the request, which material is needed, the quantities, the schedules, and where should it be delivered	The team manager will have all the necessary information: who sent the request, to which department/area, how many employees are needed, when they are required, and which skills they require
The manner for returning an answer	The person who works in the warehouse will have a defined way to return an answer to the production manager about his or her request	The team manager will have a defined way to return an answer to the service manager regarding the availability of the resources

This kind of analysis can be part of your process analysis when applying the process approach.

7.5 Documented Information

Documents and information of the QMS must be controlled. This is a key element of a QMS. The main idea is to provide control over the documented information necessary for the operation of the QMS. In order to achieve this high-importance quality objective, the ISO 9001 Standard expects a method. The new term "documented information" brings an improved and a closer reality perspective of documents and records into the ISO 9001 Standard requirements.

Documented information refers to the information necessary for the planning and operation of the QMS coming from any source and the medium on which it is contained. For the first time, the standard does not separate between procedures, documentations, and records and refers to all as documented information. The very strict definition refers to information used by the QMS or produced by it that

- Was defined by the organization as necessary for the planning and the operation of the QMS—the definition includes the media on which the information is stored and maintained. Information can be in any format and media and from any source.
- Is required to be maintained and controlled by the organization.

Documented information relates to four kinds of documentations distinguished with different characteristics related to context, use, maintenance, and media:

- Documented information needed to describe and document the QMS, for example, quality policy
- Documented information needed to document quality processes of the QMS, for example, form for management review or internal audit
- Documented information needed for the operation of the QMS, for example, work instructions, SOPs, process diagrams, etc.
- Evidence or processes or activities as records that are necessary to verify or validate results and to check and prove the effectiveness of the QMS, for example, production charts for quality control or results of customer satisfaction surveys

7.5.1 General

The ISO 9001 Standard requirements are as follows:

- The QMS documentation shall include documented information required by this international standard.
- The QMS documentation shall include documented information defined by the organization as needed to provide evidence of the effective operation of the QMS and of conformity of products and services.
- Note: The extent, scope, and size of documented information in organizations are affected by the following factors:
  - The intended outcomes and results of the QMS
  - The size of the organization
  - The level of complexity, functionality, and interrelations between its processes
  - The qualifications of the employees
  - The quality objectives

A principle that must be considered when defining your documented information is as follows: the amount and details of the documented information must be relevant to the intended outcomes and results expected of the QMS.

Documented Information Required by the ISO 9001 Standard

Throughout the standard, there are requirements for documentations needed to support and operate the QMS. The last version was relatively clear and specified, where the ISO 9001:2008 Standard expected procedures, documentations, and records. The ISO 9001:2015 Standard does not put itself anymore in the position of telling you which procedures to maintain, which procedures not to write, or which records are necessary. You will decide what is necessary for you to maintain the standard requirement. It is a harder task because it leaves more room for interpretations and debates for you and your auditor.

I propose in this chapter (as a recommendation only) a way to analyze the standard textually and assess whether documentation is required. Normally, the text of the standard implies with clear statements where documentation is expected. But sometimes you must read between the lines and figure it out. Let us look at some example statements:

- Clause 4.3 "The scope shall be available as documented information": It is clear that documentation here is required.
- Clause 8.1 "Establishing criteria for the processes": On the one side, it is not specifically stated that documentation is required, but, on the other hand, I cannot think of other ways to meet these requirements without maintaining documented information of the mentioned criteria.
- Clause 9.1.2: It is required to determine methods for obtaining and using data relating to customer perceptions of the degree to which requirements have been met—There is no need to provide documented information regarding those requirements, but there is a requirement to prove that these activities were planned and are being performed. There is also no need to develop a process or procedure that describes which methods are used, which tools are used to collect the data, and which tools are used to analyze the data. But it is required to prove that those were considered, planned, and are being performed.

Structure of the Documentation

A QMS must have a structure for the documentation it uses. The idea is to develop a structure that demonstrates the relations and interrelations between the various documents throughout the QMS and enables tracing back a process through the documented information; some documents relate to other documents, outputs of one document are the inputs for another document, and so on. The structure should be effectively planned and should present the relations between the different types of the documents in the organization. These relations are based on your organizational needs and will support your activities. In other words, let your organizational needs and operational activities dictate the structure of the documentation.

A good example for relations and interrelations of documented information is a work instruction and the references to specific forms; a work instruction describes an activity of some kind and refers to a specific form that the employee must fill out in order to provide evidence and effectiveness of that activity.

The traditional and most common method to analyze the structure of the documented information in the organization will be to arrange, classify, categorize, and set levels to the types of the documented information used in the organization according

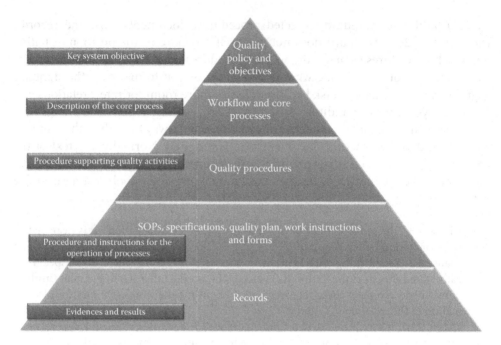

Figure 7.2   Documentation pyramid.

to their characteristics and role in the QMS. I prepared here a diagram that suggests one way of defining the structure of documentation (Figure 7.2).

This pyramid of documentation describes the operational flow of documented information in the organization and represents the levels for the types of documents that are usually used in a QMS, from the planning of strategic stages to the daily work of performing activities and filling out forms. This method clearly illustrates that with each descending level of the pyramid, the amount of required documented information will increase. This kind of pyramid provides navigation for users throughout the various documentations in the organization. In other words, when an interested party (internal like a user or an employee or external like an auditor) needs to get familiarize with the documentation of the organization, they can use this structure for orientation. Such a structure may set the basis for responsibilities and authorities regarding

- Planning and approval of documented information
- Updating documented information
- Locating and deposition of documented information
- Disposal of documented information

In the last revision of the ISO 9001 Standard (2008), it was required to maintain a quality manual and usually this structure was presented in that document. The ISO 9001:2015 does not require the maintenance of such quality manual anymore. I find the quality manual very effective because it provides a frame for the documentation.

The new revision of the standard (ISO 9001:2015) excluded this requirement, but you may still keep and maintain a quality manual, in which I suggest you document and include the description of the structure and its requirements. Let us review the different levels and see how they serve the QMS.

Strategic Level

In the first level of the documentation pyramid, we will find documentation of declarative statements by the organization that includes the key elements of the QMS, and that will set the strategic direction of the QMS. This documented information has a clear goal: to introduce and communicate the intentions, scope, and structure of the QMS in the organization. It includes elements such as QMS scope, quality policies, quality objectives or references to them, organizational structure, exclusions, general process of the organization, etc. The following table shows the types of the documented information suggested or required by the ISO 9001 Standard, which addresses this first level:

Subject	Relevant ISO 9001 Clause	Requirement for Documented Information
Description of the context of the organization	4.1 Understanding the organization and its context	No documented information is required
Description of organization and the organizational structure	5.3 Organizational roles, responsibilities, and authorities	No documented information is required
Description of interested parties and their expectations	4.2 Understanding the needs and expectations of interested parties	No documented information is required
Scope of QMS	4.3 Determining the scope of the quality management system	Documented information is required
Details of exclusions	4.3 Determining the scope of the quality management system	Documented information is required
Organization's quality policy written in conformance with the ISO 9001C Standard requirements	5.2 Quality policy	Documented information is required
Quality objectives or reference to them	6.2 Quality objectives and planning to achieve them	Documented information is required
Identification of processes included in and are necessary to the operation of the QMS	4.4 Quality management system and its processes	Documented information is required
Description of process interactions or reference to another document specifying them	4.4 Quality management system and its processes	Documented information is required

Reviewing this information mentioned in the table would provide answers to questions like

- Who are the interested parties of the organization?
- What is the scope of the QMS?
- What are the quality policy and objectives of the firm?
- What are the main products and services?
- How do the main processes flow?

Workflow and Core Processes

Workflow describes the work process of an organizational unit (an organization or a part of the organization) where each step depends on the preceding step. It depicts the business activity with step processes, their resources, their sequences (progression), and interactions that transform inputs into outputs, for example, materials into goods or information into services. Workflow can appear as a diagram or as a description (text). I prepared here an example for a diagram (Figure 7.3). It is very basic but enough to provide the main idea:

Each block describes a set of processes, subprocesses, and business activities for the operation of the QMS. The workflow should be aligned with the scope of the QMS; the workflow can include all of the organization's processes or just a certain and identified functions of the organization, certain and identified units of the organization, or one or more functions across a group of organizations. The workflow also determines the needs for documented information.

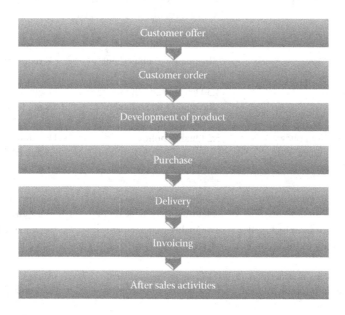

Figure 7.3   Example for a workflow.

Procedures or Process Diagrams to Support Quality Activities

On this level, you may include procedures or instructions that support quality activities that the standard describes and demands (the ISO 9001:2015 Standard requires activities not procedures). For example:

- Maintenance of documented information
- Control of nonconforming product
- Internal audit
- Corrective action

In practice, the new revision of the ISO 9001 (2015) does not require the establishment and maintenance of quality procedures as the previous revision (2008). But still you are permitted to keep these procedures, maintain them, use them, and audit them if you find that they serve the QMS, they contribute to the effectiveness of the QMS, and they initiate improvement. I certainly recommend the use and maintenance of these procedures because I find them as an effective tool for managing critical quality activities. But do not forget to

- Update these procedures according to the new ISO 9001 2015 requirements
- Remove procedures that support quality activities that are not required anymore (e.g., preventive action)

Quality Plan, SOPs, Specifications, Process Diagrams, Procedures, Work Instructions, and Forms

This type of documented information, which belongs to the fourth level in our pyramid, has the goal of supporting operations of processes and directing and instructing personnel on how to perform activities. This type of documented information has the objectives of

- Supporting the workflow in the organization
- Defining the required information and data needed to perform activities and operate processes
- Describing responsibilities and authorities of people and/or organizational functions regarding processes and activities
- Ensuring effective planning, operations, and control activities during the realization of the product
- Describing activities needed to support the workflow
- Describing the interrelations of processes or activities
- Describing methods for monitoring and measurement activities
- Describing the expected evidences and records
- Communicating information regarding processes and activities
- Assisting in training
- Reducing mistakes

Type of such documented information:

- Quality plan: A quality plan is a list of specifications or activities for the organization to follow, meet, or maintain in order to realize a product according to requirements (customer's or regulatory).
- Process diagrams: A process diagram describes the flow of several processes or activities, the required inputs and outputs, and the interactions between those activities.

- SOP: Documented procedure describing step-by-step instructions to achieve an intended process objective or result. This type of documentation is usually planned according to a template: a model, a standard, or an outline dictating the structure, format, and content of such procedure. The SOP is part of the overall process.
- Specifications: Specifications is a document stating requirements unique to a process or a product.
- Procedures: A procedure provides instructions to individuals on how to perform a specific activity; which resources are needed, which inputs are required, which activities are to be performed, and which outputs are expected from this specific procedure. Examples include capturing a customer order or transferring an order to production.
- Work instructions: Work instruction presents a detailed sequence of steps or actions to perform an activity or a task. It is a more detailed level of the procedure. The work instruction will be usually presented in the form of text but visual description can also assist. For example, instructions for entering a customer order into the ERP system will include a textual explanation of operating the ERP system accompanied by screenshots.
- Instructions for operations: This type of documented information is similar to a work instruction but refers to a specific machine or tool needed for the realization of the product. Such instructions usually bear technical details. For example, instructions for cleaning an injection molding machine between production orders.
- Test plans or protocols: Test plans or test protocols are documentations used to specify which tests must to be executed in order to demonstrate that results of an activity meet requirements previously established in a specification.
- Form: Logically structured document referring to a certain process and designed to document its execution. A form contains blank fields that indicate and instruct the users which data are required to be entered. Filled up (used) forms are usually considered records.

Regulatory Requirements for Documentation

When regulations require the maintenance of any kind of certain documented information, it should be planned, established, implemented, and maintained. For example, health records of personnel in certain positions or certain roles. The ISO 9001 Standard gives such regulatory requirements the same scale as any standard requirement for documented information when it relates to the realization of the product. In practice, you must identify such requirements and include them in the list of the controlled documented information.

External Documentation

The control of documented information shall relate in addition to external documented information. External documented information will be distinguished, registered, and controlled. The method for control shall first identify which external documented information is necessary for the planning and operation of the QMS and thus requires control, for example, a technical drawing of a product received from a customer. This is a classic external record that must be controlled for edition or version; using the older version may result in delivering the wrong product to the customer. Other examples of external documented information may be

- Information received from customers: drawings, packaging instructions, diagrams of production tools, quality requirements, design files, and customer's approvals or agreements
- Information received from suppliers: drawings, use instructions
- Regulation and standards

It is necessary to verify that external documented information follows all of the standard's requirements.

Extent, Scope, and Size of Documented Information

The extent, scope, and size of documented information in organizations are affected by the following factors:

- The size of the organization: The more departments included under the QMS or the more activities needed to realize the product, the more documented information is needed.
- The level of complexity, functionality, and interrelations between its processes: The more complex and demanding the processes, the more intense and acute the documented information.
- The qualifications of the employees: For less qualified and trained employees documented information necessary for the operation of the QMS should be more excessive and detailed.
- Quality objectives: The more the quality objectives that refer to organizational units and roles, the more the documented information will be necessary to prove the achievement of those objectives.

A good example to demonstrate this is the choosing between maintaining a process, a procedure, or a diagram. How can it be determined when a procedure is needed and when a process should be charted? The extent of the documentation of a process will be determined according to the parameters mentioned earlier. Choose a process and evaluate the following:

- How many departments and organizational units are included under the QMS?
- What is the level of its complexity? Do I need to document only the principles and main steps of the process or each activity?
- What is the level of the personnel's qualifications? Do I need to provide them with a general guideline or a detailed description of the work instruction?
- How can I measure the effectiveness of this process? Do I need to collect results of this process? Do I need to analyze the results or is it enough to just review the output?
- Does the documentation assist me in achieving quality objectives?

List of Necessary Documentation Required by the ISO 9001 Standard

Please review the following table. It specifies the ISO 9001 documentation requirements for documented information.

Clause	Description	Standard Requirement	Type of Documented Information	Requirement for Documented Information
4.2	Understanding the needs and expectations of interested parties	Determining and updating the expectations and needs of interested parties.	Document	It is not obligatory to maintain expectations and needs of interested parties as documented information but you will have to update them. Although it is not required anymore, a quality manual is an option.
4.3	Determining the scope of the quality management system	Establishing the scope of the quality management system.	Document	A document describing the scope of the QMS in the organization. Although not required anymore, a quality manual is an option.
5.2	Quality policy	Establishing a quality policy as documented information.	Document	It is obligatory to maintain a document describing the quality policy of the QMS and the quality objectives or reference to other documented information regarding the quality objectives as documented information. Although not required anymore, a quality manual is an option.
6.2	Quality objectives and planning to achieve them	Retaining documented information on quality objectives.	Document	It is obligatory to maintain the quality objectives as documented information. Although not required anymore, a quality manual is an option.
7.1.5	Monitoring and measuring devices	Retaining documented information as evidence of the fitness of monitoring and measuring devices.	Record	It is obligatory to maintain documented information as records regarding the fitness, status, and maintenance of monitoring and measuring devices.
7.2	Competence	Evidence of competence of person(s) doing work under its control that affects its quality performance.	Record	It is obligatory to maintain documented information as records regarding competence of personnel performing realization activities that may affect the product.
8.1	Operational planning and control	Determining, maintaining, and retaining documented information to ensure confidence of processes and to demonstrate the conformity of products.	Document	It is not obligatory to maintain the criteria and control of processes as documented information, but you will need it documented to perform the controls. Instructions and forms may serve here.
			Record	It is obligatory to maintain evidence as documented information that the processes were performed as planned.

(Continued)

Clause	Description	Standard Requirement	Type of Documented Information	Requirement for Documented Information
8.2.3	Review of the requirements for products and services	Results of the review of requirements related to the products and services and any new requirements for the products and services shall be maintained as documented information.	Record	It is obligatory to maintain the results of the review of requirements related to the goods and services and any new requirements for the products and services as documented information.
8.3.3	Design and development inputs	Maintaining documented information on design and development inputs.	Record	It is obligatory to maintain the inputs to development processes as documented information.
8.3.4	Design and development controls	Documented information of controls to the design and development process activities is retained.	Record	It is obligatory to maintain records as evidence for the implementation and execution of controls is required as documented information.
8.3.5	Design and development outputs	Retaining documented information on design and development outputs.	Record	It is obligatory to maintain records as evidence for the adequacy of design and development outputs.
8.3.6	Design and development changes	Retaining documented information regarding design and development changes and the relevant review.	Record	It is obligatory to maintain records regarding changes and their review in the design and the development.
8.4.1	Control of externally provided processes, products, and services/general	Establishing criteria for the evaluation, selection, and reevaluation of external providers.	Document	It is not obligatory to maintain records of these activities as documented information but you will have to provide evidence of planning and controlling them. Instructions may serve here.
8.4.1	Control of externally provided processes, products, and services/general	Results of evaluations of external provider shall be maintained as documented information.	Record	It is obligatory to maintain the results of the evaluation as documented information.
8.4.2	Type and extent of control of external provision	The organization shall determine the verification, or other activities, to ensure that the externally provided products and services meet requirements.	Document	It is not obligatory to maintain this information as documented information but you will have to provide evidence of planning and controlling them. Instructions may serve here.

(Continued)

Clause	Description	Standard Requirement	Type of Documented Information	Requirement for Documented Information
			Record	It is not obligatory to maintain records of these activities but you will have to provide evidence that performance of external providers meet the specifications.
8.4.3	Information for external providers	The organization shall communicate its requirements to external providers.	Record	It is not obligatory to maintain the records of those activities but you will have to provide evidences of performing them (e.g., purchase order).
8.5.1	Control of production and service provision	Documented information describing characteristics of the products and services, or description of activities to be performed.	Document	It is obligatory to maintain as documented information the characteristics of the products or services, or activities needed for the control of production of products.
			Document	It is obligatory to maintain as documented information the required results of the controls.
8.5.2	Identification and traceability	Documented information of unique identification of traceability.	Record	It is obligatory to maintain the outputs of tractability activities as documented information.
8.5.3	Property belonging to customers or external providers	Documented information needed to inform customer or external provider of the status of property belonging to customers or external providers.	Record	It is obligatory to maintain notices to customers or external provider regarding situation of their property as documented information.
8.5.6	Control of changes	Documented information describing the results of the review of changes.	Record	It is obligatory to maintain documented information for the review of changes and their results.
8.6	Release of goods and services	Maintaining evidence of conformity of product with the acceptance criteria.	Record	It is obligatory to maintain documented information necessary to provide evidence that products conform before releasing them and traceability of authorization.

(Continued)

Clause	Description	Standard Requirement	Type of Documented Information	Requirement for Documented Information
8.7	Control of nonconforming outputs	Documented information describing the nonconformities and any following actions taken.	Record	It is obligatory to maintain documented information for the process of identifying nonconforming product and their handling.
9.1.1	Monitoring, measurement, analysis, and evaluation—general	Process for ensuring that monitoring and measurement can be carried out.	Record	It is obligatory to maintain the results of monitoring, measurement, analysis, and evaluation as documented information.
9.1.3	Analysis and evaluation	The results of analysis and evaluation shall be used as an input to the management review.	Record	Results of analysis and evaluation are to be introduced as inputs to the management review. Although not mentioned, documented information will be needed here.
9.2	Internal audit	The organization shall retain the evidences for the implementation of audit program	Document	Planning and submitting the audit program shall be maintained as documented information.
9.2	Internal audit	The results of the internal audit will be maintained as documented information	Record	It is obligatory to maintain the audit results as documented information.
9.3.3	Management review outputs	Documented information as evidence for performing management review and of the results of management reviews	Record	It is obligatory to maintain the results of the management review as documented information.
10.2	Nonconformity and corrective action	Documented information as evidence of nonconformities, any following actions taken, and the results of any corrective action.	Record	It is obligatory to maintain the evidences of nonconformities, their handling, and results as documented information.

Bear in mind that wherever a record is mentioned there must be some kind of system that can support that record, for example, a form (which is a controlled document) or a software.

Records

Records are evidences of performing an activity and represent outputs of a process. Records provide evidence that objectives were (or were not) met. Records are logically the last level of the documentation pyramid in your QMS and are the outputs of processes, procedures, work instructions, specifications, and plans. Records may serve two main purposes:

1. Verification of execution where records are used to prove conformity to requirements or specifications. A procedure, specification, or other documented requirement demands the execution of a process or activity. With records, it is possible to verify that it was done according to the specification: resources, sequence, responsibility, and activities.
2. Evaluation of effectiveness. With the records, one can review the effectiveness of an activity and evaluate the results against criteria.

The two functionalities of a record and a form should not be confused. Once a form is filled in and filed, it becomes a record—an evidence that an activity was performed. Print a blank form—it is a document. Use it and write information in it—it is a record. A screen used to enter customer order in an ERP system may be counted as a form. The details of a certain order are the record. Types of records include

- Evidence of the supply chain: customer orders, delivery notes, invoices, credit notes, receiving slip, and certificate of compliance accepted from suppliers
- Evidence of the production phase: forms for production orders, test protocols with results, labels with production details such as serial numbers, validation forms, job release approvals, and records of quality assurance and batch approvals
- Records of maintenance of resources: records of machine maintenance, production tool status reports, and records of trainings
- Records related to quality activities: records of management review, internal audits and nonconformities, and filled-in customer questionnaires

ISO/TR 10013:2001

The ISO/TR 10013:2001 Standard guidelines for QMS documentation is a very effective tool for the development and maintenance of the documentation necessary to ensure an effective QMS. It uses disciplines that are well known in the quality management area and the standard is very easy to understand and implement. I warmly recommend using this standard while determining the structure and format of your documented information.

7.5.2 Creating and Updating

The definition of the documented information is combined from several characteristics that affect the documented information and its functionality such as identification, relevance, media, etc. All those properties must be defined and applied when designing, releasing, and updating documented information. The ISO 9001 Standard requirements are very clear:

- When creating and updating documented information, the organization shall ensure that identification and description of this documented information are defined and clear. The identification and description may include characteristics like title, date, author, or reference number.
- For each type of documented information, the appropriate format will be determined and maintained.
- For each type of documented information, the appropriate media will be determined and maintained.
- When creating and updating documented information, the organization shall apply reviews and approval needed for the suitability and adequacy of this documented information.

Identification and Description of Documents

Any documented information (internal or external) must be identified and represented with words (described) in order to make the use and function of this document clear. A document must have a name, catalogue number, or other means of identification. This is done through defining elements that can identify, describe it, and submit it to the control. Anyone in the organization that picks up the document will know where to assign it. The ISO 9001:2015 Standard requires the determination of a method for identification of documents. The following example for identification of documented information is quite basic but will hold in an audit.

The organization maintains the next operational hierarchy of documentations:

- Documented procedure that describes a process
- Work instructions that specify how activities will be performed
- Forms that document activities and deliver evidences

Let us assume that the organization manages a process for receiving customer orders. This documented procedure is numbered and identified: PR-004—Receiving Customer Order. This caption will appear on top of the documentation that describes or displays the procedure:

- PR: stands for procedure
- 004: the three digits represent the process in the workflow
- Receiving Customer Order: the name of the procedure

On the documented procedure (whether it is a diagram or a text that describes the procedure), references to lower work instructions will appear. Let us go one tier below to the work instruction. Take a look at the following number: WI-004-002: Entering Customer Order into the ERP system.

- WI: stands for "work instruction"
- 004: the first three digits represent the process
- 002: the second three digits represent the subprocess
- Entering Customer Order into the ERP system: the name of the procedure

And to it I add form FO-004-002-003: List of Open Orders

- FO: stands for form
- 004: the first three digits represent the process related to this form
- 002: the second three digits represent the subprocess related to this form
- 003: represents the third form related to this subprocess
- List of Open Orders: the name of the form

Together, they combine an identification number, maintain the interrelation between each other, but most importantly they provide identification of the documents. This kind of identification must be applied to documented information when creating or updating a document.

Identification of Records

Although the standard neglects the differentiation between documentations and records and refers to records as documented information, I would like to relate to the records because I feel that records do have special properties. Records like documents must be identified. A record must have a name, catalogue number, or other means of identification: an element that identifies it. Normally, it inherits the identification of its related document. Anyone in the organization that stumbles upon it will know where to assign it.

But important is who performed the activity. As stated before, a record is an evidence of performance. Each record must have the identity of the person that filled it or at least the function that is responsible for it. The classic way is the name, date, and a signature on the record. The identification of the person will be clear and understood. A digital signature counts as well.

Format and Structure of Documented Information

The structure and the format for each type of documented information will be determined. The objective is to constitute unity when creating a document. In other words, to create a situation where all types of documented information look alike in the organization and to avoid different formats or structures and to reduce confusion and uncertainty. The types of formats may be text, flow charts, tables, a combination of the three, or any other method that will serve the organization.

The format and structure of the document will determine the content of the document—what the document must contain. I suggest here a few elements that could appear on a document but are not obligatory:

- Title: The title should clearly identify the document. It can contain the caption and the catalog number of the document. Examples: PR-004—Receiving Customer Order, WI-004-002—Entering Customer Order into the ERP system, FO-004-002-003—List of Open Orders.
- Review, approval, and revision: Information regarding the review and approval, status and date of revision of the document must be indicated on the document.

- List of changes: Changes of revisions, the approval, and their reason could be stated on the document (when not on the document somewhere else).
- Purpose: The purpose of the document should be described on the document. The idea is to allow anyone that reads the document to understand its objective. Example: This documented procedure has the objective of defining and describing the steps that must be followed when receiving an order from a customer.
- Scope: The scope refers to the areas in the organization that the content covers. For example: "Sales" could mean the sales manager, sales personnel, and the back office. You may include which areas in the organization are not included in the scope.
- Responsibility and authority of executing the activities: Responsibilities and authorities of persons associated with the content of the document shall be defined in the document. Example: This documented procedure concerns the back office of the sales department and field sales personnel. That means that actions and activities together with the records that will be produced are in their responsibility.
- Description of activities: The activities and actions required to achieve the objective of the document are to be described. The level of detail depends on the complexity of the activities, the necessary methods, and the levels of skill and training of people that are needed in order for them to accomplish the activities. Nevertheless, a few aspects must be covered:
    - In case of documented procedure, the customer of the process and its needs must be clear.
    - In case of documented procedure, the inputs and outputs of the process must be clear.
    - Special terms and definitions related to the activities will be explained.
    - The resources required for accomplishing the objective of the process or instruction must be described (in terms of personnel, training, equipment, and materials).
    - The sequence of activities and description of the needed activities must be clear.
    - Within the activities and the sequences, it will be clear by whom or by which organizational function an activity must be performed; why, when, where, and how.
    - The controls of the process or activities and their application must be described.
    - When nonconformities are discovered, the measurements for the removal must be defined.
    - References to documented information related to the required activities shall be indicated: further instructions, forms, or other means of documented information.
- Required records: The expected records related to the activities should be indicated. When forms are to be used for these records, they should be identified and mentioned.

I prepared here an example of a work instruction designed according to the requirements mentioned earlier:

Title	WI-004-002: Entering Customer Order into the ERP System		
Review, approval, and revision	Date 01.01.2015	Revision 2nd	Approval through sales manager
List of changes	Date 01.01.2016	Change Update—Addition to Section 2—capturing of additional information necessary for transfer to production	Approval through sales manager
Purpose	Describing the steps for entering a customer order into the ERP system		
Scope	Sales back office		
Responsibility and authority	Sales representatives		
Required inputs	FO-004-002-001—Manual order from sales personnel FO-004-002-002—Order form from web application		
Description of activities	1. Terms and definitions (e.g., sales program) 2. Needed resources (e.g., user with sales permissions) 3. Sequence of activities—what is needed to be done 4. The controls of the process (e.g., activating a function that warns the user when a field is not filled out, reviewing the printout of the order)		
Reference to other documented information	PR-004: Receiving customer order (related process) WI-006-001: User guide for operating the ERP system		
FO-004-002-004: Customer order	Required records FO-004-002-003: List of open orders		

Media of Documented Information

The media on which the documented information will be saved, stored, used, or archived must be defined. The definition will refer to any kind of media used by the organization: printed paper, magnetic, or electronic. When defining the media for each type of documented information, you should consider the following aspects:

- Access of personnel to the media: When you define a type of media, be sure that all users have access to this media. For example: When you decide that forms will be saved as PDF files, be sure that all workstations have a PDF reader installed on them.
- Distribution of documented information: The media must support the distribution of documented information. For example, if you are updating a document frequently, managing it as a hard copy may not be effective because then you will need to update all the copies.
- Revision, approval, and changes of the documented information: The media must support your requirements for changes, approval of changes, and the management of revisions. For example, a word document is very much limited when it comes to managing details such as approval and revision.

- Deterioration of media: The deterioration of electronic or magnetic media on which the records are kept, for example, plays an important role. Media such as CDs or magnetic tapes do not last forever. In fact, their lifetime is shorter than you think.
- Storage: The method for storage capabilities may affect the type of media.
- Removal of documented information: The removal of media can be very complicated when documented information is scattered across many locations. Therefore it is important to decide upon where to store information.

Reviews and Approval for Suitability and Adequacy of Documented Information

Each document used by the organization must be supervised, reviewed, and approved prior to submission for use. The objective is to ensure that the document was appropriately designed, is suitable for working, and will assist the organization in meeting customer's as well as regulatory requirements. For each document, the role, function, or authority that reviews, approves, and releases it will be defined.

The function or authority that reviews and approves the document must have some degree of relevancy to the document and the activity it supports. How is relevancy determined? You need to review qualifications, experience, and background of the subject and their association to the activities. This requirement will ensure that documents will be checked by appropriate functions or roles and all important aspects will be accounted for. For example, when you design a routing card for a production process, the production manager may be responsible for reviewing that all the required fields are on the form and approving it. But if you are creating a process validation form, there will be other parties that would like to share their opinion about it: the development and the QA for example. There are situations when more than one function would need to review and approve the document. This may occur when more than one process is recorded on one document, for example, packaging instructions. Then, the storage and the production manager will need to discuss the matter and together create the optimal document. When necessary, documented information will bear a validation period.

The activity of approval and release of a document will be defined, where approval is necessary. I used to print a master copy, get it signed by the responsible party, and store it in a master documents folder. But today you can achieve this with the help of document management systems (DMS) that provides digital approvals. Usually, these systems support the standard requirements. When purchasing such a system, make sure that it answers the ISO 9001 Standard requirements. If you do not have this option, a simple table or an Excel chart describing the document and the authority for review and approval will be sufficient.

7.5.3 Control of Documented Information

After analyzing the needs for documented information necessary for operating the QMS, and developing the strategy to implement them, the organization shall establish efforts and create conditions in order to manage and control activities needed to administrate the documented information. The ISO 9001 Standard requirements:

7.5.3.1

- The organization shall ensure availability and suitability of documentation required by the QMS and by the ISO 9001 Standard.
- The organization shall control the sufficient and appropriate protection of documented information and will identify risks for its safety and integrity.

7.5.3.2

- Distribution, access, and retrieval activities necessary for the control of documented information shall be determined and maintained.
- Activities for storage and preservation, including activities to ensure that documented information remains legible necessary for the control of documented information, shall be determined and maintained.
- Activities for the control of changes of documented information shall be determined and maintained.
- The different levels of access permissions to and the authority to modify the documented information will be defined and controlled.
- Activities for retention and disposition necessary for the control of documented information shall be determined and maintained.
- Documented information of external origin needed for the planning and operation of the QMS will be identified, controlled, and appropriately distributed.
- Documented information retained as evidence of conformity shall be protected from unintended alterations.

Availability and Distribution of Documented Information

Each document will be distributed and available to the relevant parties (roles or functions) at the appropriate locations and at the appropriate point of use. Defining the availability and distribution of documents must include the following:

- User authorization: who is authorized to use a document
- Location of the document: where must a document be kept before and after use
- Form of availability: paper or magnetic media, for example

This availability should be defined and determined. I recommend tabulating that information as follows:

Type of Document	Location (a computer file path)	Media
Work instructions	Company server/department/WI	PDF files
Forms	Company server/department/forms	PDF files
Quality forms	Company server/quality/forms	PDF files

By the way, this table itself is considered as documented information.

Suitability of Documented Information

Suitability of documented information relates to the assurance that it was appropriately designed: it supports the processes and activities, is adequate for working, and will assist the organization in meeting the quality objectives and product requirements. Each document shall be approved for suitability before release for use. Which aspects indicate the suitability of documented information?

- Purpose: The purpose of the documented information must be clear to the person who designs and creates it and to the users who use it.
  - Work instruction or documented procedure: Users shall understand what is expected from them.
  - A form: Users shall understand the significant of the entries, their flow to another process, and their use as evidence.
- Properties: The properties of the documented information are defined and applied, format and structure of a document maintained, identification applied, and protection measures carried out.
- Effectiveness: The documented information achieves its objectives and the process or activity that it supports is accomplished (and not just a form that needed to be filled up in order to satisfy the internal audit)

Legibility of Documented Information

Legibility of documented information refers to its quality of being understood:

- Documents will be legible enough to be identified.
- Records will be legible, readable, and retrievable for their entire retention time.

When you plan and implement the use of documents in the organization and the creation of records, some factors that may affect legibility must be considered:

- Intactness, clarity, usability, and accuracy: The factors almost directly affect legibility. Documents and records must be handled with care and the personnel are responsible for maintaining accurate and legible records with all the required information filled out and correctly filed:
  - Documented information must bear all the required information and details.
  - I personally recommend recording the date of filling the records. People with problematic handwriting will have to work harder and provide legible data and information on the record.
- Deterioration: The deterioration of media on which the documented information is saved plays an important role. For example, media such as CDs or magnetic tapes do not last forever.
- Update of technology: Changing of technology used to save and access the documented information is a major aspect. The organization must ensure that they may access documented information in the future. A good example is the update of computer operation systems where old files cannot be read with new versions.
- Relevance: It will be possible to submit the documented information to its relevant context, process or specific product or process:
  - On documents, it will be clear to which process and product of part of the QMS they refer.
  - Records shall not carry or store information that is not relevant to them.
- Digitization: Conversion of analog records or hard copies to digital formats can result in loss of information.

- If you decide to scan records to computer files, please ensure that the scanning is done appropriately and the information on record is readable and clear after the scan. I sometimes find myself on all fours searching for old documents in cold cellars just because it was impossible to understand the information from the scanned record.

Storage of Documented Information

Storage locations, facilities, and conditions of documented information must be defined for each type of document and record.

- Documents: The designer and the users of all kinds of documents shall know the location where the document is stored and the media on which it must be saved.
- Records: The producers of records shall know where and how records must be saved; a closet, a designated folder, as an electronic file or folder, or as a software.

The objective is to allow each person in the organization to retrieve or track back documented information according to its type and context. Storage definition shall cover the archiving of the documented information as well. The organization shall consider storage factors that may create conditions for the protection and preservation of documented information. Important factors when considering the storage of documented information are

- Media of documented information: The type of media on which the documented information is saved defines usually the storage requirements and the physical conditions such as temperature, humidity, or weight. Where there are known risks to the media, precautions will be taken.
- Access and retrieval of documented information: Storage conditions must support the access requirements to the documented information; who has access to and how and when to document necessary information.
  - It is critical to avoid storing documented information on personal workstations such as computers in order to provide full access. For example, when employees leave or change jobs, the IT administration deletes the contents of their hard disk or personal files regardless of its value or significance to the QMS.
- Volume and growth rate of documented information: The storage must supply enough space according to the activity of the organization and place limitations must be accounted. If the organization is producing millions of records per year, the server capabilities must support this volume. By capabilities, I mean not only space but also the performance of the server must support access of many users.
- Workflow: Storage of documented information must support the workflow, in general, and processes, in particular. Documented information usually acts as inputs or outputs to processes and all participants of the processes must have access to it.
  - By storing documented information along the work flow, you may avoid losing it.
- Risks to the integrity of documented information: The integrity of documented information refers to the required conditions needed to preserve the content and completeness of documented information. The location or facility of the storage should not be exposed to external risks such as a flood, fire, or burglary and must prevent damages to the documented information.

- Security and confidentiality requirements: Security protection measure (physical as well as logical) shall be implemented.
- Backup systems: Backup processes should be regularly executed and the retrieval of documents and records shall be planned in case of a disaster.

Storage must take into account the life cycle of documented information in the organization where for each phase it must be defined how documented information is stored. For example, let's look at the life cycle of a customer order (a record):

Phase	Action
Creation of the records	Signed customer order—a printed/scanned form filed in a designated folder
Assigning the records to some logical system	Enter the customer order into an ERP system or print a controlled form as Word document
Maintenance of the records	Integrity of records is controlled through verifications that direct the user in entering information into the ERP system or forms with defined fields
Use of the records in the workflow	Transfer to production—record in the ERP system linked with customer order number or scheduling a manual product order based on the customer order
Disposition or archiving of records	Once the customer order is closed, it receives a certain status or it will be identified on the form (e.g., with a stamp)

When storage of records is outsourced, all of these must be discussed with the provider.

In my treatment of protection and preservation of documented information, I separate documents from records because I find that these two types of documented information have different needs and therefore set different requirements.

Protection and Preservation of Documents

The organization shall determine actions to ensure that documents of all kinds will remain safe and available for use. These actions shall refer to the appropriate maintenance conditions of documented information. What are the appropriate conditions? The ones that will prevent loss of confidentiality, improper use, or loss of integrity.

In practice, it is necessary to identify which conditions may affect each type of document in the organization. Visit the different departments in the organization, take a look, and try to review the parameters that might affect the safety of the respective documents. When conditions that may harm or pose risks to documents are found, you must define what measures are required in order to avoid them. This is a kind of small-scale risk analysis regarding the safety of a document. Parameters included here are

- Environment
- Safety
- Ability to preserve retention value
- Handling of documents
- Unauthorized access

For example, water and heat may damage a paper document and therefore in such environments work instructions or documented procedures should be laminated, placed in a heatproof pouch, and kept away from hazards.

Backup and support systems for the company's server are another example of protection. The organization is required to demonstrate that sufficient backup system is in place that allows access to documents when, for example, electronic systems crash or generally become unavailable. Another aspect is human resource behavior with documents. Define how employees should handle documents and where they should be stored. I find myself printing work instructions again and again because they get lost all the time.

Protection and Preservation of Records

Protection and preservation of records refers to the equipment and facilities that manage records and to the methods and techniques that are planned in order to protect records. Integrity of records is critical for business continuity, and therefore, records will be appropriately protected. Protection of records ensures that the information contained in the records is available and accurate for future reference and records must not be changed over time. For each type of record, you need to identify the risks (for damaged or loss) and apply appropriate methods of protection and preservation. In doing so, the following should be considered.

- Legibility: Protection of records shall support their legibility and ensure that the content in the records will remain clear and available for use over the life cycle of the record.
- Storage: It is necessary to ensure that storage of records is adequately in place according to the type of the record, for example, organizing folders in the workstations, ensuring no confusion when filing, storing, or archiving the records or verifying the functionality of data repository.
- Flow of information: It is necessary to ensure that records will reach their destination and will not be lost. This is important for the flawlessness of the workflow.
- Unauthorized access: It is necessary to ensure that no misuse of records is possible (through management of access and authorization).

The solution for protection shall extend to all areas and scopes of the organization. Backup systems are one good example of how the organization protects its records. Such a system proves the integrity of records as well as provided a solution in case of a disaster.

Basically, an analysis regarding the following issues is required:

- Who is responsible for records?
- Which conditions may pose risks to the records?
- What measures are necessary in order to ensure records are protected?
- When and where should those measures be implemented?

A Documented Method for Managing Documents and Records

The ISO 9001:2015 Standard does not demand a documented method, that is to say, a procedure, for controlling your documents and records. The last revision of the standard, the ISO 9001:2008, demanded a specific procedure for managing records (e.g., Documents

Control and Records Control). In case you already have such procedures in place, you are entitled to keep and maintain them. But do review their content once more and make sure that it covers all of the ISO 9001:2015 Standard requirements. Personally, I would adopt these procedures and update them regularly because I find these issues very critical and they need the support of a documentation that describes the methods, activities, and technologies used to implement the control of documents or records and the required evidences. In order to create an effective procedure, try to relate and include all the issues that were mentioned in this chapter 7.5—Documented information.

Access and Authorization to Documented Information

Authorization and access to documented information shall be defined and applied, and it is recommended to develop a policy in this issue. "Access" relates to

- The permission to view the documented information—users can only view documents or records but cannot change them (or at least the master document).
- The permission and authority to create, view, and change the documented information.

Management of access is required in order to protect documents from unauthorized use or modification and records from unauthorized entries or modifications. Managing access and authorization means administrating the permissions of users or personnel to access different documented information in the organization. What factors may influence the permission?

- Workflow: Access to documented information must support the workflow. In other words, you must allow employees the access to documented information based on their function in the workflow.
- Roles, responsibilities, and authorities in the organization: Access to documented information is much affected by the functions in the organization and their need to view, use, or edit documented information.
- Language: Indirectly, the language of documents and records enables access to the information in the document or record.
- Content of the documented information: Access to documented information may be affected from the content of the documented information. For example, some documents will bear customer's property and include classified information and thus must be confidential or at least restricted for changes.
- Media: Access and authorization depends on the type of documented information and the media on which the documented information is maintained.

In practice, the organization must demonstrate the ability to manage access and authority and to provide employees the access to documented information at "point of use," meaning at their place of work. Which aspects of the documented information are affected by access?

- Retrievability: Access to documented information defines the capabilities of retrieving the data. In other words, how users or employees may access the data: existence of hard copies where necessary, saved documents on a server, access through the data management systems, etc. For example, setting an indexing and filing system for records (hard copy or computer) will ensure easy retrieval.

- Storage: The storage of the documented information must support access to it.
- Security: The security measures and their degree will be affected from the access. For example, where documented information is held electronically, adequate password or other access systems will be applied. Another example is the protection through antivirus and spy systems.
- Functionality: Access to documented information will be based on its use. For example, an ERP or CRM system can manage access and authorization to documented information (forms and records) through the management of user permission where each user receives access to defined system applications.

Updates and Changes to Documented Information and Version Control

Changes or updates to documented information must be controlled and documented. The documented information will bear details regarding the status revision and changes that have occurred in the documented information. The objective of controlling versions is to ensure that individuals are using the correct information and to guide them in accomplishing activities and operating the QMS. The information and requirements that we use to operate the QMS and that appears on documented information change over time as conditions and requirements change. Changes in documented information may affect the QMS and its operation. Not controlling the versions of documented information may result in the use of obsolete or unapproved documents or distribution of incorrect records. This is why version control is critical for documented information. What are the attributes of version control?

- Version status: The revision of the version must always be identified and clear to the user.
- Version number control: Numbering versions is necessary for the tracking of the version and indicates which status has a document.
- Centralizing documents: All documents must be centralized in order to eliminate "islands" of independent documented information, for example, saving documented information on personal computers.
- Format: The format in which documents are saved and distributed must be defined.
- Distribution: Distribution as well as disposition or removal of obsolete versions must be implemented in order to ensure that only valid versions are available. Because it is not always possible for a user to verify that he or she has the latest version, you must ensure that only the most recent version of a documented information is available for use.
  - User must have access only to valid forms
  - User must have access only to valid records
- Disposal: It must be defined when and how documented information will be disposed.

When is it needed to control and document changes? And which changes are needed to be documented? Each time a change may affect the activity that the documented information supports (internal as well as external). For example when

- An employee requests for a change in a form
- A change occurs in a procedure or a process due to a customer complaint
- A customer requests for a change in product specifications
- There is a change in a requirement such as regulation or standard

You need to identify these events and requests and initiate a change, review the change and its consequences, and reapprove the document. The control of changes will determine

- Identification of events or requests for change
- Identification of the relevant documents that will be affected by the change
- Identification of parties, roles, and authorities needed for the review and approval
- Method of review: what are the inputs, where and when will the review take place, and who will participate
- Form of approval and necessary records
- Submission to a process of removing obsolete editions, and distributing and implanting the new one

Which details are expected to be recorded?

- Date of the review
- Identification of the relevant document, its editions, or revision
- The reason and cause for the change
- What was changed
- Who reviewed
- Remarks
- Consequences for regulatory requirements
- Parties present at the occasion
- Approval

The details may appear in a designated form that follows all the changes, as in the following table:

Document number	FO-004-002-001
Edition	001
Change requested by	Mr. White—Quality Assurance manager
Reason for a change	Request to add another field to the form:
	Reason for stopping a machine
	The field is necessary in order to document the reason for stopping a machine during a production batch
Approved by	Mr. Pink—Production manager
Signature	
Date of approval	01.03.2016
New edition	002
Communication of change	01.03.2016
Removal of old editions	05.03.2016

I believe that managing such a table regularly will hold an audit. Controlling the numbering of the revision is very much challenging, particularly in paper-based systems where the human factor is significant. Numbering of documented information can be done as follows: The revision will be indicated in the new form—FO-004-002-003(003) or FO-004-002-003 (01/03/2016).

Another method for maintaining version control is implementing a system for the management of documents' life cycle. With documents' life cycle, the stages that a document goes through from its creation to its eventual archival or disposal are controlled. Normally, organizations purchase a designated software for this kind of management. This type of system determines for each type of document the

- Creation of documents
- Metadata of documents
- Approval and release
- Retention time
- Storage
- Management of changes
- Distribution and retrieval
- Archiving
- Backups

Change of Records

Credibility and reliability of records ensures that the information contained within the record is accurate for reference in the future. Change of records might impact processes or products (good or services) and may affect employees at all levels of the organization. For example, if in a service-providing organization the contract details of a customer are approved without a careful review, it will directly affect the service that the customer receives. It is important to identify the reasons or events that may cause a change in records. For example:

- Updates to a database, including changing records, for example, a cross-update to product descriptions—who is authorized to make such a change?
- Change of format of documented information may initiate change in records, for example, fields may be neglected or added on forms.
- Change of technology in the organization may require change in records; how employees process and distribute records (in all forms—physical and digital). If the organization decides to implement new software, it might require more or different information and thus the records will have to change.
- Changes in regulatory requirements, standards, or laws may initiate creation of new types of records or the alteration of existing records or set new retention periods.

The organization must determine a method to ensure that the changes are identified and controlled.

When a printed record is manually altered or corrected, I recommend that

- The original entry stay visible on the record
- The person who altered the record be identified
- The modification be dated (similar to a correction made on a contract)
- The reason for the correction be stated

Protection from Unintended Change of Evidence of Conformity

The requirements for protection against unintended alteration of evidence of conformity are needed in order to preserve the accuracy, credibility, and reliability of documented information. This evidence has the very important role of proving

- That actions were undertaken in order to maintain conformity
- The conformity of products and services to the requirements of customer or regulators are met
- The ability of the organization to meet the requirements

Unintended changes may be accidental, causeless, inadvertent, or made intentionally—any change that is not part of the process of using the documented information. Which records are needed to be protected from changes or alterations?

- Records or action needed to address risks or opportunities (Clause 6.1)
- Records needed for the maintenance of infrastructures (Clause 7.1.3)
- Records needed for the maintenance of process environment (Clause 7.1.4)
- Records needed for the maintenance of monitoring and measuring devices (Clause 7.1.5)
- Records proving the allocation of resources for the realization (Clause 8.1)
- Records proving the execution of necessary activities for the realization (Clause 8.1)
- Records of review of changes in design and development (Clause 8.3.6)
- Records related to the acceptance of purchased goods or services (Clause 8.4.2)
- Records needed for identification and traceability (Clause 8.5.2)
- Records needed for the preservation of products or services (Clause 8.5.4)
- Records for review of changes for production or service provision (Clause 8.5.6)
- Records of release activities (Clause 8.1)
- Results of analysis and evaluation (Clause 9.1.3)

Retention of Documented Information

Retention refers to the ability of the organization to

- Keep documented information in a secured and protected manner that will preserve the knowledge information and data in them
- To retrieve the documented information for possible future use or application

The retention time of documented information (documents or records) will be defined and known in the organization. For obsolete editions of a documented information, it is necessary to define the retention time according to the following conditions:

- The organization and management of documented information determine the ability to retrieve it.
- When regulatory requirements set retention time for obsolete documents, or records, the retention time will be accordingly maintained.
- Normally, retention time of an obsolete edition of a document will not be less than the lifetime of records that this document bears.

Archiving of Documents

Activities for archiving old documents and obsolete editions will be determined. It is necessary to define what is to be done with old versions that are not updated, how one handles them, and whether they are to be disposed or archived:

- The storage of unupdated documents will be defined. The activities will define the location of the archiving, the retention time, and in which media.
- Invalid documents that are not disposed are to be indicated or marked. The mark will indicate the status of the documented information beyond any doubt and ensure that no one will use it.

If printed documents become invalid for use, you must ensure that they are not used:

- Destroy them.
- Mark each one with a stamp or a watermark.
- If the documents are saved on the company's server, responsibility must be assigned to someone for deleting or archiving obsolete documents from the server or replacing them with updated ones.

Implementing the awareness of not using obsolete documented information will be applied throughout the organization. Persons must understand

- How to identify an invalid documented information
- What are the consequences of using it
- What is to be done when one finds invalid documented information at the place of work

Archiving Records

Activities for archiving records will be determined. It is required to define what should be done with obsolete documented information or old records. The method shall refer to the following aspects:

- Is archiving necessary or disposal may be sufficient?
- How should the archived records be handled? Where should the archive be stored or in which media must the records be archived?
- How long should the documented information or records be retained?
- Are there any regulations or specifications that dictate requirements for archiving (very much common in the medical device industry)?

Once again, I suggest managing a table that describes the archiving parameters for each type of documented information:

Type	Test Protocols
Location	Server:\Public Quality\Quality Archive\Test Protocols (a computer file path)
Media	PDF—scan of the paper form
Retention time	5 years
Archiving approved by	Mr. White—Quality Assurance manager

Disposal of Documented Information

Disposal of documented information refers to the action or stage of a document or a record in which it is either destroyed or permanently retained. The ISO 9001 Standard expects that when a document or record is no longer needed or required, it would be removed from the QMS in order to ensure that it will not be used again:

- Users will not have the possibility to use this obsolete document
- Users will have no access to this certain record

Each type of documented information shall have its own disposal protocols indicating who will dispose it and when and how it will be disposed. The activity of disposal will relate to the following issues:

- For each type of documented information, disposal activities will be defined based on different properties of the documented information such as media, storage, and utilization.
- Authorities for disposal and approval of documented information will be determined.
- Removal from the QMS— make sure that all copies and editions are removed and cannot be used anywhere in the organization.
- It is recommended to set a schedule for disposal according to the retention time of each type of documented information.
- When documented information is converted to digital media, in some cases it will be possible to dispose the old media. But make sure that the retention time is valid for the new media.
- In some cases where documented information is transferred to outsourced archive or storage, it may be possible to dispose it.

In cases where disposal of documented information must be approved, I would set a process for request of documented information disposal. In this case, you can ensure that only approved documented information was disposed. Such processes can be easily covered with a form. In other cases, disposal may not mean complete destruction or deletion of documented information but removal from the QMS and transfer to archive.

A good example for managing authority for disposal is administrating rights to edit folders, where it is possible to allow certain users to delete files from a folder. Another example is managing user authorities on management systems such as ERP or CRM where certain users have authority only to view records, others have authorities to perform transactions or actions with records, and other have the authority to delete them permanently from the database.

Documented Information of External Origin

The ISO Standard 9001:2015 explicitly indicates that documented information that was originally created externally but is necessary for the planning and operation of the QMS will be submitted to the controls suggested in this clause. For more details regarding the determination of external documented information, please see chapter 7.5.1—External documentation (of this book).

Nevertheless, when addressing documented information of external origin, you must consider the aspects mentioned in clause 7.5.3:

- The documented information will be available at the relevant workstations at the appropriate stages of the workflow.
- The documented information will be suitable for use and remain legible.
- Storage, protection, and preservation requirements and conditions will be determined for documented information of external origin.
- Methods and activities for updating, removing, and disposition will be defined.
- Distribution, access, and use authorities will be defined and managed.

Method for Controlling Documents

As mentioned earlier, these is no standard requirement form for the management and control of the documented information, but you may follow a procedure to do so (what I would personally do). Another method for maintaining these requirements effectively is by creating charts or tables for managing all these demands where description of activities and the definitions regarding the documented information will be charted on tables according to the issues that the method will cover: chart for updates, chart for distribution, chart for editions, chart for revisions, chart for changes, chart for control, etc. (see paragraphs 7.5.2—Creating and updating documented information/format and structure of documented information and 7.5.3—Control of documented information/availability and distribution of documented information or storage of documented information for more details and examples).

Another effective method for managing documented information is by implementing a document management system (DMS) or a records and information management (RIM) system. Such systems allow the administration and control of documented information through the management of metadata of the documented information (metadata refers to details regarding the documented information—origin, creator, version, date of creation, relevance, ownership, etc.) and the application of digital tools that submit the documented information to a logical process. What is the difference between these two systems—DMS and RIM?

- DMS: A computer program used to track and manage digitally the life cycle of a document. The DMS system performs actions like saving history of documents, managing versions of documents, and managing the storage and distribution of documents.
- RIM: A software that allows the practice of controlling records—data and information in the organization covering the complete life cycle of a record. The RIM system performs actions like arranging and categorizing, prioritizing, storing, securing, archiving, preserving, retrieving, tracking, and disposing of records.

My definite suggestion when purchasing and implementing such a system is to ensure that the system covers all of the ISO 9001:2015 Standard requirements. Otherwise, you may find yourself managing parts of documented information manually.

8 Operation

8.1 Operational Planning and Control

Operational planning and control initiates the master planning for the realization of products or services with the objectives of planning, realizing, controlling, leading, guiding, and instructing all participants on the different functions and roles that are involved in the realization of a product: how to manage design and development, how to prepare for the realization, how to identify and locate the appropriate resources, which activities are needed, which controls are to be applied, which documented information is necessary, how one verifies or validates the results, and which evidences are expected. Let us review the ISO 9001 Standard requirements:

- Processes, methods, and activities that were planned according to the requirements in clause 4.4—Quality management system and its processes—and are needed to meet requirements shall be developed, planned, implemented, and controlled.
- The quality objectives at relevant functions, levels, and processes shall be considered while planning the realization of the product.
- The methods and activities shall relate the actions needed to address risks and opportunities determined according to the requirements of clause 6—Planning.
- The planning of methods and activities for the realization of the product shall include the appropriate control of processes in accordance with the defined criteria.
  - The appropriate criteria (acceptance or rejection) for these processes and activities shall be determined.
  - The appropriate criteria (acceptance or rejection) for products and services shall be determined.
- The resources needed to achieve conformity to product and service requirements shall be determined.
- The organization shall determine the documented information necessary to provide evidences and confidence that processes were performed as planned and that the requirements were met.
- The organization shall determine the documented information necessary to provide evidences and confidence that product and service conform to their requirements.
- The outputs of the planning shall be suitable for the organization's operations; the organization may be able to maintain them and they will support its operations.

- Changes shall be submitted to the controls suggested in clause 6—Planning:
  - Planned changes in the planning and the realization of the product shall be controlled.
  - Unintended changes shall be reviewed for their consequences or any adverse effects, and actions to mitigate risks shall be carried out.
- Outsourced processes that are part of the process realization shall be controlled according to the requirements of clause 8.4—Control of external provision of goods and services.

Chapter 8.1 lays out the principles with which the organization should integrate the requirements of the quality management system (QMS) in the realization of a product. The ISO 9001 Standard requires practical actions needed for the planning and the controlling. The goals of practicing operational planning and control are to develop an interface between the QMS and the realization processes and to provide a practical method to execute quality requirements. The objective is to provide sufficient instructions to persons who realize the product or provide them with the services necessary for realization.

Planning Quality with a Quality Plan

In this book, I promote the use of the quality plan because I find it to be the most effective method for planning quality. Quality planning requires a type of blueprint or guideline for the organization that states how quality requirements are to be achieved, the quality plan, its methodical approach or structure that describes and specifies all the requirements needed to be followed, met, maintained, and documented while realizing the product and consists of all necessary information that would assist the participants to manufacture a product that meets its requirements. I highly recommend the maintenance of such documentation. The quality plan integrates all relevant demands for activities, resources, and information concerning the realization of a product and makes them available to any interested party. In other words, when one carries out all that is mentioned on this plan, the product is bound to meet its requirements. The goals of the quality plan include the following:

- Identifying all the specifications and characteristics of a product including quality requirements. These must match with the expectations of the customers.
- Identifying all the required processes and their activities needed for product realization.
- Describing the necessary or expected inputs and outputs of each activity and so defining the relations between the various processes.
- Defining the required resources and conditions needed to support these activities.
- Defining the controls that will ensure intended outcomes—validation and verification.
- Defining the criteria for acceptance (or rejection) of outputs of activities (processes or products).

The activities included in the quality plan must be determined by every department or area of the organization that takes part in the realization of the product or has influence on the quality of the product: sales and marketing, development, process design, production, quality, and logistics. The idea is to let each person that uses the plan to deliver his or her inputs and to allow understanding what the other expects.

With regard to the customers, the quality plan must consider specifications, methods, and schedules that are dictated or requested by the customers, and when required, the quality plan shall be endorsed by the customer. The quality plan and its constituents are to be reviewed and updated regularly. Updating this plan shall be subject to the requirements of clause 6.3—Planning of changes.

The planning shall include the following information:

- Identification of the product—information that will link the planning with a product or a service
- The relevant process flow diagrams
- The different levels of realization—manufacturing, testing, storing, or provision of services
- The different operations needed for the realization of goods or services
- Reference to documented information like work instructions or procedures—documented information that describes which activities are needed in order to realize the product
- Quality requirements and process controls including instructions or test protocols—these will define which methods and instruments are to be used to control processes or their outputs and which process parameters are to be controlled
- Required documented information for evidence—the documented information that will provide evidence that the operations were performed and that results meet the specifications
- Responsibilities, functions, and authorities for performing operations, activities, or processes related to the realization of the product

Where purchase and requirements for external providers are involved, they will be referred to. The plan shall include or refer to other documentations indicating characteristics of processes, required activities, and corrective actions necessary to control the processes. The following are types of documented information that may be used or referred to by the quality planning:

- Customer specifications
- Drawings
- Standards and regulations
- Process diagrams
- Instructions, such as work instructions
- Test protocols

Reference to the Quality Objectives

Establishing and maintaining a quality planning means implementing the QMS and its objectives while realizing products or services. Quality planning is to promote quality objectives by describing the activities, resources, controls, and documented information that shall support the achievement of relevant quality objectives. We already know that processes are to be planned and controlled in accordance with the organization's strategy. The quality plan provides you with a practical approach to meet these objectives and implement the strategy by defining the exact tools and measures needed to achieve them, that is, by performing these activities, the quality objectives of the organization shall be reached.

For example, let us assume that the organization decided that one of its quality objectives is to reduce the number of customer complaints. After a review of last year's complaints, it was discovered that most of these complaints are related to the packaging activities and the packaging of the product. The conclusion is that further controls related to those activities must be planned and implemented in order to achieve this specific quality objective—reduce the number of such complaints. The quality plan is where you act to implement these controls.

Identifying Process Operations for Realizing Goods and Services

In clause 4.4 the standard specifically employs the process approach and requests:

- The establishment of a QMS according to the ISO 9001:2015 Standard
- The definition of processes and their interactions necessary for the operation of this QMS
- The improvement of the QMS and its effectiveness

The planning of the operations and their control is the exact point where the outputs of the process approach (the definition of processes and their interactions and controls) are in use. Identifying the processes and activities necessary for the realization of goods or services is the first step of the operational planning and control. The definitions shall relate to the following:

- Specifications of activities needed to be carried out: methods, activities, techniques, practices, responsibilities, documentations, and specific records
- Required resources necessary to support the activities: infrastructure and human resources
- Documentations such as procedures, flow charts, diagrams, routing charts, check-lists, work instructions, test instructions, or forms of any kind
- Records needed to provide evidence for these processes

At the end of this definition, the organization shall have a list of processes and activities that manage the realization of the product and thus must be planned and controlled.

The main function of the quality plan is the gathering and description of all activities, operations, and processes with which the organization realizes goods or services. The planning shall relate to activities since the earliest phase of the realization and shall cover all stages of realization. Which types of activities are expected?

- Management activities
- Provision of resources
- Activities for product realization
- Activities of monitoring and measurement including verifications and validations
- Reviewing activities
- Post-delivery activities

The above may include references to procedures, flowcharts, diagrams, routing charts, checklists, work instructions, test instructions, or forms of any kind. One may define generic procedures that relate generally to realization activities or plan specific procedures using a high level of details relating to specific process or product. The end result of this stage in the planning is that for each product, semiproduct, or service, the necessary activities for its realization are identified and defined.

An important aspect to be considered in the planning is the integration of quality principles and processes that are required by this standard. For example, documented information—the quality plan—shall ensure that all required documented information was considered during the planning; specifications are available and records are maintained. Another important aspect is customers' expectations.

Identifying Product Characteristics

The quality planning shall identify all of the product characteristics with respect to customer expectations or specifications:

- When necessary the planning shall indicate all the components, sub- or semiproducts, bill of materials, parts, and materials of the product.
- The plan will identify different characteristics of a product such as geometrical, material, or functional requirements.
- It is necessary to specify the tests that must be performed in order to ensure the adherence to the specifications of the product.
- A description of the material and function tests that are required and were determined either during the design and development phase or by the customer.
- Identification of all monitoring and measuring devices that are necessary for measuring products and process outputs.
- A test for sampling products from a process.
- Requirements for purchase and external providers will be specified.
- A reference sample of the product may be used during the planning.
- If applicable, the management and routing of materials will be described.

When it is not applicable to include all this information, reference to the documents where this information exists will be included in the plan.

Establishing Criteria for Acceptance of Products and Services

The organization shall establish criteria for

- Processes
- Acceptance of products and services

A criterion is an established set of values with reference to the outcome of an activity used for the acceptance of process outputs (e.g., products or services). In order to evaluate whether process outputs met requirements and to decide whether to accept or reject the outputs, it is necessary to have some kind of a standard on which a judgment or decision may be based. Establishing criteria for processes means setting a basis for comparison and evaluation of process outputs:

- The criteria are used for examining a particular set of issues and its results.
- The criteria are used to ensure that certain (defined) conditions are in place.
- The criteria are necessary for the evaluation of the accomplishment of the goals and objectives of a process.
- The criteria should allow you to measure a process and identify trends or changes.

Based on the criteria, a decision regarding the acceptance or rejection of a process, an activity, or an outcome could be made, and they have the instructional roles and guidelines that allow personnel to make a judgment or decision regarding process outputs, which include the following:

- Compliance with applicable standards
- Compliance with applicable regulations
- Compliance with quality objectives
- Compliance with intended outputs
- Compliance with product specifications

For example, in compliance with applicable quality objectives, when planning the criteria of a process, the relevant quality objectives may be referred to. If one quality objective is to reduce customer complaints and an analysis of the complaints show that a specific product characteristic generates the problem and the following complaints, it would be necessary to define the appropriate control and relevant criterion for this characteristic.

The following are the types of criteria:

- Criteria for acceptance or rejection of process outputs—criteria that would ensure that process outputs achieve their objectives
- Criteria for acceptance or rejection of final product—criteria that would ensure that all product requirements are realized, for example, test protocols or test instructions
- Criteria for quality of special processes—criteria for validation and approval of processes where the resulting output cannot be verified by subsequent monitoring or measurement

- Criteria for customer satisfaction—criteria that will be used to evaluate the degree of customer satisfaction
- Criteria for the evaluation of performance of external providers (suppliers or subcontractors)—criteria related to performance of suppliers
- Criteria for the release of products or services—criteria that will be used to decide whether the product meets all its requirements
- Criteria needed to evaluate changes in processes or product requirements—criteria necessary to ensure that all conditions for the release of changes in process or product are defined and to evaluate the consequences of the change
- Criteria for process environment—criteria necessary for approval that environmental work conditions that may affect processes or products were maintained
- Criteria for the handling of nonconformed products—criteria that should assist in deciding what is to be done with nonconformed products: approval for use, disposal, segregation, or rework

The following are principles for the setting up or the establishing of criteria:

- A determined method shall dictate the use of the criteria.
- The frequency of the measurement shall be defined and will be appropriate to the process or product it is measuring.
- Methods for analyzing and publishing of data shall be defined.
- The responsibility for conducting the measurement shall be defined.
- The distribution of data shall be defined.

After defining the appropriate criteria, it is necessary to bring it into effect, that is, for each activity or operation that must be measured, inspected, verified, or validated, it must be clear what the relevant criteria are. One must provide users, personnel, or employees the ability to decide upon the quality of an output: a product or a process. This link between activities and their relevant criteria may easily be executed on the quality plan—indicate to users of what they need to control and where the information is available. The criteria may be documented on

- Contracts
- Work instructions
- Test instructions
- Procedures
- Guidelines
- Designated documented information
- Standards or norms
- Regulations

Resources Necessary to Support the Realization

The resources necessary for the realization of a product must be defined during the quality planning: facilities, machinery, knowledge, information, necessary work environment, storage requirements, working tools, and human resources—including training and qualification requirements. These resources are supposed to be clear at this point and arise from the requirements and specifications of the products and services. The objectives are as follows:

- To ensure the identification and definition of resources necessary to support the operation and monitoring processes and to deploy needed activities:
  - The identification of outsourced resources shall be included.
- To promote the availability of resources needed to achieve customer satisfaction.
- To ensure the correct planning of resources, which means their effective allocation and efficient usage.

The end result of this stage of the planning is a listing of the needed resources for each activity or process related to the realization of the product. If the organization is outsourcing activities, the needed resources for those outsourced activities such as information, knowledge, expertise, technology, processes, and shared training shall be defined.

Let us review a common example—planning of human resources. In this case, it is necessary to define the required human resources for the realization of the product that will ensure fulfillment of customer requirements. Which aspects must be considered?

- Qualification, knowledge, or experience
- Training
- Availability

The above become significant in cases where only certain employees can perform certain activities. In situations where the organization must employ workers who have specific training in order to realize the product, it should be mentioned in the plan. The resources may be defined in a production plan along with the setup of machines, availability of production tools, and preparation of monitoring and measuring devices.

Controls of Processes

Process controls are needed for obtaining data and information necessary for the release of processes or products. Applying controls of processes and products supports the process approach, where the organization is required to ensure that the control of processes is effective and implemented in the QMS. These controls are the means by which the business strategy (that was translated into activities and operations) is being verified. Goals of control of processes include

- Monitoring and controlling activities of the QMS
- Monitoring and controlling the use of resources of processes
- Monitoring, measuring, analyzing, and improving operations related to the realization of the product
- Ensuring process parameter and product characteristics
- Enhancing the effectiveness of processes through the use of appropriate methods

The process controls define the steps and operations for evaluating process outputs with their relevant criteria needed to ensure that processes meet their requirements. They

- Define which activities are included in the controls (and define those that are excluded)
- Provide means to collect data regarding the performance of a process or the characteristics of a product
- Ensure that processes are performed within desired tolerances or specified scope (where a process begins and ends)
- Ensure that process outcomes stay within a desired range, which is defined through the criteria
- Enable the consistent realization of goods or services
- Achieve consistent, targeted results aligned with the organization's strategic goals and quality objectives

The controls of the processes indicate points in the operation or stages where significant factors affect the process and its outcomes. Controlling these process indicators provides significant information regarding process performance related to the quality objectives. One example is the key performance indicators (KPIs) where critical factors in the process are related to the quality objectives. Controlling of these KPIs

- Evaluates variations in a process
- Ensures that operations are being performed correctly
- Ensures that various elements in the process are functioning correctly
- Ensures that processes interact with each other (an objective of the process approach)

The controls of processes monitor their outcomes and provide the information necessary to adjust resources in order to meet process objectives. We already know that output of one process is the input to the next process. And that gives a significant weight to the controls. Each process requires its appropriate controls needed for the demonstration of the ability to achieve planned results and for the transfer of desired outcomes between processes. Outputs of the design and development may indicate which controls are needed to maintain product characteristics and process parameters.

The controls shall prove the next process aspects:

- The appropriate inputs are delivered to the process.
- The necessary activities of the processes are carried out.
- The outputs are desirable and meet the specifications.
- Undesirable outputs are prevented.

Which process elements must be controlled?

- The resources that are applied to the process
- The process environment that is supporting the process

- The inputs that are delivered to the process
- Skills and qualifications of personnel that are operating the process
- Necessary knowledge is in place
- Necessary documented information is used to ensure effective planning, operation, and control of the processes is available
- The outputs of the process

Until now we discussed the inputs for the planning of controls over processes: which aspects must be considered and addressed when planning controls. Let us now proceed to the practical area: the quality planning must outline how the organization shall implement the controls in its internal business processes and describe the activities for applying these controls, that is, what are the verification or validation activities. The main Idea: While a process progresses, methods and tools are collecting and evaluating data and comparing it to the defined criteria. Their goal is to ensure that process objectives are met through reviewing the progress of the process, evaluating its outcomes, and using the criteria to arrive at decisions. How is it done?

- A defined method for the measurement and control is implemented.
- The responsible parties for executing the activities are determined.
- The method refers to the appropriate criteria.
- It is determined where and how the method shall control the process.
- The method collects data that are relevant to the objectives of the process.
- The data are being analyzed, compared to the criteria, and reported.
- The relevant documented information is updated with the new data.

The controls should provide you with quantitative or qualitative data about the effectiveness of the process. The methods with which the performance of a process will be controlled should be practicable and appropriate to the nature of these processes and their outcomes and will be adapted to the type of the activities. The types of process controls include

- Process controls—controls that collect data regarding outputs of processes: industrial control systems, test protocols, procedures, and work instructions
- Controls of qualifications—controls that verify that the appropriate conditions for the resources are provided to the process
- Environmental controls—controls that collect data regarding the environmental parameters that influence the product and process: environmental control devices
- Regulatory controls—controls applied to verify that regulatory requirements are applied: audits and documented information
- Quality management-related controls—controls supervising that quality requirements are being met: internal audit and control of nonconformities

Figure 8.1 illustrates how the control of processes shall be implemented during their planning:

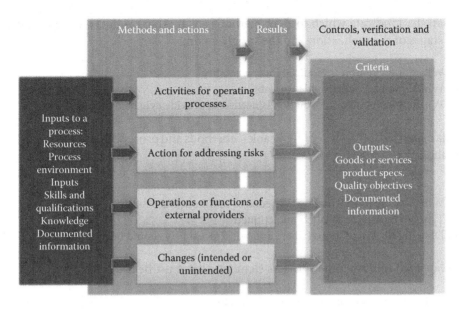

Figure 8.1   Implementing controls during the planning of the process.

Integrating Process Controls in the Quality Plan

While implementing the quality planning, it is time to determine where and how the controls will be deployed. For each activity or operation that requires a control, the plan shall indicate

- The exact stage of an activity that must be controlled—which input or output of which activity must be reviewed
- Which factors are to be controlled—which characteristics of a product or parameters of a process must be controlled
- How these factors will be controlled—what are the methods and the tools
- What type of results are expected—what type of information or data is expected

These controls may be described in

- Work instructions
- Test instructions
- Procedures

Traceability

The quality plan shall refer to the instructions and procedures related to traceability requirements. The specific requirements for maintaining traceability are dealt in Chapter 8.5.2—Identification and traceability. When traceability is a requirement, the plan shall indicate which tools or methods ensure that products will be traceable in all stages of the realization and the supply chain:

- Which information is required for traceability
- Which method will provide traceability
- Which documented information is needed to provide evidence of traceability

The plan shall indicate—where applicable—methods and procedures for accomplishing traceability:

- Work instructions
- Which product information is needed to maintain traceability
- Types of records (e.g., labels and etiquettes)
- Drawing for instructions (where and how traceability should be maintained)

The plan shall indicate other process elements that must be traceable:

- Status of monitoring and measuring devices
- Qualification of personnel
- Conditions of process environment

The plan shall indicate the necessary documented information for maintaining traceability:

- Shipping document
- Tags and marks
- Numbering systems
- Date stamping
- Barcodes
- Labels and etiquettes
- Attachments

When international standards, directives, or regulations related to the realization of the product or provision of the service that demand the use of identification methods of materials, products, process entities, or agents are applicable, the quality plan shall refer to them and indicate which identifiers are necessary.

Inclusion of External Providers

Methods and activities necessary for the control of outsources, processes, or operations shall be considered, implemented, and applied when such operations are relevant to the realization of the product and may affect its quality. The objective of this principle is to ensure that external operations that are part of the realization of the product will deliver intended results. The organization may demand that its suppliers or subcontractors maintain high-quality requirements, but the responsibility for the quality of the outputs lies directly with the organization. These will be planned with reference to the requirements of clause 8.4—Control of external provision of goods and services. The planning of related processes shall include references to the following aspects:

- Specific requirements—the planning of a process shall include or have access to the specific requirements of the external provider regarding the realization of the product.
- Interaction with the supplier—the interactions and interfaces with the suppliers shall be mentioned or referred to in the quality plan.
- Control and criterion—the controls and the appropriate criteria of the outputs received from supplier shall be defined and implemented.
- Documented information—the required documented information needed to provide evidence that external operations meet requirements is defined and maintained.

Addressing Risks and Opportunities

The ISO 9001:2015 Standard requires the evaluation of risks and opportunities, the planning of activities to handle those risks and opportunities, and their integration into the QMS in general and throughout the realization processes in particular. This is the core of the risk-based thinking and it was discussed in detail in chapter 6.1—Actions to address risks and opportunities. Successful implementation of the risk-based thinking will allow the organization to obtain a systematic method for identifying, evaluating, and controlling risks and opportunities and—where needed (and defined in advance)—eliminating or reducing risks or using opportunities to leverage the QMS. Addressing risks and opportunities will span across all stages of the realization of goods or services:

- The planning stages of the product
- Purchasing of material
- Production of the product
- Marketing
- Delivery
- Installation
- Service maintenance
- Disposal

In practice, the outputs of the risk-based thinking according to clause 6.1 are to serve as inputs for quality planning. Those planned actions, controls, and tools derived from the risk-based thinking shall be integrated in the planning of processes (e.g., will be mentioned and addressed on the quality plan):

- Reference to the responsibilities and authorities
- Identification of the relevant products or components or service activities
- Identification of the relevant risks and identification of the QMS elements that may be impacted
- Identification of any specification regarding those risks: standards, technical, safety, customer, and regulatory requirements

- Definition of criteria relevant for risk acceptability
- Definition of specific controls and actions to address such situations—the required verifications or validations
- Definitions of the different stages of realization in which these activities are to be applied and control implemented
- Definition of corrective actions when such situations actually occur

Controlling Planned Changes

Changes to product specifications or activities needed for the maintenance or improvement of the performance of the QMS are a natural occurrence that the ISO 9001 Standard promotes and therefore provides appropriate controls:

- In clause 6.3—Planning of changes in the organization—it is required to plan and control changes to the QMS methodically.
- In clause 8.1—Operational planning and control—it is required to identify planned and unplanned changes in the planning of the processes and submit them to a control.
- In clause 8.5.6—Control of changes—it is required to review and control changes for production or service provision and provide necessary evidence.

Here while planning the processes, it is necessary to integrate the method of undertaking changes to the processes or product characteristics. Since we are dealing with operational planning and control, this requirement relates to processes and activities needed for the realization of the product or their outputs. Such changes may influence the interaction between processes or activities and affect the workflow between the different related areas in the organization. This planning of changes refers to changes occurring with external providers as well.

When changes in the processes or product specifications are required, one will submit them to the method described in clause 6.3, where it is required to

- Determine which changes are necessary in order to maintain and improve the performance of the QMS
- Determine a method for managing those changes
- Evaluate and control associated risks
- Identify and develop opportunities
- Evaluate with each change the potential consequences (this is the critical part)

In practice, undertaking changes in processes or product specifications may affect several process aspects: new activities, resources, or inputs:

- Applying a new set of inputs—Where and how these inputs will be integrated in the process? Which prior processes need to be modified in order to provide these new inputs? Which documented information needs to be modified (where formats are influenced) or updated (where records are affected)? How will these new inputs be transferred forward?

- Resources—How shall the tools, instruments, and resources that are used in the process handle the new inputs? Do process elements have the capability to process these new inputs in the first place?
- Operations—What is its influence on activities? Do current activities need to be changed or do new activities for processing these new inputs need to be planned?
- Controls—Do verifications or validations need to be changed as well?
- Risk-based thinking—Are there any influences on known risks? Must controls be updated?
- Outputs—What are the new process outputs that result from these new inputs and these new processings?
- Interaction—How will the new outputs be transferred forward? Does one need a new method to hand over these outputs forward?

Reviewing Unintended Changes

Unplanned changes refer to unexpected events and their consequences that were not submitted to the controlled method required in clause 6.3 and were imposed upon the organization due to some alteration of conditions:

- A breakdown in infrastructure
- Absence of employees
- Delivery failure from suppliers

Addressing these unplanned changes is a preventive action necessary to reduce and mitigate risks due to unexpected events. By reviewing these unplanned changes, you are required to assess consequences that may result after the change has come into force. This may be achieved through the establishment of a set of proactive actions that are developed to ensure that changes do not impact your quality objectives:

- Performing your business operations
- Delivering product according to specifications
- Maintaining customer satisfaction

Addressing these unplanned changes refers to those affected business operations related to the realization of the product. In this case it is not possible to

- Determine the need for the change
- Submit the change to the method for control
- Evaluate the associated risks
- Identify the opportunities

But it is possible to evaluate the consequences for unplanned changes that may arise from its occurrence. For each change it is required to analyze and assess what are the implications and consequences and form a set of actions to mitigate these consequences. Such actions will reduce the nonconformities while business operations are required to work not as usual. Once again, the review of such unplanned changes may be through the review of several of the process aspects:

- Inputs—Does the change have any influence on process inputs? Are there any new inputs? Which documented information needs to be changed (where forms are influenced) or updated (where records are affected)? How will these new inputs be transferred?
- Does the change have any influence on tools and instruments and resources that are used in the process? Do these process elements have the capability to support the new situation in the first place?
- What is the influence on activities? Do current activities need to be changed?
- Do any of the verifications or validations need to be changed as well?
- Are there any impacts on known risks? Should controls be updated?
- What are the new process outputs that result from this change?

The review will need to be integrated into a method that will ensure the smooth integration of unplanned changes. Unplanned changes in outsourced operations shall need to be considered as well.

Changes in the Quality Plan

The quality plan is an ever-changing document and might be impacted by changes. Therefore, it needs to be submitted to control that will check its accuracy and relevancy and will update it when needed. Where processes, operations, or activities may be changed over time or where conditions are changed, their relevant documentations such as procedures or instructions as well as relevant resources may need to be modified and updated. Which types of conditions may demand updating the plan?

- Changes in the product
- Product's characteristics that must be changed
- Change in suitability and compatibility of the QMS elements indicated by the plan
- Changes in processes
- Changes in resources
- Processes that are found unstable

These changes must be approved by all relevant parties that take part in the realization of the product and influence the quality of the product. If required, changes will be submitted to customers for a confirmation. In case the quality plan is controlled, it shall be submitted to the control of documented information including as specified in clause 7.5.3—Control of documented information.

Necessary Documented Information

Documented information necessary for the confidence that the processes have been carried out as planned and process outcomes meet their requirements shall be planned, determined, defined, and maintained. The organization shall determine which documented information provides the most appropriate and effective evidence that the realization processes are performed according to the planning. The evidence shall provide the following information:

- The inputs are available.
- Activities are performed in the correct sequence.
- Process outputs are being generated and they are verified or validated.

What type of documented information is expected? The requirements for documented information are derived directly from the requirements in clause 7.5—Documented information:

- Documented information required by this international standard necessary for the realization of the product
- Documented information defined by the organization as needed to provide evidence and confidence that processes are performed as planned and the outcomes meet the requirements
- Documented information as evidence that goods or services are realized according to their specifications

The extent, scope, and size or type of the documented information will be determined by the types of activity, criterion, and controls. In other words, the type of realization activity, its criterion for acceptance, and the implemented control shall define the extent and type of documented information. If the organization is manufacturing a measurable product, most likely the documented information of controls will consist of

- A test protocol that defines which product specification or characteristics must be controlled
- Records that provide evidence that the measured product meets its specifications

But if the organization delivers an online service, the scenario looks different: the customer logs in and performs an action, and the system saves the data—other methods for control are needed here such as verification of customer details and verification of data. Such controls are not completed with the traditional documents and records but rather with codes and computerized applications, which act very differently.

One should distinguish between documents and specifications and records as evidence. By documents I refer to those specifications that are used to document, design, analyze, and measure, support interaction between processes, and train and communicate business processes and requirements. They are also used to define the required resources for a process. They are needed to describe the sequence of activities: at which point of time, in which location, reference to a process, and by which method.

They deliver details regarding the performance of processes' activities. Documents and specifications include

- Documented procedures that describe processes
- Work instructions that specify how activities will be performed
- Formats that document activities and provide evidence

Records are used to provide information and data regarding the performance of activities of processes and to ensure that the processes have been carried out as planned:

- Evidence of the supply chain—customer orders, delivery notes, invoices, credit notes, receiving slips, and certificates of compliance (COC) accepted from suppliers
- Evidence from the production phase—formats for production orders, test protocols with results, labels with production details such as serial numbers, validation forms, job release approvals, records of quality assurance, and batch approvals
- Records of maintenance of resources—records of machine maintenance, production tool status reports, and records of training
- Records related to quality activities—records of management review, internal audits and nonconformities, and filled-in customer questionnaires

A good example of a document as specification is the work instruction that assists the manufacturer in performing realization activities; it is a list of planned activities for operating a process, reference to the appropriate documentations, and a specification of the expected records. The records that are required from this work instruction and are used to prove its execution are considered as evidence.

Tables indicating the documentation requirements are mentioned in the chapter of this book—7.5 Documented Information.

Suitable Planning for the Organization's Operations: A Practical Quality Plan

The organization is required to maintain output of planning suitable to the organization's operations, in other words outputs that the organization can maintain and that will support its processes. Since I support and recommend the use of the quality plan, I will try to describe how to maintain an effective one.

The output of the quality planning, that means the format, tools, and means with which you will apply the quality plan, may vary according to the nature of the product, the nature of the activities for the realization, and the expectations and needs of the interested parties such as customers or suppliers. There is no ISO 9001 Standard requirement to document the method or to maintain any documented information applying to this method. But in an audit, you will need to prove design and implementation of such method, and therefore I suggest documentation. As a practical quality plan, I would manage on a conventional table in order to make it clear and usable. My objective is to create a table with which each user of the QMS can find his or her scope in the realization of the product, understand which actions and operations he or she must perform, which tools and information are available, and which outputs are expected of him or her. The following table is an example.

Area	Details	Responsibility	Reference to DI
Specifications of product—Identification of the product, its characteristics, and the expectations of the customer			
Product specifications	Identification of the product (distinct or general)		
	Drawing of products	R&D	File on CAD system
	Product dossier	product manager	
Customer requirements	Customer orders	Sales back office	ERP system
	Contracts	Sales back office	Contract according to customer identification
Allocation of resources—Identification of all needed resources and infrastructures necessary for the realization of the product			
Resources for realization	List of required machines	Production supervisor	Form number FO-06-01-01
	List of production tools	Production supervisor	Form number FO-06-01-02
	List of required conditions of process environment	Production supervisor	Form number FO-06-01-03
Human resources	List of required qualifications	Production manager	Form number FO-06-01-04
	List of required trainings	Production manager	Form number FO-06-01-05
Operations and activities—Realization activities, actions for addressing risks, activities for preservation of product, traceability, and post-delivery			
Specifications of realization activities	Process diagrams	Production manager	Procedure number PR-01-01-1
	Procedures	Production manager	Procedure number PR-01-02-1
	Work instructions	Production manager	Procedure number PR-01-03-1
	Assembly instructions	Production manager	Procedure number PR-01-03-2
	Checklists	Production supervisor	Form number FO-05-01-05
	Forms	Production supervisor	Form number FO-05-01-04
Controls including verifications and validations—Defines the controls, verifications, and validations, and criteria of process, resources, and infrastructures			
Control of processes	Test protocols for processes	Production manager	Procedure number PR-01-03-3
	Criteria	Production manager	Procedure number PR-01-03-3
	Forms	Production manager	Form number FO-01-03-01
Control of resources	Test protocols for productions tools	Production manager	Form number FO-01-03-02
	Maintenance protocols for machines	Production manager	Form number FO-01-03-03
	Maintenance protocols for infrastructures	Production manager	Form number FO-05-06-01

8.2 Requirements for Products and Services

8.2.1 Customer Communication

Communication with customers is considered to be one of the most common business activities in an organization. The ways to communicate with your customer may vary according to the type of activity related to the communication, the type of your product or service, or the type of the agreement that an organization has with its customers, and there are cases and scenarios where communication will be dictated and controlled by a third party. All these issues will be structured in planned arrangements for communicating with the customers. Let us review the requirements of the ISO 9001 Standard regarding communication with customers:

- The organization shall initiate communication with customers and delivery of information on various topics related to the product or the service.
- The communication will include sharing information with the customer regarding delivery or post-delivery activities including changes and updates relevant to the product or the service.
- The organization will exchange information with customers regarding enquiries, order handling, and contracts.
- The organization shall initiate communication for handling customer complaints and feedbacks.
- The organization shall initiate communication for issues related to the handling of customer property.
- The organization shall determine specific requirements for communicating issues related to contingency actions, when relevant.

The requirements for communication with customers shall be determined and established: they will be decided, will be formally enacted, and will be clear to the users of the QMS. The ISO 9001:2015 Standard does not specify the need to develop a method or arrangements for communicating with the customer. However, during an audit you will be required to prove that

- The information for communicating with customers is defined as required earlier.
- This information is managed and maintained.

This is why you will need a method to prove these points. The method does not need to be documented but defined; you are not required to maintain procedures or work instructions but to prove that there are processes that maintain communication with the customers.

Information regarding the Product or the Service

The ISO 9001 Standard requires the initiation of communication with the customer. The objective is to ensure that all necessary inputs have been accepted or transmitted and all necessary outputs are determined and are being channelized to the customer. The flow of information is bidirectional—either transmitted or received. Two aspects may determine the type of communication channel and information that will travel through it:

- The characteristics and nature of the product or service will determine the type of information.
- The types of activities in the realization process that are related to enquiries by the customer.

This information may be communicated to the customer during various activities related to the realization of the product:

- Acceptance of order or request for offers
- Information regarding development or design plans
- Enquiries of status of order
- Results of acceptance testing
- Deliveries of goods or services
- Requirements for inputs such as customer specifications
- Demonstrations and presentations of products, prototypes, or results of tests
- Information about defects or faults

Information regarding products or services may include attributes like

- The description of the product
- The usage of the goods or service
- Training necessary for the use of the product
- Instructions or warnings regarding the use and operation of the product
- Previous editions or versions of the product
- Materials, components, and characteristics of the product
- Processes and realization activities of the product
- Warnings and advisory notices regarding the product
- Changes, updates, and improvements in the product

Changes, updates, and improvements in the product or the service must be brought to the attention of the customer. According to the standard, "change" is any alterations made to the product's features or to the attributes of a contemporary or future model that was ordered and agreed for with the customer. The rationale behind this is that the organization is planning changes in the product and will deliver a modified product to its customer. The customer, on the other hand, has known expectations that after the change might not be fulfilled. Information regarding the product including changes and updates include

- Changes that may affect the characteristics of the product
- Improvements to the product
- Problems regarding the product that the customer must not know about

The same applies to improvements (which are a kind of change); the organization must communicate and announce to the customer of all improvements with documentations, e-mails, telephone calls, written correspondence, or meetings.

If one decides to maintain a method, it shall describe all the arrangements for communicating with the customer: what form of communication and through which media. One option is documenting the arrangement on customer order confirmation. Take, for example, Microsoft. Microsoft has a very effective way to announce its users of Windows Operating System of any changes, updates, or improvements: It maintains a specific component in the operation system that allows the users to view the updates, changes, and improvements to the software and decide whether they wish to accept or reject them.

Investigations, Applications, or Inquiries Regarding Orders or Contracts

The standard requires communication for managing all inquiries from the customers regarding the status of orders or contracts. The objectives are

- To make it clear to the customer how he or she may contact the organization or who is the contact person regarding orders and inquiries
- To allow access to information regarding order-related subjects that may be of interest to the customer

When defining the communication, the organization must take into consideration what information will be transmitted to customers:

- Orders or contracts—the process of submission of orders or contracts as well as inquiries or investigations before submission
- Status of orders—inquiries or investigations concerning the realization activities such as delivery dates, transportation, installation, and warranty claims
- Progress of activities that is the responsibility of the customer
- Technical support—how the technical support will be communicated and organized with the customer
- Service and maintenance—how activities of service will be coordinated with the customer
- Reports about present or future changes
- Information about nonconformities: defective products or services that were delivered or provided to the customer

If you decide to maintain a method, it shall include the organizational responsibility or a function that will manage it.

Customer Feedback

The ISO 9001 Standard requires the monitoring of customer feedback. While customer satisfaction may be submitted as a subjective opinion, customer feedback provides an objective point of view on whether the organization supplied a product according to the specifications. Customer feedback provides an early warning about potential quality problems, effectively identifies risks or opportunities, generates the inputs for improvement, and initiates necessary corrective actions. Feedback evaluation may be maintained through the following:

- Customer complaints—the organization monitors customer complaints regarding performance and functionality of the product.
- User surveys—the organization conducts surveys regarding functionality, characteristics, and performance of the product.
- Reviews—the organization initiates a review regarding the product and its functionality.
- Journal reviews—the organization researches sector and industry tendencies.

For each of the aforementioned actions, it is necessary to define the information necessary for containing all the inputs of customer feedback and distributing them to the relevant parties. A method of review of the feedback will be dealt with in chapter 9.1.2—Customer satisfaction—where determining the method for monitoring and managing customer satisfaction is required. Here in clause 8.2.1—Customer communication—we are to determine the communication channels in this method. Handling complaints is dealt with in detail in the following paragraph.

Complaint Handling

The ISO 9001 Standard regards the issue of customer complaints as being highly significant. An important aspect on this issue is the facility or means by which a customer can communicate with the organization and register a complaint. No less important is the organization's delivery of a satisfactory response to the complaint. The first concern of the organization is to define the information regarding the complaint. The ISO 9001:2015 Standard requirement refers to the definition of the information for handling complaints. As for my concern, handling a complaint should be managed using a tested method, an application process of the complaint. The information shall cover various aspects of the process, and a method shall be described in the following issues:

- Communication: The method shall describe to the customer how they may reach the organization to file a complaint. The method could be incorporated with other types of inquiries (such as ordering of products). However, the organization must prove that its customers are notified of the ways that they can submit complaints— a specific telephone number, an online form, a designated e-mail address, and an interaction center—such options for filing a complaint must be made known to the customer, and the contact details must be published (printed on a brochure, the firm's website, or on the product itself).
- Details: The method shall define the required information and details that must be recorded while receiving a complaint. The information and details should be sufficient in order to ensure effective treatment. The organization must document the information to ensure effective investigation into the nature and cause of the complaint. The details of the complaint are the inputs to quality processes and activities such as actions to address risks and opportunities, nonconformity, corrective action, and improvement. If the complaint refers to a product, then the following information must be specified: the model, identification, or traceability

(batch number or serial number), production details, contact person, delivery date, and any other information that would identify the product and support the investigation. If the complaint refers to a service, then specify when the service was given, by whom, to which product (with identification details), and the service identifier. The best way to ensure that such details are provided is to design a format (manually or electronically) and ensure that all fields are completed.

- Distribution: The method for handling a complaint shall consider the distribution of the information of the complaint to the appropriate roles or functions in the organization. It will be achieved with the definition of the interface between the organizational unit that receives the complaint and the one that shall handle and investigate it: an e-mail, form, or designated system.

- Responsibilities: The function that would be in charge of the process of accepting the complaints must be defined (e.g., back office, customer support representative, a customer relationship manager, or an account manager). As well as defining the function responsible for accepting the complaint, the function responsible for handling the complaint must also be defined. Some complaints may need to be passed on to a designated person in order for them to be appropriately handled. This person should be clearly appointed.

- Time frame: It is necessary to determine the exact time frame for the response and the treatment. The customer should not be left waiting for an answer. If the answer is taking too long, the customer should be notified and assured that the complaint is under investigation. Another aspect related to time frames is local or national regulations; where regulations require that you submit an action in a defined period of time, the organization must apply these requirements and implement them in its processes.

- Answer: After accepting, processing, and handling the complaint, it is necessary to inform and communicate the results to the customer. The method shall define the means and details. One cannot ensure that each complaint is closed to the satisfaction of the customer, but one can ensure that the results were transmitted and the customer received them.

Communication regarding Handling Customer Property

The ISO 9001:2015 Standard defines another controlled type of information between the organization and the customer—information regarding property belonging to the customer. In clause 8.5.3—Property belonging to customers or external providers—it is required by the standard to inform customers or external providers when their property is either damaged or found unsuitable for use. When such an event occurs, the organization must maintain documented information regarding the notification to the customer.

When defining communication and establishing communication channels, the matter must be considered and appropriate channels must be planned. This type of communication should be integrated into the quality processes:

- Upon receiving goods from the customer for processing
- When goods or equipment received from the customer are found unsuitable for use during the realization processes

The communication shall include information such as

- Which products were damaged
- When the damage occurred
- What happened
- Further actions or instructions

Communication regarding Requirements for Contingency Actions

Contingency actions are part of the risk-based thinking; they are a type of output from the risk management and relate to actions that must be taken as a response to unplanned events and that require the attention of the customer. When defining and implementing the information for communication with the customers, the issue of communicating contingency actions must be considered and appropriate channels must be planned. The standard states clearly that not all contingency actions must be communicated but only when necessary, that is, determined with the customer that he or she must be informed. For each contingency plan, the organization must define the following:

- Contact person
- Which information must be transferred

Effective Communication Channels

The main objective of clause 8.2.1 is to initiate effective communication channels with your customers, whether they are internal or external. Effectiveness of communication is measured when the information has reached its designated goal at a defined point of time. Effectiveness of communication channels will be achieved by introducing the next points in the organization:

- A method for accepting customer requirements is properly defined and maintained in the organization.
- The customer receives all the information and data that he or she needs and that answers his or her enquiries.
- The organization maintains an effective way to receive complaints from the customer and provides him or her with answers.
- The organization maintains an effective way to receive feedback from the customers.
- The organization maintains an effective way to communicate with customer regarding his or her property.
- The organization maintains effective channels to share information and situations of nonconformed products.
- The organization maintains an effective way to inform and receive inputs from the customer in case of contingency.

Try to understand the rational view behind this requirement—the customer and their needs should be the center of your attention. When the customers have needs, they initiate communication with the organization through defined channels. The organization is obliged to answer and fulfill these needs, which may be related to several issues and may take place at different phases of the realization processes: marketing, design and development, production, delivery, installation, and post-delivery activities like warranty claims, service and maintenance, feedback, and filing complaints. For each need or type of interaction, you must define the organizational tools with the right response and reaction: channel, responsibility, and information to be handled.

A Defined System

After the necessary information to be exchanged with the customer has been considered and identified, the organization shall plan a method to manage the different types of information. The method will be based on three simple principles:

1. How—the organization shall define by what means the information will be managed, received from the customer, saved, and delivered to the customer: notices, printed media, CRM system, ERP systems, e-mail, fax, online via website, or designated software. The definition will be bidirectional—customer to business and business to customer.
2. Where—the organization shall define where the information will be maintained and kept. This depends very much on the type of information, communication, or activity, for example, on an explicit form in an information system or on the client portfolio. Some of this information may be necessary for the planning of the control of realization of the product and thus must be controlled as documented information.
3. Whom—the distribution of the information to the relevant parties in the organization that need this information for the realization of the product: marketing, design and development, production, logistics, and technical service.

Authorities and Responsibilities for Customer Communication

In order to maintain effective communication channels, the organization shall appoint designated roles for each type of customer communication. Each role will be responsible for managing the communication channel related to a specific context (marketing, production, or logistics). The objective is to appoint a person or a function that will bear the responsibility of operating the communication with the customer and exchanging the required information. For example:

- A sales representative is responsible for delivering price quotes and inputting orders to the marketing and sales information system.
- When developing a specific product with the customer, a project manager will be designated the task of communicating with the customer, receiving and documenting product specifications and requirements, conducting meetings with the customer, conducting meetings with the internal functions to realize the product, and so on.
- When operating a call center, a representative is then considered responsible for communicating with the customers on various issues.
- Key account manager or sales person is responsible about notifying customers when defective products or services are found or delivered.

The definition may appear on the job description, the role definition, or on a work instruction. Strong emphasis is given to the treatment of complaints and the handling of customer feedback. If the customer maintains a designated system for the correspondence, then the relevant parties should receive training with instructions on how to use the relevant system.

Time Frames

Defining the precise time frames for each communication type or channel is necessary in order to allow the continuous flow of information between the organization and the customer. The objective here is to specify the exact time frame for a response, that is, over what period of time must the requested information be transferred to the customer. The time frame depends on the nature of the communication or the related activity. In principle the requisite response time for each type of communication will be defined. For example:

- An offer—the number of work days to handle a request for an offer would be defined.
- A complaint—an immediate answer would be delivered to the customer stating that the complaint is under review and that a full answer will be delivered to the customer according to the priority and seriousness of the case.
- Changes or update to the product—the response time will be determined according to the nature of the change or update.
- Information and details about nonconformed products or services—the response time shall be determined according to the impact of the nonconformity on the product.

Documentation of these time frames depends on the business activity and its specifications. These types of specifications appear usually on the contract. And so contractual business attributes such as type of customer (platinum, gold, silver) or type of service agreement (premium, normal) may set the reaction time. When this information is necessary for the planning of the control of realization of the product, it must be controlled as documented information.

Customer Records

Customer records are considered as information and data that are necessary for communication with the customer (among other activities such as invoicing or delivery of goods). Such records shall be kept according to the nature of the communication. The purpose is to define the needed inputs for initiating communication with the customer. The organization shall define which details are required in order to maintain an effective communication with the customer:

- Names and details
- Managing contact details such as telephone number, e-mail address, and postal address
- Managing details according to the activity—contact persons or activity hours
- Managing financial details—bank account details and credit card numbers
- Managing specific customer information related to the realization of the product—particular notices and remarks, examples of products, types of documents (order, delivery notes, invoices), or technical specifications

The next stage is to define where, how, and in which format customer's records will be maintained and distributed to relevant parties in the organization. These definitions will refer to the realization processes. Each realization step may require a different type of information from the customer and that means a different type of communication channel. The next table is a suggestion of how to manage this information:

Process	Customer Records
Marketing lead	The name and telephone number of the prospect plus a completed lead survey for the sales representative to communicate to a potential customer
Sales order	Full customer details, including account details
Development	The name of a contact person by the customer related to the product
Delivery	Address for delivery and a contact person in the logistics
Accounting	Details for debit

Regulatory Requirements

Applicable statutory and regulatory requirements may demand communication channels and necessary details and information with customers on various topics other than complaints or customer feedback. If such requirements regarding communication with the customer are applicable, the organization must implement them. This fact may influence the entire communication process with the customer. It might be necessary to collect specific data or to design specific processes or documentation in order to respond to these needs. This is why the organization is requested to identify these applicable statutory and regulatory requirements and integrate them in its processes. See clauses 4.2 (Understanding the needs and expectations of interested parties), 5.1.2 (Leadership and commitment with respect to the needs and expectations of customers), 8.2.2 (Determination of requirements related to the goods and services), and 8.6.5 (Post-delivery activities).

8.2.2 Determining the Requirements for Products and Services

By determining the requirements for the product and the service, the organization determines exactly what the product specifications are, in order to evaluate later (clause 8.2.3) whether it has the ability to meet these specifications. The determination enables the later verification of its feasibility and ability to supply the product according to the customer's expectations and specifications. Thus, it is necessary to review all requirements from various inputs and conditions—customer, regulatory, safety, and organizational.

The requirements of goods or services are the collection of inputs needed to set criteria for its validation. In order to accurately identify these, the organization needs to develop a method to collect and review these inputs. Here are the ISO 9001 requirements:

- The organization shall ensure that the requirements for the products and services to be offered to customers are defined.
- The organization shall identify and review regulatory requirements applicable to the product and service.

- The organization shall determine additional requirements necessary for the realization of the product or the provision of services.
- The organization must ensure and confirm that it has the capability to meet the requirements and supply the product as required or offered to the customer.

The main objective of clause 8.2.2—Determining the requirements for products and services—is to ensure that all the requirements are clear, documented, and distributed to the appropriate parties in the organization or parties that take part in the realization of the product and the provision of the service. The standard introduces us to several main types of requirements that need to be reviewed when accepting or collecting requirements related to the product. Reviewing these will bring you closer to an accurate determination of the requirements. There are no standard requirements for documenting the process. However, during an audit, you will have to prove that all of the product requirements are clear and identified: product catalogs, offers or tenders, order, and so on.

Determining the Requirements for the Products and Services

A product may be developed as part of a contract, as a product available for a market sector, as product that is embedded in another product, or as a service. The product requirements are the collection of inputs that are supposed to fulfill the expectation of customers. In order to accurately identify these, the organization must create a method that collects and reviews these inputs. The objectives are to understand fully the expectations of the customer and to prevent misunderstandings. The requirements for the product and services may be influenced by several factors:

- Requirements that are defined on a contract
- Results of a market research
- Known needs and expectations of the customer
- Requirements provided by the customer, for example, as a request for purchasing or a tender
- Necessary requirements of the organization like quality requirements or regulatory requirements
- New requirements that are the results of a change in customer expectations

All these types of requirements must be identified and known to the organization. Bear in mind that those requirements must be validated in the later stages of the realization. Thus, it is required to define a method for identifying and understanding the needs and demands related to product and proving whether the organization is able to fulfill these requirements. The method shall include all the relevant parties that participate in the realization of the product: sales, development, production, purchase, and so on.

There is no standard requirement for documenting the method; however, you will be requested to prove that this method is defined and implemented. It is difficult to direct as to how this is to be implemented because each type of product has its own manner of determining the requirements. I would plan and introduce key tools that capture the expectations of the customers. These may include

- Customer files where specific customer requirements are saved
- Meeting summaries where the customer expressed his or her expectations
- Checklists that prove that I considered critical topics while determining the requirements
- Order forms that express customers' expectations
- Management systems that manage all these elements
- A summary report of research if you ordered a market research
- A document of characterization when submitting a product for design and development
- Customer's approval of the product characteristics

The requirements shall be communicated to all relevant parties in the organization. Where product requirements are changed, the organization shall ensure that relevant documents are updated and all relevant parties are informed.

Identifying Applicable Regulatory and Statutory Requirements

Applicable statutory and regulatory requirements may affect the ability of the organization in determining the requirements related to the product or the service. Such expectations of interested parties have a potential effect on the organization's ability to consistently provide products and services that meet the requirements.

The identification of relevant regulatory requirements is critical when determining requirements related to the product. The regulatory requirements may include government, safety, and environmental regulations. These may contain provisions for product safety, characteristics, identification, and functionality; and they may affect or set the product's requirements—design and development, production, assembly, storage and transportation, handling, recycling, and the disposal of materials or products. They may apply to realization activities such as acquisition, storage, handling, recycling, elimination, or disposal of materials or products.

The objective here is to ensure that regulations are available and understood and submitted as inputs when product requirements are being determined. The process of determining the requirements will ensure that any changes or updates in the regulatory requirements are identified, gathered, and accounted for. This is how you may guarantee that the relevant product requirements will be up to date with regard to the regulations. Availability of the regulations is measured not only by their physical presence but also by the employees' awareness of the regulations and their influence and effect on the product's realization or service provision. For example, if one fixes flat tires, one may need to properly dispose of the old tires that were replaced because environmental regulations forbids one from throwing them in the garbage. This regulation shall be considered when one determines the requirements for the service to replace flat tires.

Additional Requirements Necessary for the Realization (or Provision) of the Product

The organization shall evaluate and determine other realization requirements that are not the customer's concern or that the customer need know about, but are nonetheless necessary to the realization of the processes; these are likely to affect the quality of the product and may also affect the ability of the organization to meet the customer's requirements. Such additional requirements may include

- Qualifications, skills, or knowledge of personnel needed to realize the product
- The use of special equipment, tools, and accessories (such as clothing or software) for the realization or provision activities
- The use of certain raw material or components
- The need for workshops and trainings to benefit the customer
- The need for certain resources such as workstations or infrastructure during the realization or provision activities
- Necessary processes for the realization or provision activities
- Required standards
- Activities for the disposal of products or materials
- Requirements for the operating environment of the product

The organization will ensure that all such requirements are identified and determined. Additional requirements relating to interested parties that may affect product requirements are

- Personnel—when employees must acquire certain qualifications or trainings.
- Customers—when the customer did not convey adequate specifications regarding its requirements.
- Technique—when the requirements have technical aspects that must be considered.
- Suppliers—when suppliers must ensure they are able to provide goods or services according to what was requested of them.
- Regulators—when regulatory requirements may affect the realization of the product. For example, if one is delivering dangerous substances (where transportation is part of the delivery of the product), one may need the appropriate transport certifications.

Ensuring the Ability to Provide a Product or a Service

The output of the determination of product and service requirements is the proof that the organization is able to supply the product that will be offered to the customers. In other words, to transform the determined requirements into a set of desired operational terms that will assist the organization in deciding whether it has the ability to provide those requirements or not:

- Characteristics of the product such as functionality, intended use, safety, and security are clearly defined and unambiguous.
- Resources (like infrastructures and knowledge) are available.
- Feasibility of realization—the organization maintains processes and operations that shall enable to achieve the requirements and the requirements are prioritized and approved.
- Traceability is established between the requirements of the product and the customer's requirements.

8.2.3 Review of Requirements for Products and Services

After determining the requirements for the product or the service, the organization is required to develop a method to review customers' requirements. The review of the requirements for products and services refers to the examination of the operational conditions that may affect the ability of the organization to provide a product according to the requirements specified by the customer and determined by the organization. The ISO 9001 Standard requirements are as follows:

8.2.3.1

- The organization shall review with the goal of ensuring its ability to meet the requirements for products and services to be offered to customers.
- The review of requirements shall be performed prior to the engagement by the organization to provide the product to the customer.
- The organization shall review product requirements specified by the customer. The determination shall refer to delivery and post-delivery activities as well.
- The organization shall review product requirements that are necessary for the realization of the product.
- The organization shall review a product's requirements that are not expressed or forwarded by the customer, but are required for obtaining the intended use of the product.
- The organization shall review regulatory or statutory requirements applicable to the product and service.
- The organization shall ensure that differences between the specifications conveyed by the customer and the actual requirements submitted to the customer are settled.
- The organization shall ensure that a product's requirements are defined, clear, and understood, including the delivery and post-delivery requirements.
- When the customer does not provide documented requirements or specifications, the organization must obtain an approval prior to acceptance (forwarding for realization).
- In case the customer does not deliver its specific requirements, the review will be based on the product or service information available to customer.
- Note—in case the review is not practicable, the organization shall inform the customers of the product's specification prior to acceptance of order (like product catalogs in Internet sales).

8.2.3.2

- The results of the review shall be maintained as documented information.
- Where and when new requirements arise from the review, they shall be maintained as documented information.

The objective of these requirements is to provide an internal approval that the organization can realize and supply the product according to the requirements and expectations of the customer. The review will be conducted before the submission of a commitment to the customer. In other words, it must be clear before receiving the approval of the order what is to be done, by whom and when and whether the organization is capable.

Accepting a Request from the Customer

Review of requirements related to the product begins with a method for receiving the requirements and coordinating them with your organization. The objectives of the review are as follows:

- Capturing the requirements of the customer
- Translating the needs into organizational terms—product specifications, processes and activities, risks or opportunities, and regulations
- Reviewing, ensuring, and approving the ability of the organization to realize and deliver the product or to provide the service

The organization is required to incorporate these steps in its process of managing customers' requests or orders. I suggest here some simple basic principles:

- Creating a tool for the presentation of product or service specifications to the customer, where the customer is made aware of the product and its characteristics, customer offer, products catalog, and Internet site, participating in tender, or identifying a need in the market—the organization is aware that it has the ability to provide goods or services to interested or potential customers.
- Accepting a request from the customer for delivery of products or a service—creating an interface between the organization and the customer to receive, exchange, and transmit information related to the expectations of the customer regarding his product.
- Submitting the request for review and performing the review—the different organizational responsibilities shall share their inputs in the review and assess the ability of the organization to provide the product according to the specifications and expectations.
- Delivering a proposal to the customer—the assurance of the organization that it is able to supply the products or the services according to the requirements or informing the customer of differences.
- Settling differences—settling of differences is required if during the review the organization identifies deviations between the requirements of the customer and its abilities to provide the product or the service.
- Approving the review—the approval of the organization that it is able to supply the products or the services according to the requirements or informing the customer of differences and settling them.

I have listed here a few methods that are considered ways for capturing and determining requirements related to the goods and services:

- Customer offer
- Tenders
- Customer order
- Order from Internet site
- Contracts
- Product catalog

Reviewing Ability to Meet Requirements

The organization must evaluate whether it has the abilities to provide the customer with the product according to their requirements and specifications. In order to review the ability, certain information must be available during the review:

- Product specifications—how the organization perceives the product.
- Design specifications—specifications that indicate the characteristics of the product (e.g., drawings).
- Customer-specific requirements—how the customer perceives the product.
- QMS requirements—what verifications or validations are needed, which quality objectives are related, and which quality requirements are requested by the customer.
- Relevant standards—they may set additional requirements.
- Regulations—they may set additional requirements.
- Risks—risks related to the realization of the product must be taken into account.
- Terms for purchase—additional requirements for purchase activities related to this order:
  - Certain suppliers
  - Certain materials
  - Additional outsourcing
  - Capacity or availability of suppliers
- Terms of order—additional details regarding the order:
  - Quantities—are the quantities achievable? Does the organization have enough capacity and resources to supply these quantities?
  - Commitment for schedules—are the schedules achievable? Does the organization have sufficient capacity and resources to meet schedules?
  - Prices and payment conditions—Do the terms and payment conditions satisfy the organization and the customer?
  - Environment and conditions—the environment in which the product must function or the service must be provided will be made known.

The review can be carried out by a certain function or a group of roles related to the realization of the product. A good example is the internal Monday morning meeting where people from various functions in the organization sit together and discuss the work plan for the coming week. During the review, all quality requirements related to the product must be available and all parties that participate in the review must be aware of them.

Resolving Differences between the Organization and the Customer

The review of requirements related to the product or service occurs during the interaction between the customer and the organization. One of the objectives of the review is the confirmation that the organization can deliver a product to the customer according to its needs and expectations even if disagreements, gaps, misunderstandings, or differences between the customer and the organization may arise related to the requirements of the product. According to the standard:

- Product requirements must be clear and unambiguous in a way that would allow effective product validation during later phases of the realization of the product or provision of services.

- Customer's order should not be submitted for realization before all disagreements or differences between the customer and the organization are resolved.
- Resolving such differences is settled by accepting an approval from the customer of the terms for supplying the product or receiving the service.
- In case the organization is not able to provide the customer with a product according to his or her specifications, the organization must inform the customer before the approval of the order.
- The customer shall confirm that he or she accepts the product as offered by the organization.

Differences between the organization and the customer refer to the fact that the customer wants a specific product or service but the organization is not able to provide it exactly as the customer expresses and expects. This is a natural occurrence. For example, differences may appear as price disparities, functionality, attributes, and logistic issues like storage, packaging identification, delivery, and disposal or delivery schedules. These differences must be resolved before the organization's commitment to supply goods and services to the customer or before the customer approves the order. The organization shall undertake measures to eliminate those misunderstandings or gaps. Through the process of accepting and reviewing an order, the organization must make it clear at which point of the review the customer agrees and approves the terms of the organization, for example:

- Signed form of the order confirmation
- An e-mail with affirmation
- The action of accepting terms and conditions on the Internet website
- Signing a service contract

Here are some ways to identify and eventually resolve such differences:

- Appointing a designated authority or function for receiving and understanding customer requirements will reduce the number of opinions and interpretations.
- Identifying and contacting functions and roles by the customer. Such functions may assist and support the organization in understanding the requirements: quality control, purchase, design, and development.
- Ensuring that technical or professional terms and abbreviations are understood, well interpreted, and explained.
- When the requirements are in a foreign language, the organization shall ensure that it has the capacity to have these translated.
- Creating a communication channel designed to receive specific inputs will ensure the submission of definite inputs.

Approval of Requirements

The approval of the requirements related to the products or the services is the last step before forwarding the order for realization and thus may be critical. The objectives of the approval are as follows:

- Ensuring that the customer accepts the specifications of the product and the terms of delivery
- Avoiding nonconformity
- Eliminating waste of resources (and thus causing more nonconformity)

When documented requirements or specifications regarding the product are not provided by the customer, a confirmation of the same should be sought before acceptance, for example, a signed order conformation or an approval via e-mail.

Results of the Review as Documented Information

The results of the review shall be maintained as documented information and will serve as evidence. The evidence is to prove that all of the earlier mentioned methods have been carried out: accepting the request for the customer, reviewing the ability to meet requirements, resolving differences, reviewing changes, and approving the requirements. There is no specified format for such a requirement in which details may be documented. The results of the review have two main objectives:

- To ensure the ability of the organization to provide the product according to the specifications
- To define and determine product requirements and specifications necessary for subsequent verification and validation

Documented information for new requirements for products and services has the following objectives:

- Make sure the new requirements and expectations of the customer are accepted by the organization.
- The organization has reviewed its ability to provide such requirements.

As I see it, the evidence of the review are divided into two types:

- Documented information that describes the review: procedure for contract review (not required but certainly recommended), formats for documenting meetings, formats for order details, and ERP system for managing orders
- Documented information that proves the review itself: minutes of the meeting, evidence of communication with the customers, like e-mails, records out of a CRM system, signed order conformations, or offers, internal orders, production plans with reference to orders, records of consignation, signed tenders, and contract agreements

Informal Review

In some cases customer requirements are based on a constant fixed list that represents all the product characteristics, for example, a shop that sells products (either the traditional shops or a web shop); the customer goes in, selects a product, makes a payment, and leaves. When accepting an order in those types of businesses, the traditional review, as mentioned earlier, is not practical:

- The organization offers its catalog of products with prices and delivery terms.
- The customer selects a product and accepts the terms by approving the conditions of the order.

With those two steps, the entire process is completed. In such cases, the review can cover other relevant information available to the customer. This note refers to information regarding the specification of the product that would be delivered to the customer in case a traditional review of requirements was conducted:

- Information regarding the product—information regarding the functionality and characteristics must be available to the customer. If it is a food product, then the ingredients, and if it is a product combined of multiple items, a list of the supplied items.
- Necessary conditions for fulfilling or operating the product—if it is a software, then the required operation system and technical specifications needed to install the software. If it is a food product that must be kept cool, it must be mentioned on the product.
- Necessary operating instructions—information regarding the operation of the product. If it is an electric device, instructions for the operation should be provided.

8.2.4 Changes to Requirements for Products and Services

The ISO 9001 Standard requirements include the following:

- When changes of customer requirements occur
  - The relevant documented information will be updated or amended.
  - The changes will be distributed and communicated to the relevant personnel.

Acceptance of Changes

The organization must manage and control changes in customer requirements. The ISO 9001 Standard is aware that customers may change their minds and needs. This is why manufacturers and service providers must maintain control over the updating and validity of the requirements. The objectives are as follows:

- Accepting the change—from the customer, a supplier, or an internal party such as employees from the logistic or production division
- Reviewing the change and its implications
- Informing the relevant parties—customer, suppliers, and organizational departments
- Accepting an approval that the information was delivered and the change is feasible
- Updating relevant documented information

The review of changes required by the customer shall determine whether the organization is still capable of providing the product or the service. Thus, several contexts that may affect the products or the services must be reviewed:

- Order details like prices or quantities
- Product characteristics or technical specifications
- Implications on risks or opportunities
- Compatibility with regulatory requirements
- Technical abilities of the organization
- Availability of resources

The first stage in the process is accepting a request for a change and submitting the change to the relevant parties for a review:

- Changes received from the customer
- Changes received from other interested parties that are involved in the realization of the product like a supplier or regulator
- Changes initiated by the organization—changes in conditions that were agreed upon with the customer and to which the organization is committed: materials, packages, processes, human resources, and transportation

This action of acceptance must be integrated into the method of the review. How to do it depends pretty much on the process of order management—the characteristics of the process will determine how a change will be reviewed and submitted forward. But the next issues must be determined:

- Who is responsible for the evaluation?
- What is to be evaluated?
- Which documentation needs to be reviewed and updated (diagrams, specifications, instructions)?
- Who approves the changes?
- To whom must the changes be communicated?

By evaluation I mean the determination of the extent of the change—where would the change have implications:

- Identification of product specifications
- Identification of activities related to the realization of the product
- Identification of resources that might be impacted by the change
- Identification of the relevant organizational parties that must be informed

The outputs of these processes are as follows:

- Communication of the change to relevant interested parties that are involved: personnel, suppliers, and regulators
- Updating relevant documented information that delivers inputs necessary for the realization of the product according to the specifications of the customer

Let us take, for example, a simple case of updating order confirmation; the customer has made an order, and after a few days he or she has decided to add or subtract some items in that order and to change the delivery dates. The review may be conducted thus

- The customer initiates communication through defined communication channels (see clause 8.2.4—Customer communication).
- The sales person updates the order confirmation on the order form or in the ERP system.
- The sales person performs an availability check with the relevant parties either through the system or by communicating directly with them.
- After clearing the information with everyone, he or she approves to the customer the change.

The objectives were apparently achieved; the customer delivered his or her wish, everyone involved in the realization was informed, the documented information was updated, and the customer received his or her approval. The issue becomes complicated when the product is more complex, and more parameters must be evaluated, more interested parties are involved, or where the product is a specially tailored one. In this case, one may need to update a whole set of documentation and to review the change with various departments or functions.

Updating Documented Information after a Change

The next objective is ensuring that all relevant documented information is updated with the changes to the requirements. Such assurance depends very much on the characteristics of your processes and the related documented information. One is required to identify each type of documented information that is involved and be affected by the change in the requirements. Which information may be changed?

- Order details—information related to the delivery of the product or service: prices or dates of delivery and address for delivery
- Specifications of product—specifications for product, quality, packaging, production, or purchase
- Requirements for resources—specifications regarding the resources that are used to realize the product
- Requirements for delivery—transport or geographical requirements
- Technical requirements—specifications regarding new technologies used for the realization of the product

Take, for example, a product that involves technical plans or diagrams. The customer wants to make a change to the technical specifications. This means the plans and diagrams related to the design and development should also be changed or updated and these changes must be forwarded to production.

8.3 Design and Development of Products and Services

Design and development should be carried out according to a disciplined method to prevent or minimize the occurrence of quality problems. Hence, the organization should ensure that products or services are developed in compliance with the customer specifications and requirements and in accordance with design and development planning or quality planning.

Managing and controlling the design and development activities in the organization defines the quality tasks necessary for design and development. Those activities shall cover the transfer of customer specifications and requirements into a designated process plan, translation of those requirements and specifications into technical details, and producing outputs necessary for the realization of the product.

Objectives of managing and monitoring the design and development under the QMS include the following:

- Identifying the necessary design and development tasks and setting the design and development activities as stages
- Ensuring delivery of a product or service according to the defined specification
- Defining the required inputs for the design and development activities
- Eliminating potential quality problems as much as possible
- Ensuring that the design and development progress according to regulatory and customer requirements
- Managing changes in design and development

8.3.1 General

The ISO 9001 Standard requirements include the following:

- The organization shall plan and determine the design and development of activities as a process.
- The stages of the process shall ensure the subsequent provision through processes, operations, and activities for the realization of products or services.

Process for Managing the Design and Development

The organization is required to plan and develop its design and development activities as a process. The goal of this process is to ensure that the realization of the product or service will be according to their specifications. As we know each process under the QMS has distinct properties:

- The different process stages, their sequences and interactions, as well as the activities and operations for design and development shall be planned.
- Inputs for design and development will be defined and planned.
- Responsibilities and authorities will be defined and delegated.
- Specification of the product or the service as the customer expects—the requirements will be delivered to design and development activities clear and complete and will answer the expectations of the interested parties in the organization.
- Design and development output shall take into account the resources in the organization that support the realization processes:
  - Personnel
  - Infrastructures
  - Processes environments
  - Monitoring and measuring resources
  - Knowledge
  - External providers
- Support activities—design and development activities shall be planned in accordance with the support operations in the organization for
  - Competence
  - Awareness

- Communication
- Documented information
- The controls of the process stages shall be defined:
  - Verifications
  - Validations
  - Appropriate criteria
- Changes of the elements relevant to design and development will be managed.

Ensuring Subsequent Provision of Products and Services

The outputs of the design and development have the role of serving the planning of the realization. In other words, the role of the outputs of design and development will be to formulate the activities that will be used for the realization processes of the product or the service. This subject will be dealt in detail in chapter 8.3.2—Design and development planning—under paragraph Requirements for Subsequent Provision of Products and Services. The process will cover the following aspects:

- The stages or milestones of design and development
- The inputs for design and development
- The expected outputs from design and development
- The controls, verifications, and validations on the progress of design and development

8.3.2 Design and Development Planning

The design and development processes, operations, activities, and controls shall be planned in accordance to several requirements (the ISO 9001 Standard requirements):

- The nature, duration, and complexity of design and development activities
- The necessary process stages of design and development, including activities for design and development reviews
- Verification activities necessary for the control of the progress of design and development
- Validation activities necessary for demonstration and conformation of design and development
- Definition of the responsibilities and authorities in the design and development process
- Determination of internal and external resources needed for design and development activities
- The need to control interfaces between persons involved in the design and development process
- The degree of involvement of customers and users in the design and development process and the relevant process stages that they will be involved in
- The need for subsequent provision of products and services
- The expected degree of control on behalf of customers and other relevant interested parties in the design and development process
- Documented information of the design and development activities necessary to demonstrate that design and development requirements have been met

Nature, Duration, and Complexity of the Design and Development Activities

In determining the stages and controls for design and development, the organization shall consider the nature, duration, and complexity of the design and development activities. This requirement is important for identifying the process environment in which the design and development will be planned and developed and for identifying the interrelations between activities of design and development or other activities of the QMS. According to those distinctive methodologies, frameworks and tools for the design and development activities shall be defined and determined. Let us discuss the three:

- Nature refers to the essential qualities, constraints, or characteristics of the design and development activities. It refers to the type of activities, the type of inputs and outputs, the type and amount of data that those activities require or generate, the personnel, their capabilities and experience, the process environment in which they will be performed, and the related risks and their controls.
- Duration refers to the planned period of time in which the activities must be accomplished.
- Complexity refers to the quality of process elements and their combinations. It is important to assess the complexity of the activities in order to be able to evaluate whether the organization can perform these activities. Another goal of understanding the complexity is the ability to develop processes or activities that will serve this complexity. The complexity in turn will help in determining the requirements for inputs, personnel, or documentations and the interrelations between design and development activities and other activities of the QMS.

Process Stages of Design and Development

Process stages describing the design and the development process must be defined and determined. Process stages represent the main steps that the organization must follow and complete while designing and developing. They are the basis for the method to manage design and development. Defining and mapping the process stages

- Is an important stage necessary to evaluate whether design and development has achieved its objectives
- Creates a clear picture of design and development progress
- Allows authorities in design and development to release the outputs to the next stage
- Assists in prioritizing the activities of design and development

When a group of activities is considered to be a stage of design and development, contributes to the progress, requires inputs, and generates outputs, it should be counted as a process stage. For each process stage, you must determine

- Its relevant activities
- The associated resources (internal as well as external)
- The associated controls and actions for addressing risks
- The necessary development tools
- The necessary monitoring and measuring devices
- Its verification and validation activities

- Its responsibilities and authorities
- The need for involvement of customers and users

Examples of process stages include the following:

- Marketing reviews
- Concept and planning
- Research
- Prototype design
- Development realization process
- Development of realization controls
- Test of batch manufacturing
- Changes and modifications
- Launching the product
- Control after launch
- Updates and modifications

The most simple way to determine and define the process stages is using a process chart that illustrates the progress of design and development in the organization. After defining the process stages, you may derive the design and development activities and assign them to specific process stages.

Design and Development Review

At suitable process stages of design and development, methodical reviews of design and development shall be planned and performed. This review has the following objectives:

- Evaluating the ability of the results of design and development to meet requirements
- Identifying any problems and proposing necessary actions

Verification Activities

Verification of design and development activities refers to the examination whether the development activities are advancing as planned. While planning the design and development process stages, the organization shall describe how one will examine whether design and development is advancing according to plan and when the verification activities will occur during the design and development process stages.

In practice you must verify the progress of the development process by controlling it through attributes such as availability of inputs, transfer to development phases, performing activities, responsibility, time frame, risks, and acceptance of required outputs. The application of the verifications will be discussed in detail in clause 8.3.4—Design and development controls.

Validation Activities

Validation of design and development activities refers to the examination and confirmation that inputs of the activities are competent and that the outputs of the activities achieved the required results.

In practice you must include validation activities during the design and development process stages to ensure that

- Inputs transferred into design and development are adequate and suitable
- The resources invested in design and development are adequate and suitable
- The outputs of the design and development activities meet their objectives

The test and assurance will be done with predefined criteria. The application of the validations will be discussed in detail in clause 8.3.4—Design and development controls.

Responsibilities and Authorities

The organization is required to identify the organizational personnel responsible for conducting, performing, and accomplishing design and development and quality tasks. The organization must refer to any role that performs any kind of activity in design and development: engineers, scientists, programmers, lab technicians, and so on. The bottom line is defining and determining which functions or roles participate in the design and development activities, and each of the participants in the design and development must have a clear and defined understanding of his or her responsibilities and the activities that he or she is required to perform, how these activities are integrated into the development cycle, to whom he or she must report, and what the outputs expected of them are.

Each development and design activity has corresponding responsibility and authority assigned to it. This is also part of the process approach promoted by the ISO 9001 Standard. The responsibilities shall include

- Design and development tasks
- Design and development inputs—make sure that the inputs are available and complete
- The interrelation with other design and development roles
- Communication with the customer or his or her representative in the organization
- Conducting design and development reviews, verifications, and validations
- Approving changes to design and development

The definition may be documented on a job description, a development plan, or a procedure. You may also use a development team structure (like an organizational structure) with interrelationships outlined between the participants. In cases where certain tasks require training that was not included in the training program, such training must be identified, planned, and given to the appropriate parties.

Internal and External Resources

It is necessary to define which resources are required for the design and development activities or any development tools or instruments that will be used during the activities. The purpose serves to define the appropriate resources, accessories, and process environment used for the design and development activities and to eliminate the use of irrelevant or invalid resources. The types of resources include

- Human resources
- Infrastructures like labs
- Tools and equipment like software for design and development
- Monitoring and measuring devices
- Feasible materials like parts, components, and raw materials
- Documentations

I find it effective to define and allocate resources and their capacity to certain activities. Thus it will be clear which resources are necessary to each activity. Please review the following table:

Process Stage	Activity	Resources
Prototype design	Test of prototype design	Technician—20 work hours Testing lab—20 work hours Software for testing—20 work hours
Test of batch manufacturing	First test batch	Technician—24 work hours Production manager—24 work hours Machine of type A—16 work hours
Test of batch manufacturing	Test of packaging	Order from Supplier XYZ of 200 packages

Mapping the resources needed for design and development will assist you in allocating organizational resources better. When it comes to external providers, it is necessary to define what products, components, or services are to be purchased in order to achieve the development, and the requirements for purchasing shall serve as input for the design and development process.

Control of Interfaces

The interfaces and interrelations between different design and development entities need to be defined. Interfaces refer to the overlap where more than one area, functional group, subject, or field involved in design and development may affect each other or have links to each other. It refers also to interfaces with external interested parties. The plan should specify those interfaces. The objective is to ensure effective interrelations between the different parties involved in design and development: communication, clear assignment of responsibilities, ensuring that outputs are verified and suitable for the next phase, and the assurance that information and data can flow seamlessly. The success of design and development depends on the degree of communication and interrelation between all the functional groups that are involved.

In order to make the interface more effective, I recommend describing the interface on a flow chart as a process explaining the relations and interrelations between the entities throughout the development stages—how documents relate to other documents, how the outputs of one document are the inputs for another, and so on.

For example, if one is developing a product that has mechanical as well as electronic components, both developed by different teams, one must ensure both teams have efficient communication between them.

When defining the relations that serve the design and development process, one should refer to the interface with external objects, bodies, or parties that may have an effect on the process or may take part in any of the design or development activities. These may be customers, suppliers, authorities, or other developing teams or organizational departments.

The organization needs to define controls of those interrelations in order to ensure that they are effective. Which aspects may be controlled?

- Transmission of inputs
- Acceptance of outputs
- Exchange of information
- Sharing of risks and actions to address those risks
- Common reviews

One way to define the controls of interfaces is to initiate a review that will involve areas of design and development. In this review the topics that were mentioned earlier will be discussed.

Involvement of Customers and Users

Involving the customer during design and development increases the chances that the objectives of the product and the customer's expectations will be properly achieved. In this way one may ensure that exchange of data and information with the customer is guaranteed. This involvement of customer may be performed as joint reviews that may be scheduled on a regular basis or at defined process stages. Which issues may be reviewed?

- Information regarding the product
- Progress of the development
- Conformance of the design and development outputs according to the customer's specifications
- Conformance of design and development outputs according to regulatory requirements when such are applicable
- Demonstrations of the product and its features
- Results of validation activities
- Progress of activities expected by the customer
- Issues and problems that might have arisen during design and development
- Opportunities that occur during design and development

In practice the organization must define

- Where and when (in which design and development stages) will the customer be involved
- Which inputs must the customer bring at each stage
- Which outputs must the organization provide the customer with at each stage

Those definitions can be included in the development plan (in case one is managing the same).

Requirements for Subsequent Provision of Products and Services

The process of the design and development shall ensure that the outputs can be integrated in the realization processes and activities once the development is released. This requirement emphasizes the relation between the outputs of design and development and quality planning in the organization. There are two approaches to managing the link between design and development and quality planning:

- You may ensure that the products are developed in accordance with the quality planning implemented in the organization. This will ensure smoother integration of the new product in the realization processes.
- You may redesign quality planning along with the progress of design and development, and items concerned with each stage of design and development shall be updated.

Control on Behalf of Customers and Other Relevant Interested Parties

When the customer or other interested parties are involved in design and development, their role, responsibilities, and authorities must be defined in detail. Which other interested parties other than the customer may take part and which tasks might they fulfill?

Interested Party	Role
Customers	Approval of specifications of the product or service
	Delivering inputs
	Approval of outputs
	Test of product functionalities
	Approval of product before release
Employees who are not involved in design and development	Test of product functionalities
Regulators	Test of product functionalities
	Approval of product before release
Suppliers	Delivering inputs
	Approval of outputs
	Test of product functionalities

Documented Information of the Design and Development Activities

The organization is required to retain documented information that demonstrates that design and development requirements have been met. The records have the following goals:

- To prove design and development progressed according to the planned process stages
- To prove that all the expected inputs were delivered
- To prove that all the expected results were accepted
- To prove that results of the design and development activities meet their objectives
- To prove that certain required conditions were fulfilled while conducting the design and development activities

The records shall be designed in a format that will enable them to be delivered as inputs to subsequent stages in design and development. The required records shall cover the next areas of the design and development process stages:

- Design and development inputs
- Records of design and development reviews
- Design and development outputs
- Design and development changes
- Records of reviews of changes to design and development

The documented information shall be submitted to the controls suggested in clause 7.5—Documented information.

8.3.3 Design and Development Inputs

Design inputs serve as the basis for product development. The inputs for design and development processes and activities must be determined in order to ensure their availability to those processes and activities. The ISO 9001 Standard requirements include the following:

- The organization shall determine the essential requirements of inputs to the specific types of products and services to be designed and developed.
- The next issues shall determine the type and extent of the inputs to design and development activities:
  - Inputs relevant for the intended use and performance of the products or services
  - Data and information derived from design and development of previous relevant products or services
  - Statutory and regulatory requirements
  - Standards or codes of practice that the organization has committed to implement and act upon
  - Potential consequences of failure of design and development to meet requirements or specifications derived from the nature of the product or service
- The inputs shall be
  - Appropriate to the objectives and expected results of design and development
  - Complete
  - Unambiguous
- Conflicts that may arise from different types of inputs will be resolved.
- The organization shall retain documented information on design and development inputs.

Determine the Essential Requirements for Design and Development

The inputs serve as essential requirements necessary for the progress of design and development. Those requirements must all be defined as corresponding to the purpose of the product or service. The inputs should be determined according to the planned activities of the design and development processes and the intended use of the product. The main concept of defining one's design and development inputs is simple: the inputs must relate to the product's intended use, functional performance, and quality and regulatory requirements. Within the design and development of inputs requirements, the organization must refer to the issues that will be discussed in this chapter. The inputs may be derived from

- Development of prototypes
- Need to modify earlier developments
- Request for changes
- Problems or failures with earlier versions
- Failure to comply with the acceptability criteria
- New customer requirements
- An iterative development

Clarity, Adequacy, and Completeness of Design and Development Inputs

When defining and introducing the inputs to design and development, you must ensure that they serve their purpose, or in other words, a design input requirement must be measurable, verifiable, and objective. The requirements for defining adequate, complete, and unambiguous refer to the ability to verify those requirements of the product in later phases of the design and development. The standard demands of us to assess the quality of the inputs through several quality parameters:

- Adequacy—adequacy of the inputs refers to the ability of the inputs, their types, and content to meet the needs and expectations of the involved parties. Here it is required to ensure that requirements for the related product are appropriate and address the intended use of the product and the needs of the customer.
- Clarity—unambiguous inputs refer to the quality of having a single clearly defined meaning that leads to only one conclusion and that prevents misunderstandings or conflicts. Design input requirements that are ambiguous cannot be objectively verified.
- Completeness—completeness of inputs refers to the quality of having all the data and information that are required, and when something is missing, failures may appear down the road.

Example of an ambiguous and incomplete input will be as follows: "The application shall process adequate data in a considerable time." No one can objectively verify or validate this specification because "adequate data" and "considerable time" are open for discussion. Instead, I would define "The application shall process 60 MB of data in 30 seconds." I suggest considering and defining how the design and development inputs will be verified before they are submitted to the activities.

Avoiding Misunderstandings and Conflicts

While reviewing the design and development inputs, one must review any impractical requirements (requirements that cannot be fulfilled or that the development team might have problems fulfilling or understanding) or inputs that might conflict with each other. The goal is an early detection of any future difficulties, conflicts, or obstacles that may arise during the development phase. Resolving conflicts regarding the inputs refers to the review of the inputs and the approval that the inputs do not contradict each other and are suitable—the requirements of several inputs can be appropriately integrated in the developed product. It is possible to conduct the review together with the customer or with the responsible party for these requirements in order to obtain their points of view. Examples of conflicts or misunderstandings that may occur are as follows:

- Requirements that cannot be validated or verified
- Incomplete information or lack of information regarding the user or the intended use
- Lack of information or details regarding the user's environment
- Contradictions between requirements

Inputs Relevant to the Intended Use and Performance

The organization shall determine product requirements for performance, functionality, and intended use in order to allow the identification of all necessary design and development inputs. The objective here is to ensure that the product will acquire all its expected features and functionalities. The intended use of the product or service refers to its purpose, operation, and utilization and relates to user expectations. Intended use refers to information and outputs derived from market surveys or research, risk-based thinking outputs, customer requirements, and a review of requirements related to the product. The definition shall refer to situations where it is inadvisable to use the product or provide the service. Let us review some aspects of the functional and performance requirements:

- Physical characteristics of the product such as plans, sizes, dimensions, diagrams, drawings, samples, and prototypes. The definition shall refer to tolerances and limits.
- Requirements for handling and behavior with the product and specifications of aspects such as packaging, storage, labeling, operation, handling, and maintenance requirements; anything that might affect the quality of the product or its intended use must be reviewed during the development stages.
- Definition regarding the operating environment of the product or the environment in which the service shall be provided. If the product is to operate under a controlled temperature, the design and development activities must take that into consideration. This requirement is also valid when a software must be installed and there are technical specifications for installation.
- Interface with other products where the product is to be combined or installed with other products or pieces of equipment and/or accessories; this shall be referred to as inputs to design and development.
- Service requirement shall be taken into account. When it is already known that the product will require service activities in the long run, it will be reviewed in the design and development activities so that certain characteristics of the product may be planned appropriately.
- Compatibility of the product's components shall be considered when the product is constructed or assembled from various components and materials; it is necessary to

examine the materials used in order to ensure that the different components compete or match one another and the risk of contamination, error, or failure is prevented.

- The organization will specify any safety requirements that are needed to serve as inputs for the development and which may affect the design of the product and its characteristics and intended use.
- Appropriate considerations to the integration of environmental protection aspects shall be delivered as inputs to design and development as and when they are applicable.

In practice one has to demonstrate that for each product under design and development, one has defined the required inputs and that the inputs are available. It is difficult to say how these inputs will be introduced to design and development because each type of input has its own characteristics and each will be documented in a different way. In general, I recommend building a product file that includes all the required inputs. The format and structure will determine the types of expected inputs and their content.

Data and Information Derived from Previous Design and Development

Where applicable, the organization shall refer to outputs and experience of earlier, similar, or parallel developments that have been performed by the organization. The outputs of such developments may serve as inputs to future developments. The goal is to allow the knowledge and experience of the design and development to be independent of persons in the organization. Using information derived from previous similar design and development activities may assist in preventing quality problems in future development; these problems may have already been prevented before and may help save resources.

When planning a new product that is based on a former product, the organization may (and should) use the experience and knowledge acquired in previous developments. This type of information refers to

- Experience from former developments or designs
- Knowledge acquired during former developments or designs
- Techniques developed from former developments or designs
- The use of development tools
- Outputs of former developments or designs
- Documentation of applicable earlier developments
- Complaints, failures, or other events regarding previous products
- History of the organization (a knowledge center)

In practice, the organization should create a database that is available to development teams, which they may access and fetch data and information of previous developments. Ways to transfer the experience and knowledge of previous developments include the following:

- Design and development manual
- Data and information regarding analyses of failures and nonconformities (like FMEA)

- Outputs and techniques of production optimization (that will be used in design and development)
- Results of experiments
- Reports about materials, components, techniques, and processes

Statutory and Regulatory Requirements

Many products must be realized under local or international regulations and external requirements. These are applicable to the design and development stages as well. Both statutory and regulatory requirements are usually nonnegotiable and must be complied with. They were planned in order to assure that some conditions (e.g., regulatory and safety) are met while developing a product for use. So it is critical that the organization will identify those that are relevant to its products and areas. The relevant statutory and regulatory requirements should be identified while understanding the needs and expectations of the interested parties (see clause 4.2).

When planning the inputs for design and development, one shall identify, indicate, and refer to requirements such as regulations, regional directives, or statutory requirements and particular requirements that serve as inputs and which controls are to be incorporated into one's design and development activities.

Standards or Codes of Practice That the Organization Committed to Implement

The organization must identify and introduce standards or codes of practices that it has committed to implement during the design and development activities. Standard or codes of practices may suggest rules, mythologies, views, ideas, opinions, life cycle models, or concepts related to design and development. Those types of requirements present restrictions, conditions, or constraints that must be integrated into the activities and that may influence its performances, functionality, intended use, and safety.

These requirements are to be identified and made available to the relevant parties in the appropriate manner, with which they can understand those requirements and integrate them into the design and development activities. Under standards or codes of practices, you may find

- External procedures
- Policies
- International standards or technical requirements
- Environmental standards
- Technical specifications
- Industry standards or specifications
- Packaging requirements

Those types of standards or codes of practices are usually defined in the contract or purchase order, and customers may demand that applicable standards or codes of practice typical in their industry shall be referred to while developing products for them. They can be introduced to design and development in the form of procedures, instructions, verifications, or validation controls and tools (like software or measurement tools).

Potential Consequences of Failure of Design and Development

Potential consequences of failure due to the nature of the product or service are supposed to be addressed during the design and development activities. The detection and identification of such consequences and situations related to the use of the product or provision of service should be completed as early as possible in its life cycle and that is in the design and development phases. One may reduce the probability of the occurrence of potential failures through the identification of situations where the product might fail, the analysis of related risks and their causes that might affect the objectives of the product, the implementation of controls, and the mitigation or elimination of known hazards during the design phases. The result is the inherent assurance that the product will act as expected. Addressing these potential consequences during the design and development stages has several goals:

- To introduce risk management activities into the design control process and the life cycle management of the product or service
- To implement the appropriate control measures in the final design of the product or service
- To reduce the probability of potential consequences by improving those potential characteristics of the product or service
- To focus and refine the controls that will be implemented during the realization of the product

The organization shall also give its attention to environmental product aspects and should seek to minimize environmental impacts over the full life cycle of its products. This consideration shall relate to the design and development activities, the realization and delivery of the final product or service, as well as the disposal of the product.

In practice I recommend conducting before the design and development begins a sort of review to identify hazards, evaluate risks, and implement risk control measures, specify the activities and responsibilities, and refer to the required documentation standards, regulations, and records that may support the conclusions of the review. The conclusions of the review will be introduced as inputs, for example, as data and information derived from previous design and development.

Documented Information of Design and Development Inputs

The organization is required to retain documented information on design and development inputs. This documented information has the purpose of demonstrating that the inputs submitted to design and development meet the requirements in clause 8.3.3—Design and development inputs. Examples for documented information include the following:

- Documented product specifications
- Customer specifications of the product
- Minutes of development meetings
- Offers, orders, or contracts
- List of relevant statutory and regulatory requirements
- List of relevant standards or codes of practice

8.3.4 Design and Development Controls

Clause 8.3.4—Design and development controls—defines the type of controls that the organization is expected to apply to the design and development activities. The ISO 9001 Standard requirements include the following:

- The organization shall apply the controls on the design and development activities.
- The controls shall ensure that
  - Results for design and development are defined
  - Reviews necessary to assess the ability of the results to answer their requirements are performed
  - Verification activities necessary to ensure that the results conform to their requirements are performed
  - Validation activities necessary to ensure that the results meet the requirements of specified application or intended use are performed
  - Actions that are results of problems and events detected during the design and development reviews, verification, or validation activities are initiated
- Documented information that demonstrates the carrying out of these controls shall be retained.
- Note—reviews of verification and validation are activities with defined purposes. They can be conducted separately or in any combination, as is suitable for the products and services of the organization.

Controls on the Design and Development Activities

The organization is required to apply controls to the design and development process to ensure several objectives. But first I would like to understand the meaning of setting up of control for design and development. Controls of design and development refer to techniques and tools including the qualifications necessary to ensure that the intended results of the activities are accepted. The objectives of the controls are as follows:

- Controls related to the design, with which the organization should ensure so that the product meets the acceptance criteria
- Setting up of the controls to ensure the appropriate progress of the design and development
- Eliminating undesired results and effects on the design and development activities
- Ensuring conformed inputs and outputs of design and development activities

For each design and development phase, process stage, or activity, the appropriate control will be identified and determined. The controls shall be adjusted to the activities that they are reviewing. The controls shall be planned in accordance to the criticalness and complexity of the activity and its related risks of failure. When planning the controls, one must take into account the different conditions in which the activities are performed. Those conditions may set the concept of the control. When one feels it necessary, one may document the controls in procedures or include them in design and development procedure or instructions. Principles in setting of controls include the following:

- Identifying the design and development activity of process stages to be controlled
- Defining the purpose and objective of the controlled activity (what in the process stage is required to be controlled)
- Defining the control activities—how the control will be installed
- Defining means with which the controls are performed: related documentations like instructions or standard and monitoring and measuring devices
- The required records that will prove the control was performed and the results of the control

Clause 8.3.4 sets a list of design and development aspects that must be controlled. They will be discussed in the next paragraphs. The standard adds a note to the requirements of clause 8.3.4—Design and development reviews—verification and validation have distinct purposes. This note suggests that the organization may determine whether these controls will be conducted separately or in any combination that is suitable for the developed products and services.

Results for Design and Development Are Defined

For each process stage of design and development and activity, the required outputs are supposed to be defined. A control must be applied to verify that these outputs are defined and clear to the relevant parties participating in design and development (external parties as well). This type of control will enable the interrelation and a smooth flow of information between the activities. The idea is to ensure that the design and development does not proceed to the next stage without all the necessary outputs of the previous stage or without the necessary verification of an activity or validation of results. Types of results include the following:

- Results of activities (the outputs)
- Results of verifications
- Results of validations

Once it is detected that the defined results are not acceptable, an action must be initiated.

Planning such a control, one must review all the design and development activities (whether as a procedure or a flow diagram), identify all the activities, and ensure that for each activity an output is clearly defined. Implementing such a control can be challenging. You may use a procedure or a form that implies which results are needed and whether they are defined in all process stages of design and development. But bear in mind that each type of product may have other process stages that require a different arrangement of the procedure or the form.

Design and Development Review

At suitable process stages of design and development, methodical reviews of design and development shall be planned and performed. This is an opportunity for an overview of the progress of the design and development when participants from different areas with different skills and points of view are brought together to discuss the design and development. Such brainstorming activity can provide solutions to many problems. This review has the following objectives:

- Evaluating the ability of the results of design and development to meet requirements
- Identifying any problems and proposing necessary actions

The type and manner of the review shall be appropriate to the complexity of the activities, the complexity of the product, its quality requirements, and the degree of risk associated with the specified use of the product or service. Preparing for the review, the necessary criteria for evaluating the progress of design and development will be provided. It refers here to progress on various aspects:

- Progress of the design and development according to the process stages
- Availability of inputs and their quality
- Availability of resources
- Performance of external providers
- Results of verifications and validations
- Achievement of product objectives (design and development outputs)

When determining the review, the next issues must be referred to:

- The topics or issues of the review shall be determined.
- The type of review, how the review will be conducted, and demonstrations of results, experiments, proof of correctness, or inspections.
- The participants of the review must be identified. Here you need to refer to functionalities of design and development with the goal that each party will give its inputs.
- How the review will be conducted including activities or use of techniques or tools like software.
- What are the outputs of the review in terms of records and documented information.
- Reference to the standards or codes of practices that are to used or implanted during design and development activities. The review will assess their applicability and whether their objectives are met.
- Reference to the regulatory or statutory requirements that are to be used or implanted during the design and development activities. The review will assess their applicability and whether their objectives are met.
- Follow-up activities necessary to ensure that issues and problems that were identified during the review will be addressed.

As I see it, there is no simpler way to define the arrangement of the review rather than defining a form that covers all the issues mentioned earlier.

When nonconformities are detected during the review of design and development, the organization shall submit them to the process of handling nonconformities as required in clause 10.2—Nonconformity and corrective action. This way, it is ensured that nonconformities that occurred during the design and development stages are submitted to a controlled process and will be resolved. Further development activities shall be approved and preceded only after all nonconformities have been eliminated or reviewed.

Verification Activities of Design and Development

The organization shall define and implement activities for verifying the design and development of the product or service. The objective of the verification is to confirm that design and development outputs meet the design input requirements. The verification activities are to be included in the design and development plan. The verification activities are not about studying the results of the design and development but to review its progress. What needs to be verified?

- All the expected inputs were delivered and are in the correct format.
- Design and development tasks were performed and accomplished.
- Knowledge required for the design and development activities is available and valid.
- Human resources who participate in the design and development activities are able to perform the required tasks; they are qualified, trained, and have the required experience.
- The required documentation is available and applicable, and the needed information and data are available.
- All the required outputs were accepted.
- The outputs conform to the design and development inputs.
- The design and development controls were applied.
- The design and development processes were reviewed.

The definition of the verification shall refer to the

- Methods used for the verifications
- The personnel responsible in the organization that will perform the verifications
- The approval of the verifications

For example, when designing a product, the organization generates a document with specifications that include

- Customer expectations of the product
- Product characteristics such as measurements and tolerances
- The description of the intended use
- The necessary human resources for the development
- The required tools for the development

For each type of input, a certain documentation of the specification is expected:

- Customer expectations will be backed up with a document that describes them.
- Product characteristics will be backed up with documented descriptions of the product.
- Product measurements will be described with diagrams.

The verification will ensure that all those documentations are available and are appropriate; that is, they deliver the required information or data and can be used for later design and development activities. An effective way to implement such verifications is to maintain a product dossier for design and development that will define exactly which inputs are expected. Release of the design and development will be done only when the dossier is completed.

Validation Activities of Design and Development

The organization is to define validation activities to design and development. The validation is a critical test intended to prospectively ensure, with objective evidences, that the output of design and development (the product or the service) satisfies its users' expectations in terms of functionality, performance, safety, and intended use. Validation in the context of design and development is the confirmation, through the provision of objective evidence and using supporting data, that a requirement, that is, an expectation for an intended use, or product characteristic that can be fulfilled.

The objective of the validation is to ensure that the design and development outputs, the product, its characteristics, and the intended use, conform to the specifications that were described in the design and development inputs. Another objective of the validation is to prove that the requirements will be consistently met throughout the entire life cycle of the product. Validation test will be practiced on a process output. It may span all the process stages of design and development. For products with several intended uses, for each type of intended use, a validation will be defined.

While planning the design and development validation, it is required to refer to conditions that will represent the use and the use environment of the product or the service. For example, if you are developing a software product, the validation activities shall be performed in an environment, where the software shall run and work: hardware specifications, version of operating system, and so on. Outputs of risks analysis may serve as the list of validations. Planning the validations shall refer to the following issues:

- Which product characteristics must be validated
- Which methods will be used to validate
- How much is needed to be validated
- What are the basis, criteria, norms, or standard that will serve as comparison
- How will the results be documented

The validation of design and development is to be applied to all elements and activities of design and development:

- Inputs to design and development
- Process stages of design and development
- Resources allocated to design and development
- Outputs of design and development

In order to simplify the matter, I divided the validation requirements into the following:

- Design and development inputs—validation that the inputs do deliver the information and data that are expected.
- Process stages and activities—validation that the design and development activities will achieve the results that are expected.
- Design and development environment—validation that the environment in which the product will be designed and developed is suitable; safety measures are implemented; conditions of the development conform to the predefined conditions (where the product will function or the service will be provided).
- Tools and equipment—validation that the tools and equipment that will be used for the design and development activities are intact, calibrated, used correctly, and maintained as planned.
- Design and development outputs—validation that the expected outputs are adequate to the inputs and that the results conform to their objectives.

An important aspect in the validation of design and development is the identification of the appropriate measurement devices and equipment monitoring tools. These must be able to provide with appropriate and reliable data in order to validate processes. When a tool or piece of equipment is required for validation, it is necessary to define what it is and to verify that it is available. Say, for example, that one is using a certain material during the development and are required to monitor its behavior in certain conditions. First, one needs to check how the data will be obtained. Second, one needs to define which monitoring equipment will obtain the necessary data. Next, one must see whether the equipment is available and ensure that it will provide the desired data.

Actions Taken on Problems

The next requirement refers to the assignment of controls over necessary actions that were initiated after identification of issues and problems or events that occurred during the design and development stages, reviews, and verification and validation activities. This requirement suggests that the organization shall deal with deficiencies or nonconformities that occurred during design and development activities and were identified during the review.

The progress of design and development should proceed only after the consequences of all known deficiencies or problems are resolved or at least the associated risks are acknowledged. During the review, the following problems from different aspects may be detected:

- Problems regarding the progress of design and development according to the process stages
- Availability of inputs and their quality
- Availability of resources
- Results of verifications and validations
- Achievement of product objectives (design and development outputs)

A problem may be best detected when an aspect of design and development was evaluated against a defined criteria and the assessment indicated that the objectives were not met. For example, verification problems indicate that inputs were not submitted to design and development as expected or outputs were not acceptable. To each detected problem, an action is supposed to be initiated as a reaction in order to solve the problem. The standard demands that such actions will be reviewed for their conduct and effectiveness. When actions or solutions to problems are detected, it is required that you also review their effectiveness, that is, whether they achieved their objectives and, more importantly, whether the actions or modifications met design and development requirements.

Documented Information of Design and Development Controls

Documented information needed to prove that the controls mentioned earlier were performed and their results are satisfying is to be retained:

- Documented information of reviews of design and development activities and the handling of problems detected during the reviews
- Verification results
- Validation results

The organization is required to determine the expected records for each of the controls mentioned earlier. But in order to maintain those records, one is required to design the appropriate means that will gather the records. The purpose of the records then is to demonstrate the carrying out of the control and to present the results.

8.3.5 Design and Development Outputs

The outputs of design and development shall demonstrate that the activities were carried out in accordance with the plan through a traceability to the design and development inputs. In other words, the outputs will allow adequate evaluation of conformance to design input requirements. Design and development outputs shall be documented, reviewed, and approved before release. The ISO 9001 Standard requirements include the following:

- The organization shall define the design and development outputs.
- The outputs shall meet the design and development input requirements.
- The outputs shall reflect and describe the activities necessary for the realization of the product or the service.
- The outputs shall include or refer to monitoring and measurement requirements and to relevant acceptance criteria.
- The outputs shall specify the characteristics of the product that are essential for its intended use and safe and proper provision.
- Documented information of the design and development outputs shall be retained.

Definition of Design and Development Outputs

Design and development outputs are the results of design and development activities, and they represent the specifications for the product or the service:

- Characteristics of the product or service
- Specifications for the realization (manufacturing of a product or provision of a service)
- Monitoring and measuring activities to ensure conformity to the product or service
- Acceptance criteria for expected outputs of the realization activities

It must be clear to the design and development team which outputs are required, in which form and format, and what are the expected details. This definition of the outputs shall assist in controlling all the expected outputs that were accepted after each design and development activity and will allow traceability to the inputs. In the next paragraphs, I will go into details.

Compatibility between Design and Development Outputs and Inputs

The first requirement refers to the verification against input requirements. The design and development outputs shall be realized in terms that can be verified, validated, and proved for compliance against design input requirements. The verification of the outputs shall demonstrate that inputs to each design and development process stage are correctly reflected in the outputs. In other words, it will allow traceability between specifications (inputs) and results (outputs) and prove the accomplishments of the requirements.

This requirement demands that at the end of each design and development process stage, a verification between the output and the relevant inputs will be performed. Let us review how this type of verification may be integrated in the next sequence of activities:

- Input—a requirement to be designed and developed
  - Area or scope of the requirement
  - The nature or characteristics of the requirement
  - Description of the requirement
  - The required resources
- Output—the realization of this requirement
  - Evaluation of the requirement
  - Description and analysis of the requirement in terms of design and development
  - Description of the solution—suggesting a solution and proving its feasibility
  - Performing the solution
  - Delivering the output—product, process, documentation, and so on
- Verification or validation—proving that the output is consistent with the requirement
  - Description of the requirement
  - Description of the verification—what has to be checked: availability of characteristics, test of performance, and the use of certain resources
  - Performing the verification
  - Specifying the results
  - Evaluating the results—is the output compatible with the input?

In practice I would define for each output documentation its verification against its relevant input.

Activities Necessary for the Realization of the Product or Service

One of the main purposes of the design and development outputs is to prescribe the necessary requirements and activities necessary for the realization of the product:

- To demonstrate how the organization's QMS applies to the realization of the product
- To demonstrate how realization activities may meet customer specifications
- To demonstrate how realization activities may meet regulatory specifications
- To describe how resources shall be used during the realization processes including reference to responsibilities and authorities
- To describe the controls that shall be applied on the realization processes
- To describe how risks shall be addressed during the realization processes
- To describe which documented information shall be maintained during the realization processes: used documentation and expected records

These requirements and activities shall serve the next issues:

- The output shall indicate the workflow and processes needed for the realization of the product.
- The outputs shall specify the required materials (raw materials and/or components) and where applicable the specifications or standards to which the materials have to necessarily conform to for the realization of the product.
- The output shall include quality management requirements:
  - Relevant quality objectives
  - Addressing risks and opportunities
  - Submitting changes to control
  - Control of nonconforming product
- The output shall specify which resources are needed for the realization of the product including reference to personnel and their training requirements necessary for the realization of the product.
- The outputs shall indicate which tools, models, and methods are to be used when realizing the product.
- When required the organization shall define characteristics and traceability measures such as serial numbers, validity dates, or batch numbers, when such are required, to allow control over the realization of the product.
- The output shall provide appropriate information for maintaining effective working relations with external providers such as suppliers or subcontractors:
  - Which product components, materials, parts, and so on must be purchased
  - Which suppliers are selected for the purchases and which criteria are applicable when selecting a supplier
  - Which information regarding the realization of the product must be communicated to the suppliers such as description of the product or service, delivery date, and conditions
  - Which controls must be applied to those external providers or must be applied by the external providers
  - Determine the verification, or other activities, necessary to ensure that externally provided processes, products, and services meet requirements.
- Determine and define controls such as verification, validation, or other activities necessary to ensure that the processes, products, and services meet the requirements.

- Where applicable the controls required by customers or regulatory bodies shall be defined. The definition shall include the following issues:
  - Issues in the product shall be communicated
  - The means to be used for communicating with the customer or regulatory bodies
  - Responsibilities for the communication
  - Relevant and necessary records
  - How the feedback of the customer shall be integrated in the realization processes
- The outputs shall refer to required customer property and its handling.
- The outputs shall define which activities are necessary for the preservation of the product: handling, storage, packaging, and delivery.
- The output shall refer to risks and their subsequent nonconformities that may occur during the realization and how they may be reduced or managed.
- When appropriate, the output shall include generic or specific documented information that is necessary for the realization: procedures, instructions, plans, blue prints, diagrams, quality tests, and so on.
- The output shall specify which records are required to be maintained while realizing the product.

Establishing a quality plan as an output of design and development may be an effective way to describe these requirements and activities necessary for the realization of the product, and these outputs may serve as a basis for the quality planning. You may refer to the template and specifications of the quality planning in order to understand which outputs for the realization are needed.

Monitoring and Measurement Requirements of the Product

Outputs of design and development shall identify and determine the necessary monitoring and measuring activities and the controls to be carried out when realizing the product. Monitoring and measurement activities shall provide the means by which evidence to conformity and meeting of specifications will be obtained. This requirement refers to the quality instructions of a product that define which process inputs (materials, components, qualifications), process (activities or operations), or process outputs (products or services) shall be measured. For each of the requirements mentioned earlier, it will determine

- At which stages the monitoring and measurement activities shall be performed
- Which characteristics will be monitored and measured
- Which parameters of realization processes will be controlled
- How the expected characteristics of the product will be measured
- Which methods are to be used to monitor, analyze, and evaluate process performance
- Which methods are to be used to monitor, analyze, and evaluate process outputs
- Which monitoring and measuring devices shall be used for monitoring and measuring activities
- Which criteria are to be used for the evaluation

In some cases, it may be necessary for the customer to provide those requirements. In other cases, a third party like a supplier or a regulatory body may need to perform these activities.

Acceptance Criteria

One output of the design and development activities shall be the acceptance criteria for the product. The acceptance criteria are used as a basis for comparison or as a reference point against which the product can be compared, evaluated, and then released or rejected. A good example is the determination of quality specifications or characteristics of a product. These characteristics are to demonstrate the features of a product or a service such as measurements of the product, performance or functionality, and tolerances and limits. These acceptance criteria shall be used during the realization of the product in order to determine the quality of the product and eventually enable the release of the product.

The acceptance criteria shall demonstrate traceability to the design and development inputs. For example, one of the inputs is customer expectations of the product—a specific feature. The acceptance criteria shall include an examination that this feature exists and answers the expectations.

Characteristics of the Product

The outputs of design and development shall specify the characteristics of the products and services that are essential for their intended purpose or use and their safe and proper provision. These specifications shall include the entire life cycle of the product: realization, sales, delivery, service, and disposal. Here is the reference to information of the product that will ensure that users and end customers use it properly:

- Information regarding the proper use of the product or the service
- Information regarding the environment where the product will be used or service provided
- Information regarding safety measures necessary to preserve the characteristics of the product

A good example is from the food industry; when an organization is developing a product, it is required to specify

- How the product will be consumed
- What are the ingredients of the product
- How the product will be packed
- How the product will be stored
    - By the organization before delivery to the customer
    - During transportation
    - By the retailer that distributes the product to the end customers
    - By the end customer that purchases the product

Examples of such outputs may be

- Warnings
- Labels
- User instructions
- Safety instructions
- Health notifications

Documented Information of Design and Development Outputs

The organization is required to retain documented information on design and development representing each of the outputs. For each type of output, you must define with which format or method it will be documented: a list of specifications, a textual description, a form, a computer application, or a tangible model. It will also be defined on which media the records must be maintained (e.g., in electronic and hardcopy formats). The outputs are usually documented as

- Documented instructions, information, and discipline to be used as guidance for the realization and use of the product or service:
  - Work flow chart
  - Process map
  - Project plan
  - Procedures (generic or specific)
  - Matrices
  - User guides
  - Operator documentation
  - Work instructions
  - Checklist
  - Training material
  - Maintenance documentation
- Outputs in a form or a pattern that will demonstrate and represent the characteristics of the product or service and based on which the product shall be realized:
  - Prototype
  - Models
  - Drawings
  - Engineering analysis

Those records must be submitted to the control of documented information as required in clause 7.5—Documented information.

8.3.6 Design and Development Changes

When changes related to design and development of the product are required or requested, they must be submitted to certain controls that will ensure the objective and expectations of customers and other third parties will be maintained. The objective of controlling the changes is to provide an overview of the requirements related to the product, its characteristics, and its status before and after the change.

Changes to design and development are to be identified and reviewed to assess and evaluate their potential consequences. The changes then shall be verified and validated, when applicable, and authorized prior to their implementation. The control will cover the entire life cycle of the product or service. The review of design and development changes refers to the identification, review, and control of the effect of the changes on products that are in the design and development phase or products that were already delivered to customers. The ISO 9001 Standard requirements include the following:

- The organization shall review and monitor changes in the design and development:
  - Changes may occur during the design and development activities
  - Changes may occur to the design and development activities subsequently
- The changes will be reviewed with the purpose to ensure that there is no adverse effect on the conformity of product as a result of the change.
- The organization shall retain documented information on
  - The changes to the design and development of the product
  - Results of the reviews regarding the changes
  - The authorization of the changes
  - Action taken in order to avoid or eliminate adverse impacts

Identifying the Changes

The first step in controlling changes is to detect in which phase of the life cycle the product is situated:

- Preproduction phase—the change occurs before the product has been released where changes are required due to design modifications, or in cases where designs have failed the verification or validation tests and cannot provide the required outputs, and, as a result, the design cannot be realized.
- Postproduction phase—the change occurs after design and development have been completed, and the product was released for realization and is being delivered to the customer. Sometimes more features and characteristics need to be added due to a customer's request, the updating of a product, market fluctuations, or new regulatory requirements.

In case of changes in a postproduction stage, there will probably be need to implement more changes due to the fact that they would have to be implemented on existing realization processes that were already released, are already operational, and these would involve many QMS elements.

Possible reasons or factors for change include the following:

- Results of design review: results of verification and validation activities, errors, or failure to provide satisfying results (e.g., in calculation, material selection)
- Failures or nonconformities detected on the product after release and the need for corrective measures
- Improvements or updates to the functionality or performance of a product
- Difficulties during the realization processes: production, installation, or service
- Changes in regulatory or safety requirements
- Change requested by customers or external providers
- Postmarket reviews and experience reports
- Changes required for corrective action
- Changes that are results of risk-based thinking

In practice you must identify those business cases where changes related to design and development occur, submit them to the control, and establish a method for identifying modification requests from the relevant interested parties of the organization. Let us review the following cases:

- One has reviewed one's design and decided that the validation activities generate unsatisfying results. In response, the designer suggests using another type of material or component and then to try to validate it again. This is regarded as a change in design and development where a factor in the product was modified.
- One has decided to replace a component in the product with a similar component that performs the same role, but from another supplier. This is regarded as a change where a factor in the product was modified because the conditions of the components (other supplier) were changed.
- One has performed a validation activity and realized that the results deviated from the limits. The developer checked and detected that the method for the calculation is not correct. He or she has suggested correcting it and performing the test again. This case is not regarded as a change—you did not change anything; you only followed standard requirements and validated the design. It is not necessary to submit the correction to change control.

Characterizing the Change

In order to understand the scope and extent of a change, it is required to characterize it in terms of processes and products:

- The version, edition, or revision of the product—a distinct identification of the product.
- The status of the product—the status may refer to the lifecycle of the product or to its usability.
- The reason for the change:
  - Improvement of the product
  - A solution for a problem
  - A change in the specification or the requirements (from the customer or from a regulatory body)
- The impact of the change.
- The scope and extent of the change—to which product characteristics the change applies and accordingly which realization processes it concerns.

Regarding the scope, the organization must assess its abilities; it may be that the organization has the knowledge and experience to handle the change or may be the change is new and requires new competence or knowledge in the organization.

Reviewing Changes in Design and Development

Before a change is forwarded for realization, it must be reviewed. The goals of the review are analyzing the need for the change and its potential impact on other characteristics of the product and ensuring conformity of the product after the change has been implemented. The reason is that the change might not occur only in the product but might concern other QMS elements that affect the product and its realization: design inputs, verification tests, validation activities (product and processes), work instructions, quality tests, expected records, reference to competence of employees, and purchase requirements and verifications. Therefore, one of the results of the review is the controls that will be applied for evaluating the change. The review shall refer to the following issues:

- The effect the change has on the product specifications
- The effect the change might have on other realization processes related to the functionality of the product
- The effect the change has on the intended use of the product
- The effect the change has on other components of the product
- The implication for the realization processes
- The implication for the relevant controls of the realization processes (verifications or validations)
- The requirement of new certifications or licenses
- The implications for existing documentations

Based on the review of the changes and its results, the organization shall plan actions for implementing the changes:

- The organization shall determine which design and development inputs such as documentation, processes, resources, component, or materials need to be modified.
- The objectives of the change will be defined.
- The activities and operations for carrying out the change will be defined.
- The necessary controls such as verifications or validations shall be defined with the appropriate acceptance criteria.
- The change shall be communicated in an effective way to all interested parties: employees who are developing, designing, or realizing the product, customers and end users, external providers, and regulatory bodies.
- All types of documentation related to the realization of the product that must be modified as a result of the change shall be identified.

After a design and development change is approved for implementation, you must control its progress, results, and effectiveness. This approval shall be documented as one of the results of the review. For more information and techniques about implementing a change, please visit chapter 6.3—Planning of changes again.

The organization may develop a process using a format or a system for transferring the request for a change to the review. A format for a change request is a good example. Managing changes in design and development with a change request assists in controlling how requests will be submitted and handled. The details expected are as follows:

- General data such as date, employee details, and department that initiate the request
- Identification of the product, process, or other QMS elements
- Identification of the relevant characteristics
- Details of the requested change
  - Why is the change required
  - What will be the change or how will it be realized including details of related processes
- Review of the effect and impact of the change on other QMS elements
- How the implementation of change will be verified
- The results of the verifications or validations
- Who is authorized to approve or reject changes

Review for Adverse Effect on the Conformity

Reviewing adverse effects on the conformity of a product is the second part of reviewing change. An adverse effect is considered as such when any nonconformity that may be related to the use of the product, or on the environment in which it is active, may occur as a result of the change. Reviewing adverse impact on conformity after changes refers to the analysis of the possible risks, assumptions, dependencies, and problems associated with changes to design and development. In other words, changing one characteristic of the product may have an unforeseen adverse influence on another aspect of the product directly or indirectly or may adversely affect the expectations of the relevant interested parties: customers, regulatory bodies, or external providers. The objectives are to evaluate and assess whether a change may cause the product or the service not to meet its specifications or requirements and to allow the organization to understand how change may affect the products or services. Issues that may be referred to after the implementation of the change include the following:

- Deterioration in function or intended use of the product
- Failure to achieve the agreed specifications
- Deterioration in integrity or compatibility of components
- Disturbance in interface with other products or systems
- Problems that may occur during realization or installment
- Lack of feasibility of verification or validation
- Conformity to regulatory requirements
- Safety of the product

After reviewing the issues mentioned earlier, you may

- Conduct a risk analysis where the organization perceives how the change may adversely affect the product
- Identify process parameters, process activities, or process outputs that are affected by a change
- Establish a method to evaluate the level of the effect
- Define the controls that may indicate whether the change has adversely affected the process or the process output
- Initiate actions to eliminate adverse effects when such are detected
- Evaluate the effectiveness of those actions—are the adverse effects reduced or eliminated

All of the bullets mentioned earlier may be integrated in the change request.

Documented Information of Design and Development Changes

The organization shall retain documented information on

- Design and development changes
- Results of the reviews
- Actions taken to prevent adverse impacts

The objective of the documented information is to provide evidences that a change was implemented under a controlled method and was identified, reviewed, verified, validated, implemented, and approved. It is necessary to document

- The requirements for the change (changes request)
- The review and subsequent analysis of the request
- The results of the review regarding the QMS elements that may be affected and potential adverse impact on conformity to requirements
- The necessary activities for implementing the change
- The approval and authorization for the change:
  - The person that authorized the change
  - Date
- Actions that were planned in order to prevent adverse impacts as a result of the review

The results of the review shall be sent to the appropriate authorities for information, guidance, and action. As mentioned in the last paragraph, I support the maintenance of a format or a system that will serve as change request—it is an effective way to manage change in design and development according to the requirements of the ISO 9001 Standard.

8.4 Control of Externally Provided Processes, Products, and Services

An organization uses the services of suppliers because either it lacks the resources to do the work itself or decided that it would be better to allow an external provider to produce the products or provide it with services. This is a natural process. But nevertheless, the ISO 9001 Standard requires structure and control when using services of external providers. The goals of this clause are to create trust in the abilities of a supplier to deliver goods or services that meet requirements in the long run, to manage better the relations with external providers, and to ensure that they will deliver according to their requirements. ISO 9001 Standard requirements are given in the following sections.

8.4.1 General

- The organization shall ensure that externally provided processes, products, and services conform to specified requirements.
- The organization shall determine which controls are to be applied to the external providers and the externally provided processes, products, and services depending on the following circumstances:
  - The products, goods, or services provided by the external provider will be incorporated into the end product or one of its components.
  - The products, goods, or services provided by the external provider will be delivered directly to the end customer of the organization.
  - A process, part of a process, or a function needed for the realization of products provided by the external provider is used in the realization activities.
- The organization shall evaluate, assess, and select external providers according to their performances and ability to supply suitable and appropriate products and services.

- The organization shall determine and establish the criteria for the external providers' evaluations and selections. The criteria shall assess the performance of external providers against the organization's requirements.
- A reevaluation shall be performed.
- Documented information of the external providers' evaluation should be maintained and kept.

Identifying the External Providers

An external provider is an interested party in the organization, though independent from the organization's QMS. This also includes cases where a supplier may be part of the organization but maintains a separate QMS. For example, in some corporations, business unit A may receive services and goods from business unit B; though no financial transaction was carried out during their interaction, and both are paid by the same boss and have the same brand name, they manage two different QMSs. And so business unit B is regarded as an external provider to business unit A, and they maintain "organization–supplier" relationships. The term external provider refers to external business units that provide the organization with

- Physical elements related directly to the product like parts, components, or raw materials
- Products that are needed for the realization of a product such as infrastructure like software or machine maintenance services
- Process services that are required to realize the product like assembly of parts, coating, painting, or cleaning
- Services like consulting or training

All of the earlier mentioned have one distinctive feature in common—they may influence directly or indirectly the characteristics or quality of a product or the ability of the organization to meet predefined requirements. Including the control of external provision of products and services in the QMS extends the influence and effectiveness of the QMS to the performance of suppliers.

Ensuring Conformity of Externally Provided Products and Services

Outsourcing a process or purchasing goods or materials for incorporation in the product is a very common situation where the organization has chosen to perform a certain process or activity by an external organization (external provider), that is, the organization delegates the responsibilities of this activity to another organization. This process or purchased product is part of the realization of the final product and has a direct effect on the quality of the product or its conformity to requirements. The external provider in this case may be a supplier that is totally independent from the organization or another part of the same organization that is not subjected to the same quality management. The external provider can

- Deliver the organization finished goods that it has processed
- Perform realization processes for the organization
- Perform services on the organization premises

The ISO 9001:2015 Standard expects that those goods, activities, tasks, processes, or assignments with influence on the quality of the product and its conformity with requirements, although performed by an external provider, will be under control of the in-house QMS. Assigning the task to the external provider does not absolve the organization of its responsibilities for conformity to all customer, statutory, and regulatory requirements. Apart from identification of those outsourced processes or purchased goods, the necessary verifications and validations will be determined and applied. These are submitted on three levels:

- The controls that the supplier shall implement and perform during the realization of its processes
- The controls that the organization has to implement over the supplier
- The controls that the organization has to perform when receiving product, processes, or services from a supplier

The controls will be set in accordance to the complexity, importance, and effect that the purchasing process has on the final product. Planning the controls will include reference to the risks related to the product or service. In order to address those risks, the required controls will be defined and implemented by either the supplier or the organization.

In addition, the external provider must have a certain level of technical or professional knowledge, skills, and abilities that are

- Related to the processes and services that it provides
- Needed to control and approve the processed goods or services that the external provider delivered

Determining Which External Providers of Processes, Products, and Services Will Be Controlled

When the organization purchases, uses services, or integrates in its product goods of external providers that affect the product, a set of controls will be determined and established to ensure that externally provided processes, products, and services do not adversely affect the organization's ability to deliver conforming products and services to its customers. The type and extent of those controls will be discussed in chapter 8.4.2. But in clause 8.4.1, the standard defines three distinctive business cases in which products or services of external provision shall be integrated into the realization processes:

1. The purchased products are incorporated into the product that the organization is manufacturing (final or semiproduct). For example, in the industry of machine engineering, most parts of a machine are purchased.
2. Goods or services that the organization sells to end customers are provided by a third party—the external provider. This is a common feature among handlers or trading companies; customers purchase the goods from the company, but the delivery is effected by a third party.

3. Outsourcing of processes that are part of the realization of a product—in the electro industry, the organization may decide that it does not have enough knowledge and resources to perform the coating of parts and send those parts to a third party—a supplier of service that executes the coating and returns the parts to the organization. Also there are subcases here:

    a. The organization delivers two parts, and the supplier assembles them—the purchase of a service.

    b. The organization delivers one product, and the suppliers add another part to the product—the purchase here includes a service and goods.

        The location where these processes or activities take place does not affect the definition; the supplier may perform the work on its own premises or in the organization's premises.

Evaluation of External Providers

External providers shall be approved on the basis of their abilities to deliver products or services in accordance with predefined requirements. The approval will be given upon an evaluation. An effective and objective evaluation considers the significant parameters regarding the purchased product.

The evaluation is to be conducted on a periodic basis, and the frequency of the evaluation must be determined. The objective is to establish an ongoing control process over the external provider in order to foresee events that might become nonconformities or quality problems. This is done normally once a year, but there are cases of purchased products that require a more frequent control, depending on the risk and effect that the purchased product has on the product.

The parameters for the evaluation should represent the ability of the supplier to deliver products or services that meet the requirements:

- Quality of goods or services
- Credibility and reliability of delivery
- Quantities
- Commitment to prices
- Maintaining conditions such as QMS (ISO 9001, ISO 14001)
- Parameters like willingness to solve quality problems or the ability to train personnel

These parameters will constitute the criteria for the evaluation. The criteria will situate the supplier or subcontractor with regard to their performance.

The controls that are applied to the external providers must deliver results and indicate whether a supplier or a subcontractor met its requirements while participating in the realization of the product.

The evaluation shall cover the following parameters related to the relationship with the external provider:

- Evaluation of realization processes
- Evaluation of quality processes
- Evaluation of delivered products or services
- Review of risks related to the supplier or subcontractor

In order to determine which parameters the criteria shall cover for an external provider, I suggest the next approach—ask relevant questions regarding the conduct of the external provider:

- Does the supplier maintain stable realization processes?
- Does the supplier comply with regulatory requirements?
- Does the supplier maintain supporting documentation?
- Does the supplier maintain appropriate controls over its processes?
- How does the supplier handle nonconformities?

Answering these types of questions will create a status report regarding the supplier and its performance. I recommend evaluating each supplier on its merits; each case or supplied product or services has its significant characteristics and therefore may require different parameters for evaluation, although it is very demanding.

The controls applied on external providers like delivery records, quality control reports, agreements, complaints, and nonconformities will provide objective evidences, information, and data regarding their performances. These results will be reviewed and compared to the criteria, and conclusions will be drawn.

Setting the Criteria for the Evaluation of Suppliers

Effective evaluation of external providers must use criteria. Based on the criteria the organization can assess the performance of a supplier or a subcontractor and a decision regarding the approval of an external provider could be made. The objectives of the criteria are to rank the supplier in a quantitative or qualitative manner and to assist with deciding whether the supplier can deliver under controlled conditions and meet requirements. The criteria should allow you the assessment of the supplier's performance in a defined range, period, or area. For example, over a period of one year, the reliability of delivery on schedules is assessed. And so the evaluation according to the criteria allows you also to identify trends or changes in the performance of the supplier. The criteria shall relate to the products, processes, or services delivered by the supplier and will be based on three main issues:

- The quality and intended use of the purchased goods or services
- The conditions in which the product was delivered or the service was provided
- The effect that the purchased goods or services will have on the subsequent product realization or the final product

Those issues shall determine which parameters must be controlled. The derived parameters will be used to set the criteria. For example, a supplier delivers you packages that protect the product. The completeness and integrity of the package is critical for the progress of the packaging processes and the intended use of the product. When packages are not delivered as required, there is a threat that realization processes will be disturbed and that the product will fail. Let us review the three issues again:

- The quality and intended use of purchased product of the packages—integrity and completeness requirement.
- The conditions of delivery—delivery on time, the provision of the agreed amount, and protection of the goods.
- The effect—the intended use might fail if packages are not delivered according to the requirement.

According to the determination of the criteria, the tools and techniques shall be applied enabling the relevant parties to obtain the relevant data and information relevant to the performance of the suppliers. And these data and information shall be compared with the criteria. Principles for setting or establishing criteria include the following:

- The availability of the required or expected data must be checked.
- The frequency of the measurement shall be defined and will be appropriate to the type of process or product it is measuring.
- Methods for analyzing and publishing the data and reports of the evaluation shall be defined.
- The responsibility for conducting the measurement shall be defined. Different responsibilities and roles from different areas in the organization should be involved in the setting of the criteria in order to provide their inputs.
- The distribution of the data shall be defined.

I have given here some example criteria for evaluating the performance of external providers through common parameters.

Reliability of delivery—data will be extracted from the ERP system and purchase order module (requested date against date of receipt).

Criteria	Note
No delay	10
1–2 days' delay	7
More than 2 days	5

Reliability of delivered quantities—data will be extracted from the ERP system and purchase order module (requested quantity against received quantity).

Criteria	Note
No deviation from the ordered quantity	10
Until (−5)% from the ordered quantity	7
More than (−5)% from the ordered quantity	5

Quality of products—data will be obtained from quality assurance (test protocols or control reception notes).

Criteria	Note
Quality A	10
Quality B	7
Quality C	5

Quality of delivery—data will be obtained from the warehouse (delivery protocol).

Criteria	Note
Quality A	10
Quality B	7
Quality C	5

Quality of packaging—data will be obtained from the quality assurance (test protocols or control reception notes).

Criteria	Note
Quality A	10
Quality B	7
Quality C	5

Quality of services—data will be obtained from the maintenance team (maintenance protocol).

Criteria	Note
Quality A	10
Quality B	7
Quality C	5

Certification to ISO 9001 Standard—data will be obtained from the purchase department.

Criteria	Note
Acquire Certification to ISO 9001 Standard	10
Does not acquire Certification to ISO 9001 Standard	5

Ranking External Providers and Setting the List for Approved Suppliers

The objectives of ranking or classifying external providers are

- To approve or disapprove suppliers and subcontractors from participating in the realization processes
- To create list of approved suppliers
- To drive suppliers to improvement

This ranking represents the supplier's performance during the realization processes and shall allow you to decide whether they are suitable to provide the organization with conformed products or services. There are many ways to rank external providers. I will focus on a simple method where the output of the evaluation is a note or a grade—a relative position of value on a scale of quality for each supplier or subcontractor. The next step will be to classify these data on the predefined scale. For example, Classification of Suppliers:

Final Grade (Points)	Classification	
81–100	A	Strategic
61–80	B	Preferable
51–60	C	Transactional
1–50	D	Declined

One may determine that one's organization is willing to incorporate in the realization of the product only suppliers with B classification and above. External providers that were ranked C will need to prove improvement measures and external suppliers with ranked D are to be discouraged from participating in the realization processes. When the evaluation failed and the supplier was found to be inadequate, one must define the necessary actions to be taken:

- Update the status of suppliers in the ERP system according to their rank: inactive for low ranked suppliers that no orders could be created.
- For open orders from low ranked suppliers, tighten the controls of incoming goods or provision of services.
- Notify all relevant parties in the organization of the rank of suppliers.
- Inform the suppliers of their rank:
  - Initiate a discussion with the suppliers about their performances.
  - Demand corrective action or an improvement plan from the supplier.
- Select another supplier that has achieved an acceptable rank.

However, the emphasis shall be on establishing an effective action, and it is required that one assesses its effectiveness over a defined period of time. If one has initiated a corrective action or an improvement plan from the supplier, it is required that one assesses its effectiveness over a defined period of time. In case the organization and the external provider decided to initiate a plan for improvement of its performance, the organization may submit it to the requirements of clause 10—Improvement.

The outputs of the evaluation process are to be recorded and maintained as documented information. During later stages, one will have to retain this information and compare results of old evaluations with reevaluations. I suggest planning a simple format (digital or analog), and bear in mind that this is not a one-off action, but an ongoing one. This record will need to be updated. The standard does not demand a documented process specifically for the evaluation process but only a defined method to ensure that purchased products conform to specification. I believe that such documentation may assist one; however, when documenting the method on a procedure, there are some issues that must be considered:

- The type of suppliers that are included under the evaluation
- The parties responsible for conducting the evaluation
- Reference to the criteria
- The frequency of the evaluations
- Inputs required for the evaluation
- The type of evaluation or a description of the evaluation method—performances compared to the criteria
- Outputs of the evaluation

Reevaluation of External Providers

After one has evaluated the supplier and has provided them with one's feedback, one should, in a defined period of time, reevaluate their performances again. This is an ISO 9001 Standard requirement. The frequency of reevaluation will be determined in accordance to the effect that supplied goods or services have on the product and in accordance to the effect wtih that of the associated risks.

The reevaluation shall refer to the last evaluation, and the results of the two should be compared. In cases where measures or actions were applied to the supplier due to the last evaluation, the reevaluation should indicate their effectiveness.

List of Approved Suppliers

The output of the evaluation and reevaluation process shall provide one with a documented list of approved suppliers. This list indicates which external provider is approved to participate in the realization of the processes, which external provider must improve, and so on. The requirement is to provide the results of the evaluation, and the most effective and logical way is a controlled list. One will probably be asked to present such a list during an audit.

If one maintains an ERP system or any other systems that manages one's purchasing processes, (naturally) one's suppliers will be entered in this system. One need only define that approved suppliers are documented on your ERP system. Using the status of one's supplier on that system (e.g., whether they are active, inactive, suspended, etc.), one may declare and document the approval. One may also produce a report from the system: a list of approved suppliers and the date of the approval. This option may allow one to keep the list up to date. I propose here a format for the list:

- Supplier's name
- Supplier's details
- Status (approved or not approved)
- Rank or classification
- Date of approval
- Last rank or classification (from the previous evaluation)
- Remarks or reference to relevant documentation

Maintaining Documented Information

The next type of documented information is expected in the context of evaluation, selection, monitoring of performance, and reevaluation of external providers:

- Criteria for evaluating the external providers
- Records of evaluation and reevaluations
- Rank of the external providers
- A list of approved external providers

8.4.2 Type and Extent of Control

The main challenge with external providers is the ability to achieve control over actions that are not directly under the organization's supervision. This will be achieved by developing the appropriate controls. A consequent control of delivered products or services shall reduce nonconformities and complaints from the end customer and will enhance the quality of the end product. The ISO 9001 Standard requirements include the following:

- The organization shall ensure that externally provided processes, products, and services do not adversely affect the organization's ability to consistently deliver conforming products and services to its customers.
- Outsourced processes or realization processes that are being performed by external providers shall be included within the control of the QMS.
- The organization shall define the controls it intends to apply to the external provider.
- The organization shall define the controls it intends to apply to the resulting outputs.
- While defining the controls, the organization shall consider
  - The potential impacts of the provided products or services on the organization's ability to consistently meet customer and applicable statutory and regulatory requirements
  - The effectiveness of the controls applied by the external provider
- The organization shall define controls, verifications, validation, and other activities necessary to ensure that the externally provided processes, products, and services meet requirements.

Ensuring the Ability of the Organization to Consistently
Provide Conforming Products and Services

Before the ISO 9001 Standard introduces us to the practical requirements for controlling external providers, it presents us with the main objective of those controls: ensuring the ability of the organization to consistently provide conforming products and services. In other words through applying appropriate measures, the organization shall ensure that deliveries of processes, products, or services are well controlled. In order to reach this objective, the organization must have clarity on the following main issues:

- It must be clear which processes, services, or products are externally provided. Inputs to this issue may derive from the following quality elements:
  - Context of the organization—where generally external issues of the organization are determined.

- Process analysis—the analysis of processes indicates which processes are provided externally.
- Requirements for resources—where needs for services from external providers are defined.
- Requirements for knowledge—where needs for external knowledge are defined.
- Product specifications—in the specifications it is defined which products or parts of the product will be delivered by external provider.
- Planning the controls for external providers refers to the following aspects:
  - Outputs of the actions to address risks and opportunities—where it is identified where and how the organization may fail to provide conformed products or services and which actions must be initiated in order to eliminate such situations
  - Relevant quality objectives—where it is defined and what is expected of the organization and its processes
  - Controlling changes—where management of changes may refer to changes of externally provided processes, products, or services

After understanding the issues, one may advance to developing the controls. For those required controls, the standard introduces clear requirements.

Including the Processes of External Origin in the QMS

Processes of external origin are to be included in the QMS. Inclusion of those processes in the QMS means that the organization must plan, determine, implement, control, measure the effectiveness, and improve those processes. The reference applies the following types of activities:

- QMS activities, for example, system audits
- Provision of goods or resources, for example, provision of components or outsourced personnel
- Performance of processes that are part of the realization of the product including manufacturing activities, handling, activities of measurement, and analysis
- Performance of services, for example, transport services or consulting services

These processes or business activities provided by an external provider are supposed to be identified as part of the determination of the QMS, its processes, and interactions (see clause 4.4). In clause 4.4 it is required to maintain documented information concerning the processes—that means the list of the QMS processes. Thus, list of the externally provided processes should be available. This determination should indicate to one exactly which processes are to be included under the controls suggested in clause 8.4. Which aspects of the process approach are to be considered?

- The following are responsible parties and authorities in the organization that participate in the processes of purchasing and may affect the quality of the purchased goods or services:
  - Purchase managers
  - Carrier
  - Warehouse workers
  - Production managers

The definition shall relate to the interrelations between the participants—what the role of each participant in the process is and how they interact (how do they exchange information regarding purchased goods or services).

- Necessary inputs required for the process are as follows:
  - What are the inputs
  - Where the inputs are located (data or materials)
  - Who is responsible for their delivery
  - Reliability and integrity of inputs (e.g., the quality and correctness of suppliers' master data)
  - What is the expected format for types of information like addresses of suppliers or conversion units of products
- Interactions with other processes of the QMS
- Expected records (outputs) to support the process:
  - Information regarding the processes or products like purchase orders or acknowledgment of acceptance
  - Outputs of purchase activities, for example, reports from the management system
- Reference to other procedures, such as acceptance procedures and formats in the process
- Addressing potential impacts related to purchased products—identifying the risks and applying controls to manage them:
  - Risks related to the ordered processes, products, or services
  - Risks related to their delivery or provision
- A description of needed controls over the process—where and how will the purchased products or services be controlled:
  - Incoming control over quantities and prices
  - Control over quality
  - Reviewing performed services

The acceptance processes and activities shall be planned in order to support those defined requirements for approval.

- Required outputs—documented information that the supplier must deliver:
  - Delivery note
  - Certification of quality
  - Results of quality tests

In clause 8.4.2, there are no requirements to document the matter. However, clause 4.4.2 demands the maintenance of documented information to support the operation of processes. In that case, you will be expected to present some kind of documented definition. The quality plan may be a good place to document this definition.

Eliminating Potential Impact Related to Externally Provided Products or Services

When purchased processes, products, or services are introduced into the realization of the product, there are risks that these might be nonconformed. Introducing nonconformed goods or services is considered as the potential impact of the product, may cause critical consequences, may have a strong effect on processes or the quality of the product (final or semi), and thus may adversely affect the ability of the organization to consistently deliver conforming products and services to its customers. Such issues, situations, or risks that may bring about those potential impacts must be identified and managed. Here are some of the issues to be reviewed when identifying risks:

- Complexity of the provided product or service
- Ability of the supplier to continually supply
- Ability of the supplier to manage and control complex production processes
- Liability and stability of processes
- Reaction to failures and nonconformities
- Liability of products
- Qualification of personnel
- Awareness of quality
- Availability and cooperation of the supplier or subcontractor
- Legal status
- Capital investment and financial viability

Reviewing these issues (or risks) as mentioned earlier will assist in developing and applying the appropriate methods or controls over the suppliers: when, where, and how one must invest one's resources when controlling the supplier. After identifying those issues, situations, or risks, one is ready to apply the controls at suitable stages of the purchase process necessary for addressing them:

- Selecting the supplier
- Ordering
- Accepting goods or services
- Incorporating goods or services in the product (semi or final)

In order to manage those issues effectively, I would

1. Include this review of risks and their impact already in the risk-based thinking phases of the planning of the QMS (where you are required to address risks).
2. Evaluate the extent and degree to which the control of an externally provided process is shared between the organization and the provider.
3. Persuade the external providers to initiate actions for addressing the output of the risk-based thinking throughout their relevant realization processes. In this way one promotes the control one step forward and apply controls already during the realization of the externally provided goods or services.
4. In case it is required to submit for realization unreleased delivered goods or services, a special release must be planned.

The training of personnel of the supplier or subcontractor is another issue that may directly affect the goods or the services. You are required to ensure that the personnel realizing the products are qualified and have the appropriate skills and knowledge before they take part in the realization of the product.

Identification and labeling of the delivered goods or services is needed in order to allow a distinction among approved products, processed products, and services from those unapproved. This distinction will prevent the use of nonconformed products or assist in case nonconformed purchased products were submitted to the realization processes. According to the outputs of risk-based thinking, the appropriate means for identification and traceability will be defined and communicated to the supplier: batch number, lot number, or number of production order.

Controlling Changes of Externally Provided Products or Services

When one's external provider changes the product or the performance of a service, it must notify in advance and receive one's approval when the change may have an impact on the product. The requirements for the communication will be discussed in chapter 8.4.3.—Information for external providers.

Before accepting the change, the organization needs to evaluate the requested change and its implication on the final product and its quality. In case the changes do affect the product, it must be ensured that the intended use of the product will not be affected and conformity to the requirements is guaranteed. When required the changes will be reviewed by the appropriate teams where each team evaluates the matter from its aspect:

- A review of the development team
- A review of the production team
- A review of the quality team

In order to implement an effective change control, I propose here the next sequence of activities that will enable you to control changes:

1. Initiating a change, documenting the need for a change, and describing the change
2. Communicating and notifying the organization about the requested change and the reason for it
3. Delivering data to demonstrate that changes do not cause any modifications to the product
4. When the change does cause a modification of the product, information and details shall be delivered to the organization
5. Defining a responsible party or parties from the organization to review the request
6. Establishing a review committee consisting of delegations from the supplier and the organization
7. Communicating with the supplier and announcing that the request has been reviewed and accepted or rejected
8. Documenting outputs and decisions of the committee
9. Conducting validation of the change
10. Issuing a documented approval for the change
11. Updating the relevant documentation
12. Implementing

Planning and Applying Controls to External Providers

The controls applied to external providers shall ensure the capability of suppliers to deliver a product that meets the requirements. The objective of the controls is to eliminate any potential failures where the organization does not have a direct access to processes or their outputs. The type and extent of the controls should be relevant to the activity of the supplier. The following issues must be reviewed:

- Preliminary or introductory conditions for collaborating with the supplier (if required):
  - Maintaining quality system requirements
  - Maintaining sufficient skills and competence of personnel
  - Maintaining special work environment conditions
  - Maintaining minimum inventory levels
  - Revealing financial strength
- Preliminary or introductory requirements like regulatory requirements (if applicable):
  - Maintaining social conditions for employees
  - Controlling the use of restricted or prohibited materials
  - Maintaining environmental requirements
- Implementing controls related to processes:
  - Audit on the supplier's premises
  - Process audit by the supplier
  - Control of further tier suppliers
  - Controls over equipment and machinery
  - Control over monitoring and measurement tools used by the external provider
  - Verification of competence of personnel of external providers through control of qualification and training

This topic will be discussed during the contract stage in order to ensure the supplier's engagement with the matter. Which quality instruments may assist you?

- Results of supplier's evaluation
- Supplier's performance history
- Supplier's questionnaire (delivers an overview concerning the requirements or precondition of suppliers, delivered products, or services)
- Maintenance of management systems—quality, environment, and safety
- Accreditation of processes
- Training of personnel
- Protection of data
- Management of documents
- Protection of customer's property

Planning and Applying Controls to the Resulting Output

The controls that will be applied to the resulting outputs shall be proportionate to the effect that the purchased product or service has on the finished product or final service. The controls must be performed according to a defined plan. The goal is to prove the functional and performance requirements of the purchased products before submission to further processes in order to ensure the final intended use of the product. Products or services may also need to be validated against the acceptance criteria.

Applying controls of products will span from the moment materials begin arriving from the supplier until it is submitted to the supply chain in the organization or from the moment a service is provided until it is approved by the organization. The definition shall refer to properties of the product that must be controlled. The inspections, controls, tests, and expected results that the external supplier must perform and deliver shall be defined and clear. The definition may include the following:

- The type and scope of the tests
- In which point of time or process must the test be performed
- What is the amount of tested products
- Who is responsible to perform the tests
- What is to be performed in the tests
- What are the criteria

Which types of controls may we encounter?

- Realization requirements—how to ensure that product or service will be realized as required and which verification or validation activities are needed:
  - The organization shall describe which validations are required when products or services are not able to be verified (e.g., products where the resulting output cannot be verified by monitoring or measurement activities).
  - Material handling requirements—which evidences are expected to prove that materials, parts, or components were handled as required. When needed, the definition shall refer to the identifications of the delivered products or services.
  - Handling of nonconformed products or services.
- Product requirements—how to ensure that the product is provided as expected:
  - Configuration of the product
  - Functionality tests of the products or parts
  - Quality specifications and requirements—what quality or acceptance tests are necessary
  - Traceability
  - Measurement and monitoring of parts or components
  - Control of stored goods
- Process audit for conditions during realization—which evidences are required to prove that necessary process conditions were achieved during the realization processes.
- Requirements for delivery—when purchasing a product or service, a consideration must be given to how these will be packed and transported and under what conditions.
- Handling property, data, or information of customers—which evidences are expected to prove that property, data, or information of customers was handled as required.
- Measurement requirements—which measurements and analysis activities are needed to be performed during the realization processes or at the process outputs and the results expected.
- Monitoring and measuring devices—which tools and instruments must be used.

The earlier mentioned may be included in procedures, work, or test instructions that the supplier must follow and implement when realizing the product or the service. Defining these arrangements will give validity to test results and will be considered as a precondition to the acceptance of the products or services. How and when this information will be communicated to the external provider will be discussed in chapter 8.4.3—Information for external providers.

According to the risks related to the product, the organization shall manage traceability of purchased products or services. The objective is to enable the organization to identify end products whose quality was adverted by supplied products or services which were nonconforming. A clear identification of goods or services must be practiced in order to separate conforming from nonconforming delivered product or services. It must be clear to the users of the QMS what the status of the delivered products or services is. This will be practiced in the premises of both the suppliers and the organization. It can be achieved with managing batch numbers or order numbers. Pay attention that such requirement of one's product will require the inclusion of related records in the control of documented information as required in clause 7.5—Documented information.

Delivery of evidences of tests and quality certifications will indicate which tests were conducted and what are the results. Definition of the relevant documents and records delivered from the supplier that are defined include

- The type of documented information
- Their content

And the organization must ensure their availability on reception of goods or services from the external provider. Delivered evidences by the external provider must relate to a specific delivery. When it is required that only a certain supplier must conduct the tests, the supplier must deliver evidences that it is the case.

Evaluating the Effectiveness of Controls Applied by the External Provider

The effectiveness of the controls applied to the external providers will be expressed by the ability of those controls to deliver products or perform services or processes according to their specification. The organization shall assess the performance of its external providers:

- Examination of the conformity of products or process outputs
- Handling nonconformities
- Proving improvement of processes

For example, if you are noticing that there are many nonconformities concerning a supplier, you might need to examine its controls. But it does not end by the nonconformities. You must assess how the supplier reacts to the nonconformities and which actions it initiates in order to reveal the root cause and eliminate their recurrence. A system audit is a good way to investigate the processes of your suppliers and its relevant controls.

Activities Necessary to Ensure Requirements

One main objective of clause 8.4.2 is to ensure that only released purchased products will be used in the realization. Although purchased products or services are delivered from a third party, the responsibility to ensure appropriate provision of purchased products or services lies with the organization.

Information and data regarding the performance of the external provider shall be gathered and monitored over time. This requirement relates directly to the requirements of clause 9.1.3—Analysis and evaluation—where it is necessary to analyze and evaluate the performance of the external providers. The objective is to maintain a continual control over the performance of the external provider in terms of product realization through activities of monitoring, measurement, analysis, and evaluation. Which parameters may be evaluated? The parameters that influence the quality of the delivered product (much of those who will be used for the evaluation of the supplier as required in clause 8.4.1—General), include the following:

- Credibility and reliability
- Quality of delivered products or services
- Adherence to specifications
- Quality of goods or services
- Quantities

After deciding which parameters will reflect the performance of the external provider, one will need to include these measurements of performance when planning activities for monitoring, measurement, analysis, and evaluation:

- Decide which methods for monitoring, measurement, analysis, and evaluation will be used.
- When will the monitoring and measuring be performed.
- How and when will the results be analyzed and evaluated.
- Retain the appropriate documented information as evidence of the results.

8.4.3 Information for External Providers

The organization is required to ensure with a method that all the necessary information regarding the purchased product is identified prior to their communication with the external provider. The goals are to

- Ensure that all the requirements regarding purchase are identified, including approval of the requirements and definition of controls
- Develop the ability to transfer to one's supplier clear specifications regarding the product
- Ensure that the supplier receives all the information it needs in order to verify its ability to deliver the products or services according to the requirements
- Ensure that all the required information is received from the supplier

ISO 9001 Standard requirements include the following:

- The organization shall ensure the adequacy, quality, and clarity of specified requirements of purchased products or services prior to their communication to the external provider.
- The organization shall communicate to the external provider the requirements of the services or the products including important information. The information shall include the description of the products and services to be provided or the processes or activities to be performed.
- The information shall include requirements for approval and release activities necessary to ensure that externally provided processes, products, and services will be delivered as expected.
- The information shall refer to methods, processes, procedures, and the use of tools and equipment needed for the realization of purchased products or services.
- The information shall include necessary release activities—activities for the verification that all requirements were met.
- The information shall include competence, training, and qualification requirements relevant to the realization of the purchased products and services.
- The information shall describe the methods and content of the interactions between the organization and the external provider.
- The organization shall determine and implement activities and controls and monitor the performance of the external providers.
- The organization shall determine the verification or validation activities that the organization or its customer intends to perform at the external providers' premises.

Method for Ensuring Adequacy, Quality, and Clarity of Specified Requirements of Purchased Products

Information regarding the purchased product or services must be reviewed and verified before submission to the external provider. Thus, it is required to determine a method for reviewing the purchasing information before it is submitted for realization. The objective of the method is to ensure that the content of the information and the means with which it was communicated are used as planned. The information submitted to the external provider must be

- Correctly communicated—the organization shall use the appropriate communication channels to convey the information. Using the appropriate communication channel shall ensure that the data reaches its destination.
- Clear—all the required data are understood by the supplier.
- Sufficient—the information is sufficient to accomplish the task and the external provider has all the details that it needs.
- Approved—a responsible party (from the organization) has reviewed and approved the content.

Such method for ensuring adequacy and clarity depends much on the type and nature of the purchased product. Authorization of persons who approve the transmission of the information to the external provider can be defined on a job description. There is no requirement to document the method of transmitting the information to the external provider, but you will be required to produce evidence that such a method exists.

In simple cases where the information is easy to transmit, a format for the purchase order may serve here as a control tool—filling all the fields means that all the requirements are there. It becomes complex when the products or the required services are complex and influence the amount and extent of the necessary details, for example, in a project where many aspects must be considered:

- Technical issues
- Customer specifications
- Project with long-term assignment
- Special, not ordinary, or out of the standard requirements

Communication and Interaction with the External Provider

One issue this method will cover is the definition of the interaction with the external provider and the relevant communication channels for that purpose. The communication with the external provider is a critical process that ensures that data and information or goods (components or materials) are properly transferred from customer (the organization) to supplier and from supplier to customer. This is why a method for communicating with the suppliers is essential: what, how, by whom, and when will the information be submitted to the supplier.

For each stage of the purchase process, the designated communication channel will be defined. It is critical for both supplier and customer to know how they interact with each other along the supply chain:

- How orders are being transmitted
- How orders are being conformed to
- How changes will be communicated
- How goods or services will be delivered
- How nonconformities will be managed
- How financial transactions will be preceded

Arrangements for communication with external suppliers may be managed through personnel or through systems. Communications and interactions can occur such as sending printed orders and e-mails, installation of supplier's terminal at the organization's premises, or maintaining EDI between the organization and the supplier.

Exchange of Documented Information with the External Provider

The next aspect of interaction with external providers is the provision of the relevant inputs to each stage in the process of purchasing and the acceptance of the expected outputs in terms of documents or records: information or data that are required to operate the purchase activities and that the supplier or the employee of the organization must be aware of. There is no standard requirement for maintaining these documents or records as documented information. The types of documents or records related to the purchase processes and supports of these processes must be defined. The format and extent of the documents or records is influenced by the information that is needed to be conveyed to the supplier as well as the information that must be received from the supplier. In order to define clearly these different types of documents or records, it must be clear which information as inputs or outputs is expected for each type of interaction with the organization.

I suggest analyzing the purchase process in order to effectively identify all the required documents or records. Let us review the (traditional) purchase process:

Stage	Type of Documented Information
The organization sends to one or several suppliers a request for an offer.	Request for an offer with description of the purchased products or services including quantities and schedules
The organization orders goods or services from the supplier.	Purchase order with quantities, schedules, purchase declarations, instructions regarding delivery or transportation, required tests and inspections and payment conditions, discounts, or surcharges
The supplier approves the acceptance of the purchase order and its conditions.	Order confirmation from the supplier
Change of purchase order—the supplier or the organization initiates a change in the details of the purchase order.	Request for a change and the approval of the organization that the change has been accepted
Receiving goods or services—the organization receives goods or services from the supplier.	Receiving note—which identifications or labeling of the product are required
Control of goods or services—the organization performs internal controls of received goods or services.	Test protocol—documented information as evidences of quality control
Approval of goods or services.	Which documented information is required as approval of release of the product and of inspections or tests that are conducted (signature or a stamp with details of the approval)
Return of goods—the organization sends products back or rejects a service after control.	Delivery slip—which documented information must be attached to the returned goods
The organization delivers money in return for the goods or services that it received from the supplier.	Supply invoice

Each stage of the purchasing process may include or demand other types of information from other sources in the organization. For example, the content on the purchase order (which is a document that is being submitted outside the organization) will differ from the content on the receiving slip, which is normally an internal document.

Description of the Products and Services

A detailed description of the purchased product or service will be transmitted to the external provider in order to describe the expectations of the organization. Which information is expected to be detailed?

- Identification of the process, product, or service—all information that identifies the required product for purchase: name or description of the product or service, catalog number (internal or external), edition, version or revision of diagrams, configuration, and model. When needed, a reference to supplier's internal description of products or service (supplier product number) shall be mentioned.

- Information regarding the purchase process—required quantities, delivery dates, prices, discounts, and addresses for delivery and billing. This information shall relate to the internal processes or arrangements of the supplier that will ensure the correct delivery—for example, the conversion of unit for measure (when the organization manages the stock in units but the supplier delivers it in kgs).
- Information regarding operational requirements, technical specifications, instructions, and guidance that the supplier must follow—packaging requirements, delivery requirements, transportation conditions, development environment, operational environment, work instructions, procedures, and diagrams.
- Information regarding quality requirements—test instructions, inspection instructions, quality records, quality protocols, tools and equipment for control, quality certifications, and statistical data.
- Information regarding any supplementary requirements—regulatory requirements, safety documentation, and so on.

Requirements for the Approval of Purchased Products and Services

The organization must plan, define, and communicate to the external provider which activities for approval or release of products or services must be conducted by the external provider prior to delivery to the organization. Those requirements for approval and release must be defined as part of the purchasing requirements. Such activities may be performed by the supplier or in some cases by the organization upon receiving the goods or approving the services. The information delivered to the supplier shall cover the following issues:

- Processes for the approval of products or services
- Procedures specifying the controls that have to be applied during the realization of goods or services
- Monitoring and measurement activities that will indicate how processes or their outputs must be controlled and will ensure that processes deliver intended results
- The use of tools and monitoring and measuring devices that is needed to approve and release process outputs
- Quality control tests including final tests that must be performed on outputs of processes necessary to ensure that the finished product complies with specifications

Communicating effectively the requirements for approval and release of products or services and ensuring their implementation during the realization stages to the external provider will reduce the need for quality control of incoming products or services. Outputs of risk-based thinking and design and development verifications and validations of the product may serve as inputs for planning the activities for approval and release of purchased products. This information will be used by the external provider to

- Plan its realization methods, processes, and activities
- Plan which equipment must he or she use during the realization of the product
- Plan its controls over the realization processes
- Deliver to the organization the necessary evidences that delivered products or services meet the specifications

It is important to convey to the external provider the relevant instructions that will define expected activities, the extent of the tests, and criteria that will allow him or her to perform those activities and will enable a comparison to expected process results:

- Work instructions
- Test instructions
- Inspection procedures
- List of monitoring and measuring devices

If it is a manufactured product, ensure that the conditions for allowable tolerances are defined and known to the supplier. In case of service, the specifications of acceptance must be defined and clear.

Controlling Changes of Externally Provided Products or Services

When your external provider changes the product or the performance of a service, it must notify you in advance and receive your approval when the change is significant and may modify the product. The process of communicating and approving changes shall be defined and agreed upon on an agreement or contract level. Changes may occur to each parameter of the realization process:

- Material
- Production processes
- Assembly processes
- Human resources and personnel
- Equipment or machinery
- Test equipment and tools
- Documentation
- Packaging
- Storage conditions
- Transport conditions

Competence of External Personnel

When certain competence of personnel and training regarding the realization of the purchased product are needed, they must be identified by the organization. When such requirements are applicable, they must be communicated to the external provider. And it is the responsibility of the organization to prove that these personnel were trained according to the requirements. One may train the personnel oneself or demand that the supplier do so. Training of external personnel may be applicable in the following cases:

- Special processes
- Regulatory requirements
- Changes in production processes
- New contracts
- New parts
- New suppliers
- Nonconformities
- Rework

Control and Monitoring of the External Provider's Performance

Information and data regarding the performance of the external provider shall be gathered and monitored over time, and this information shall be communicated to the external provider. This requirement relates directly to the requirements of clause 9.1.3—Analysis and evaluation—where it is necessary to analyze and evaluate the performance of the external providers. The objective is to inform the external provider which aspects or issues are important and may affect the quality of the purchased processes, products, or services.

Verification or Validation Activities at the External Provider's Premises

A conducted inspection by a representative of the organization or its customer may be needed in order to verify that certain requirements are in place and are being implemented and that products or services being realized are consistent with the specifications submitted before. The need for inspecting processes or products at the premises of the supplier is determined from the required verifications or validations of the externally provided product. This inspection may be independent, periodical, or in the nature of follow-ups (the outputs of one inspection serves as inputs to the next). The nature and frequency of the visits depend on the quality capabilities of the supplier or the complexity of the product. In which case a review may be required

- In cases where processes are outsourced and the organization would like to audit them:
  - Inspect the quality assurance
  - Inspect process conditions
  - Inspect personnel
- In cases where the supplier uses a third party to realize the products or services that it delivers to the organization; the organization may want to review this third party.

When visits to the premises of the external supplier are needed, it is necessary to define and communicate

- The purpose of the visits—what product or processes will be evaluated
- The interval of the visits—how many visits per period
- The participants of the visits—from the side of the organization, the supplier, and when applicable the end customer (of the organization)
- The activities that will take place during the visit—the verification or validation activities that the organization will perform during the visit
- The required inputs to the visit—which inputs shall each party prepare in advance
- The outputs of the visit—what will be documented and in which format

An effective way to document these visits is by planning and introducing some kind of visit report to the QMS, but there is no requirement for documented information.

8.5 Production and Service Provision

8.5.1 Control of Production and Service Provision

Implementing production and service provision under controlled conditions refers to the assurance of variables that affect the performance of an element used in the realization of the product or service. The control is achieved by applying and enforcing a set of principles and conditions that will guide and accompany the realization processes. The control allows the organization to monitor the different variables during the realization that may affect the performance of the QMS and the quality of the product and to extract reliable data and information regarding the realization. These data will be used to verify the results. Let us review the ISO 9001 Standard requirements:

- The organization shall implement production and service provision under controlled conditions.
- The organization shall ensure the availability of documented information that defines
  - The characteristics of the products to be produced or the services to be provided
  - The activities to be performed
  - The expected results of those activities, products, or services
- The organization shall ensure availability and use of suitable monitoring and measuring devices.
- The organization shall implement monitoring, measurement, analysis, and evaluation operations and activities at the appropriate process stages. These operations and activities shall ensure that criteria for control of processes or outputs and acceptance criteria for products and services have been met.
- The organization shall ensure the use of suitable infrastructure and process environment for processes and operations.
- The organization shall ensure the availability of competent personnel and necessary qualifications for the operations and activities.
- The organization shall ensure that activities for validation, and periodic revalidation, of the ability to achieve planned results of the processes for production and service provision, are performed, where the resulting output cannot be verified by subsequent monitoring or measurement.
- The organization shall initiate activities for preventing human errors.
- The organization shall implement activities for release, delivery of product, and services as well as post-delivery activities.

Production and Service Provision under Controlled Conditions

Before beginning with the production of the product or providing the service, the organization shall define and determine the conditions necessary for the realization of the product or the service. It will be done by organizing and ensuring a set of conditions, processes, and activities that will transform requirements into a product. The implementation of realization processes includes specification of process inputs, parameters, and conditions that will ensure conformed process outputs, for example, under which conditions may the intended outputs be achieved or the requirements of the process and the operating environment that are needed for achieving the intended outputs.

The following principles are clearly defined:

- The organization shall define the condition necessary for the realization.
- The organization shall define the controls over these conditions.
- The controls shall ensure and provide evidences that the conditions are maintained during the realization processes.

Chapter 8.5.1 describes the different requirements and suggests practical ways to implement them in the QMS.

Availability of Documented Information Describing the Product Characteristics

The standard requires that the organization shall make available all product characteristics with documented information. The objective of this documentation is to provide the employee, supplier, subcontractor, or any other party that participates in the realization processes full access to the required product characteristics. With this documented information, they can assess and evaluate the conformity of process outputs with the product specifications.

An effective documentation will be used for the verification and validation of the product. This can be used twice: once during the process in order to evaluate the performance of the process and once when the process ends to verify and validate the process outputs.

For example, the assembly of products, when a product is assembled, specific assembling instructions, data forms, and inspection instructions—including all the required documentation, such as diagrams and criteria—will be available at the workstation. The employee has all the information that it requires to perform the assembly. Another example refers to the quality tester that while using documented information knows which process outputs he or she has to validate when testing a product; the quality tester has a test instruction that identifies the part to be tested, the machine that realizes it, the interval of the test, the sequence of the validation (the test), and the criteria for acceptance. All the required data and information are defined, documented, and made available at the appropriate process stage and locations.

What are the objectives of the documented information?

- Defining the different characteristics of the product/service
- Defining its intended use
- Defining the expectations of the customer regarding the product/service (aside from the product characteristics—delivery schedules, special packaging instructions, special payment conditions)
- Defining the materials, parts, or components that assemble the product
- Defining the conditions of the environment that may influence the product or its characteristics
- Defining the known associated risks and their controls

- Defining the processes, operations, and activities that are needed or expected for the realization of the product, or shall I say how these operations and activities produce the product characteristics
- Defining the quality requirements of the product or product/service including the known quality problems of the product/service

When producing such documented information, you shall make sure that the participants of the realization know

- Where the documented information is located
- How to use it

When a regulatory requirement needs certain documented information regarding product characteristics, it will be made available to the employees.

Types of documented information include

- Requirement specification—documented requirements for material, design, product, or service
- Functional specification—documented specifications that describe the functionality of a product using a method such as block diagram or flow chart
- Design or product specification—documented specifications detailing designed solution or final produced solution

Availability of Documented Information Describing the Required Activities

The standard requires that the organization shall make available all essential activities necessary for the realization processes with documented information. At any stage during the realization, documented description and instructions or specifications for production, assembly, storage, installation, service, or support are to be available for anyone who is involved in the processes. The objective here is to provide the employee, supplier, contractor, or any other party that participates in the realization processes with full access to information such the required tasks specifications and their influence on the product characteristics. This information will be used by the relevant parties to do the job correctly: carrying out activities, measuring performances against specifications, and verifying and validating the process outputs. If any questions, uncertainties, or issues regarding the realization arise, the enquirer may use this documented information as a reference. Which common documentation is already used and may serve as documented information describing the following required activities:

- For the overall planning, the organization may create a documented plan for the realization, which will ensure that customer requirements are met. The plan will provide all the data and information necessary for the realization of the product and will link activities of the realization to the order of the customer (its needs and expectations). For the manufacturing organization, it may be the production plan and for the service provider a service plan. This plan will control tasks, resources, and schedules.
- For quality assurance, a test instruction is required, which identifies the part to be tested, the machine that realizes it, the interval of the test, the sequence of the

validation (the test), and the criteria for acceptance. All the required data and information are defined, documented, and made available. Through this documentation, an employee is in a position to control process output.

- For the assembly of a product, specific assembling instructions, data forms, and inspection instructions including all the required documentation, such as diagrams and criteria, will be available at the workstation.

The documented information will cover the following realization issues:

- Overall planning of the realization in order to ensure that customer demand will be met
- Process or process flow that persons or parties that participate in the realization must be aware of
- Tasks or activities needed to be carried out
- Description of responsibilities for tasks or activities and their required qualification (when such indication may influence the quality of the product and the organization finds it important to mention)
- The use of infrastructure and behavior in the process environment
- Relevant equipment to be used during the activities
- The expected results of the activities

This documented information will instruct the employee on

- How to behave in the process environment
- How to handle materials, parts, or components
- Which tools and equipment are to be used
- How to perform certain activities such as assembly, construction, installation, packaging, labeling, or storing
- How to perform tests, inspections, verifications, or validations

When an acceptance or quality approval is required for parts, materials, or components in order for them to be used in the realization process, the approval will be documented and made available to the worker, and he or she will use these parts only when an approval has been given.

When training for employees, contractors, or suppliers is necessary, in order to ensure that they follow the instructions, this training will be documented and controlled. For example, when the documented information includes technical details, you ensure that the personnel that use this documented information understand the contents and objectives.

When outputs of risk-based thinking specify the need for documentation such as control requirements or safety measures, these will be made available at the workstations. Where appropriate, training will also cover this topic.

When a regulatory requirement requires certain documented information for the realization, for example, at a workstation, it is to be made available to the employees. For example, material safety data sheets (MSDSs) are the internationally standardized way for documenting the hazardous properties of chemicals and other agents.

In several countries, occupational safety and health regulations require the availability of such charts, when they are used during the realization.

When the documented information is exposed to environmental conditions that may harm and wear it, it is necessary to implement protective precautions and to locate the documentation in a safe area in order to ensure its use (i.e., a pouch, in laminated folder or covers, etc.).

Types of documentation include

- Documented instructions
- Standard operating procedures
- Specifications
- Work instructions
- Test specifications
- Blending or mixing procedures
- Step-by-step procedures
- Flow charts
- References for measurements and monitoring
- References for measurement procedures
- Routing cards
- Operation instructions
- Packaging procedures
- Service manuals
- Process control charts
- Diagrams
- Fabrication drawings
- Assembly drawings
- Subassembly drawings
- Technical (engineering) drawings
- Label drawings
- Package drawings
- Samples of finished products
- Samples of assemblies
- Models
- Reference material

Availability of Documented Information Describing the Expected Results

The organization is required to ensure that documented information that defines the results to be achieved will be available during the realization of processes. The records shall provide evidences that results were achieved during the realization. Results to be achieved refer to the

- Required conditions and inputs necessary for the realization:
  - Availability of suitable inputs for the realization
  - Allocation of resources
  - Availability of knowledge and competence
  - Availability of required documented information for the realization
- Process outputs meeting their specifications:
  - Availability of product characteristics and customer requirements

In order to plan effectively a format that will gather all the necessary data and information regarding the progress of the realization, I suggest the following:

1. Get a clear description of the realization process: a process flow chart or the quality plan. It is important to have a detailed description because each process stage might need to be monitored.
2. Decide for each process stage what are the critical conditions—conditions that must take place in order to achieve product conformity:
   a. Specifications of the product
   b. Operations and activities
   c. Controls including verifications and validations
3. Describe for each process stage the expected results.

One way is to plan this format according to the quality plan. This type of documented information may be attached to the product file, as attachment to the customer file (when it is a customer-specific product) or as instruction.

Infrastructure and Process Environment

The infrastructures needed for the realization of the product or provision of service must be used under controlled conditions. The goal is to monitor parameters and aspects that may affect the performance of the infrastructures or the process environment and the quality of the product. The aspects that will be controlled are specified in clauses 7.1.3—Infrastructure and 7.1.4—Environment for the operation of processes. It is defined there how one must plan and implement the infrastructure and the operating environment of processes. Now, while realizing the product, we must ensure that those plans are implemented effectively.

The performance of infrastructures and process environment is affected when they are not functioning under conditions that were set and determined as needed for the quality of the product or not properly used. When the performance is poor, it affects directly the conformity and the quality of the product. The critical parameters must be identified, and suitable controls must ensure that they are maintained before and during the realization.

The organization shall identify the key processes, activities, or operations that affect the quality of the product and the relevant infrastructures and process environment and make sure that they are intact before and during the realization. The organization must provide the necessary resources as well. For that, an effective and preventive system must be developed, planned, and implemented. The system will refer to the following issues:

- Identification of all relevant infrastructures and process environment and their contribution to processes or activities
- Definition of proper or appropriate methods for maintaining the infrastructure and process environment
- Planning of activities needed to maintain the infrastructures and process environment

- Planning of maintenance and preservation of accessories and equipment related to infrastructures and process environment
- Definition of responsibilities
- Availability of spare parts

Documented information related to infrastructures and the process environment is necessary for its operation to be maintained. The goal is to ensure the correct use of these during the realization processes; it must be clear to the users which parameters of the process environment or how the use of infrastructure can affect the conformity and quality of the product or the service and which controls are required:

- Instructions for proper operation and the use of tools and equipment related to the process environment and the infrastructures.
- Machine maintenance procedures.
- Test instructions and schedules for activities (maintenance and service).
- Troubleshooting procedures.
- In cases where there is a risk for the safety and intactness of the infrastructure and work environment, an appropriate measure of precaution should be implemented in order to ensure their protection.

This documented information shall be available for persons who use the infrastructures and process environment for realizing the product as well as for persons who are responsible for maintaining and servicing them. It is necessary to identify the departments or individuals that use the infrastructure and process environment while realizing the product and to verify that they are trained, qualified, and have the proper skills to use them.

When the organization finds it necessary, it must provide evidences that these maintenance activities were undertaken before and during the realization. When is it necessary? It is necessary when the process environment and the infrastructures directly affect the quality of the product. In practice you may use forms or procedures that describe

- The activities for the maintenance—it is important to identify activities that affect the performance of the infrastructure or process environment that affects the quality of the product.
- The period in which the activities shall be performed.
- The responsibilities.
- The approval.

Examples include

- Test protocols for production tools
- Maintenance protocols for machines
- Maintenance protocols for infrastructures

Take a look at the following example of a format for weekly control of a production machine.

Activity	Date	Name	Signature	Remark
Control of hydraulic power unit				
Control of safety installations				
Control of pneumatic system				
Testing the automatic central lubrication				
Testing all lamps and signals				
Controlling cleanliness				
Testing the general function of the machine				

Availability and Use of Suitable Monitoring and Measuring Devices

During the realization process, monitoring and measuring devices are used for the validation of products, processes, infrastructures, equipment, and the required environmental conditions. The selection of such devices and their implementation will be done in accordance with the nature of the product. The use of those monitoring and measuring devices plays an important role during the realization process.

For each product, process, infrastructure, piece of equipment, and work environment that must be monitored and measured, the organization will identify, allocate, and make available the appropriate monitoring and measuring device, for example, scales, caliper gauges, software, or particle counters.

The organization will identify the roles that are responsible for the use and operation of these devices or are responsible for their activity, namely, quality testers, maintenance technicians, laboratory technician, and assemblers. They each use different tools to control their activities and process outputs. One must ensure that

- They are using the appropriate devices related to the processes they are realizing
- They are qualified to use the monitoring and measuring devices
- The appropriate operation and user instructions are available
- Documented criteria, such as tolerances, limits, weight, and the allowed deviations required to evaluate results (whether the output is validated or not), are available
- Any training necessary is provided

Documented information is required according to the context of use of the monitoring and measuring devices. In other words, where documented information related to the use of monitoring and measuring devices is necessary to prove the conformity of the product, you may need to document

- The availability of the monitoring and measuring devices during the realization
- The methods for using monitoring and measuring devices
- The results of the measurements

For example, if you measure a process output and the results are needed for the release of the product, the identification of the monitoring and measuring device that was used to perform this measurement will need to be documented besides the results, as evidence of fitness of the monitoring and measuring device.

Labeling Activities

The label has an important role in the use and functionality of the product; it provides information regarding several parameters in the product. Labeling refers to issues such as

- Identification of the product itself and when necessary the traceability and unique identification of the product (with serial numbers or batch numbers)
- Identification of components in the product
- The status of the product regarding its readiness for use, that is, not released, ready for use, and disqualified
- The product's safety, functionality, performance, and intended use
- Special requirements such as for the automobile industry or the food industry

Labels are actually outputs of different realization processes and are part of approval processes:

- Design and development processes—model, edition, components, parts, ingredients, and intended use
- Realization processes—identification, dates, and traceability
- Risk-based thinking—alarms, warnings, and safety instructions

The labeling activities also refer to information and data published on the package, that is, the name, model, pictures, warnings, and user instructions (the next paragraph, packaging activities, will refer to the packaging of the product). The label is considered to be documented information. There are three distinct types of labels, and each is influenced by different needs:

1. Labels that indicate the status of the product during the realization processes
2. Labels that are attached to the product that provides information regarding its components or intended use
3. The information of the product that is provided on the package

While defining labeling processes and labels, the following issues are to be considered:

- The label should be intact and resistant to the ravages of time, the operation environment, storage, delivery, and use.
- The label will be protected from damages during the realization processes, namely, production, assembly, packaging, storage, delivery, and service and maintenance. When it is known that an activity damages the label, replacement of the label will be planned (e.g., when servicing a product and the label must be peeled off).
- It is required that one ensures that the use of the product will not damage the label.

- It is important to ensure that the label will not damage the product or components and will not affect their functionality or performances, that is, electrical currents, sterilization, or chemical reactions.
- The location of the labels on the product must be defined.
- Graphics, diagrams, and pictures will be controlled and maintained as documented information.
- Labels and labeling activities are to be planned in accordance with standards, technical specifications, sectoral or regional and regulatory requirements.
- When the label is a part of the package, you may need to perform a compatibility test to ensure that the material that the label is made of does not affect in any manner the product's components or processes.

The details on the label will be appropriately planned according to the nature and use of the product and will include information and data such as

- Identification of the product and the components (i.e., name and model)
- Edition or version
- Dates that are relevant to the realization of the product, that is, production, assembly, and the next service.
- Dates that are relevant to the characteristics of the product, for example, expiry dates
- Identification of production and traceability (serial number and batch number)
- Process approval
- User instructions
- Performances of the product (e.g., electrical)
- Alarms, warnings, and safety instructions
- Storage instructions
- Delivery instructions

Packaging Activities

The operation and activity of packaging the product may be a critical issue and may affect the quality of the product, and the packaging processes are reviewed as any other realization process and thus must be controlled. The handling of packages and materials will be controlled in order to preserve them and their characteristics during storage, processing, handling, and delivery. The controls shall cover issues like sealing of packages (where required); the use of appropriate materials for a package; performing the right packaging activities (which will not damage the product); ensuring the safety, functionality, and performance of the product; and the preservation of packaging materials while handling or storing them. The control of the packaging activities shall also include validation of equipment, tools, processes, as well as materials. Outsourced packaging processes are taken into account as well.

In some cases, the package affects further realization activities of the product. For example, in the automotive industry, the packaging and quantity of delivered parts is planned according to the requirements of the assembly line. In other cases, the package may represent or support a feature in the product, or carry important information regarding the product and its use.

The package is considered to be one of the product specifications and shall protect the product from

- Handling, processing, and storage
- Delivery activities until the package is opened
- The environment and possible contaminations

The packaging operations and their controls are planned and designed during the design and development stages. During the realization, those operations and controls are to be implemented and are needed to be verified or validated.

Environmental factors are to be taken into account when storing and realizing materials for packaging. The environmental parameters that may affect the package are to be defined and controlled, for example,

- The permissible level of moisture and humidity
- The pemissible storage temperature range
- The lifespan of packages and materials
- Hygiene control

When the package and the packaging activities may have an effect on the quality of the product and the risk of performing the activities incorrectly or not using the packaging as specified exists, validations shall be planned and implemented in order to ensure that the quality of the product is maintained. For example, if the product must be sealed, there are a lot of critical issues to attend to that may affect the product and its characteristics: physical, chemical, as well as biological. The controls may include quality assurance tests on samples of packed product and process validations of the sealing itself.

When applicable significance will be given to statutory and regulatory requirements, standards, and technical specifications related to the product that describes packaging requirements. When such are relevant, the organization will control that the specifications, activities, so that the required controls are implemented and practiced.

Information regarding the packaging will be communicated to distributors. The objective is to give instructions concerning the risks related to the package and the handling and safety of the package.

One may use procedures, work instructions, drawings, pictures, or examples of packaging at the workstations. The documented instructions shall relate to the packaging operations and provide a specific description of the activities to be carried out. When documented information is required, it will cover the following issues:

- Identification of the relevant product that will be packed.
- Identification of the different types of packages and the required packaging and materials necessary for the packaging process (such as adhesives and sterilizers).
- Definition of target groups that will perform the packaging activities.
- Storing and handling of packages and related materials.
- Definition of required tools, equipment, or machinery, including setups and required maintenance and their operation.

- The physical parameters of the package are to be defined (size and dimensions, form or shape, weight, consistency, color and graphics, and power requirements).
- A description of the packaging activities (e.g., the materials used, the packaging activities, and the operation of tools, equipment, or machinery).
- The required controls for validation (where applicable).
- The required controls for quality of the package.
- Required delivery specifications.

I used to photograph each product as to how the packaging of the product should be with important remarks on the photo itself and attach the photo to the work instructions.

Release Activities

The organization shall implement planned arrangements for the release of the product or service, at appropriate process stages and locations, needed to verify that the product and service requirements have been met. During the realization processes, one must ensure that those requirements are implemented and practiced. The standard requirements for the release activities are detailed in clause 8.6—Release of products and services. Please refer to chapter 8.6 for detailed information.

Delivery Activities

Transportation and delivery may be critical processes and may affect the quality of the product and thus will be defined and controlled. The objectives are to ensure the product's functionality and safety and to prevent any nonconformities during the delivery. The scope of the definition will refer to the entire realization process:

- Delivery of goods from external providers
- Transfer of goods to manufacturing
- Delivery of products to subcontractors for processing like coating or furnishing
- Delivery of the final product to the final customer

The delivery conditions are to be planned and defined during the determination of requirements related to the product (see clause 8.2.2). Now is the time to implement the controls that will ensure their implementation. When initiating controls over delivery processes, the following issues are to be considered:

- The controls will ensure that the delivered products are safe and secured during the deliveries and that all transport requirements are met.
- The products will be protected from environmental factors that may affect them, for example, heat, cold, humidity, moisture, dirt, and light.
- When there are requirements for controlled temperatures, the necessary conditions will be provided.
- The identification and labeling of the products will be intact and safe.
- The products are protected from damage such as spillage, shaking, collision, or other accidents.

- There should be protection against contaminations from substances of other products and contamination that may be caused by damage to the products themselves.
- There should be a protection against pests or microorganisms.
- When there are requirements for special vehicles, containers, packages, or their labeling for the transportation, the necessary means will be provided.

There is no requirement for a documented procedure, only a definition. But there is a requirement to provide documented information that necessary activities were performed and that the product was preserved under appropriate conditions throughout the delivery process. Then one will be required to provide evidence that the delivery conditions were imposed and practiced. So I recommend one define the documented information that is needed to control the delivery conditions throughout the transportation activities. These records and evidences will serve you during audits, regulatory controls, or customer complaints. The documented information is divided into two categories: the instructional records and the evidence of performance:

- The instructional documented information are the definitions of controls to be implemented during the delivery process, which are needed to support the delivery activities. They include
  - Packaging instructions
  - Delivery or transportation specifications
  - Specifications for protection
  - Documented precautions or warnings
  - Customers' delivery instructions

  This documented information will be made available to the relevant parties when required.
- The evidence is a record of performance, and the objective here is to prove that delivery specifications were met:
  - Labeling and identification of the product
  - Delivery notes, including details of delivery like dates and carrier
  - Reports of transportation or storage conditions during the delivery processes, such as data registration of temperature or moisture
  - Pictures taken as evidence

The organization will appoint a responsible party for the verification that the delivery activities are performed according to specifications. Normally it will be a person in storage who prepares the goods for delivery. If distributors are receiving and delivering the product to the final customer, the organization is responsible for dictating the delivery requirements to them. The distributors will have to provide the manufacturer with evidence of performance.

The organization shall define what is to be done with the product once a deviation from the delivery specifications has occurred and there is a danger that the product is no longer meeting the requirements. In this case, a link to the process of controlling nonconforming process outputs must be established and the employee will know how he or she shall report this nonconformity.

Installation Activities

Installation is regarded as a realization process, and the organization shall ensure that all the required inputs, conditions, and resources are defined and available at the time of the installation. The purpose is to ensure that the product will operate properly after the installation in terms of functionality, performance, safety, and intended use:

- The organization shall ensure the proper installation of the product and determine the necessary installation activities, steps, inspections, verifications, and approval.
- The organization shall document and provide all necessary data related to the installation of the product, such as instructions, specifications, procedures, or any other documentation that is necessary for the execution of the installation.
- The organization shall make sure that all the necessary inputs are available, for example, parts and components.
- Tools, equipment, and the qualifications of the personnel necessary for the installation shall be defined and available.
- The organization is to define what the necessary conditions for the installation are and ensure that they are available at the time of the installation.
- Access to the location of the installation shall be granted.
- Outputs of risk-based thinking shall be considered and the necessary controls shall be implemented.

Emphasis will be given to the required controls, inspections, and testing and its relevant acceptance criteria, that is, the activities that will verify and validate that the product is adequately installed, that its requirements are met, and that the intended use is accomplished. The common example is designing a checklist that verifies that all the steps of the installation were performed and all the results after the installation were accepted.

Conditions for the installation are important because they may influence the quality of the product. The organization shall ensure that the conditions necessary for the installation that are necessary to achieve conformity of products and services are available. The conditions may be

- Environmental conditions like cleanliness
- Weather conditions
- Hardware requirements
- Infrastructural requirements like power grid
- The availability of certain locations or access to locations

There are cases where installation is performed by a third party, for example, the customer or a supplier. In these cases, the organization must provide them with the installation instructions, inspection, and criteria for acceptance, must define the responsibilities (between the organization and the third party—who is responsible for what), and must train and qualify the installer. Where applicable the third party shall provide documents and records that prove that the installation was performed according to specifications.

The stages and activities of the installation may be enclosed on a quality plan (or any other installation plan) and may include

- A definition of responsible roles for each activity
- Documented customer specifications
- A specific description of the installation activities and the required results
- Definition of the necessary conditions for the installation
- A reference for tools and equipment and their use
- Reference to the operation environment, customer facilities, and other devices that may affect the installation
- Supporting documentation required to perform the installation (including trouble-shooting documents during the installation)
- Tests, controls, validations, and verifications—including criteria
- Reference for schedules and time objectives
- Expected records

Post-Delivery Activities

The realization of the product or provision of the service may include post-delivery activities. During the realization processes, you must ensure that those requirements are implemented and practiced. The standard requirements for the post-delivery activities are detailed in clause 8.5.5—Post-delivery activities. Please refer to chapter 8.5.5 for detailed information.

Preventing Human Errors

Preventing human errors during the realization of a product or service refers to design and formation of the realization environment and conditions that will proactively prevent persons from making errors during their work. This will be achieved through analysis of risks at the workplace associated with human actions and the initiation of actions that should create understanding among workers of what are their quality objectives and how should they perform their work. The following are examples of actions that can be initiated:

- Maintain effective qualification and certification activities when incorporating personnel in the organization.
- Plan training and repetitions of training.
- Standardize best practices in the work area.
- Maintain order in the workplaces or adopt (appropriate) frameworks for orderliness in the workplaces.
- Reduce chances of being disturbed with unnecessary items or events in the workplaces.
- Perform maintenance activities as planned to prevent machinery and equipment deterioration.
- Promote awareness of contribution to the effectiveness of the QMS, including the benefits of improved performance.
- Make relevant documentation available and accessible at the appropriate workstations or at the appropriate process stages.
- Implement troubleshooting documentations.
- Segregate nonconformed process outputs, tools, or equipment to eliminate their use.

- Clarify safety measures and instructions.
- Define access to the different work areas.
- Obtain feedback from workers.
- Set quality objectives related to the work areas.

In practice one may use the process of improvement to initiate such actions. For documentation purposes, one may include preparations for these actions on the quality plan.

Validation and Periodic Revalidation

Before I begin with the issue of validation, I would like to discuss a few terms and definitions (just to toe the line between us):

- Verification: Confirmation, through the provision of objective evidence, that a process delivered its specified requirements. The verification will be achieved through activities that control process outputs, such as calculations, measuring, or tests, and compare them to specifications.
- Validation: An approval, through the provision of objective evidence, that a process delivered its specified requirements for a specific intended use or application. The validation is achieved through the collection of data and information during the realization, which establishes an evidence that a process is capable of consistently delivering qualitative product and that requirements for a specific intended use or application have been fulfilled.
- Revalidation: A reapproval of validation in case it is evident that a process does not deliver its expected outputs or any changes occurred within the process—a change of equipment, a change in human resources, raw material, and so on—that may affect the product.

The organization is required to validate realization processes and their outputs where their results and outputs cannot be verified—monitored, measured, and evaluated. These are processes whose outputs cannot be controlled at the end of the process or as a finished product and thus cannot provide evidence, through verification, that the product meets its specification. Before the release of the product for use, the organization shall establish a high degree of assurance in the performance of the manufacturing process that defined inputs were submitted to the process and certain conditions were controlled during the realization and so ensuring that characteristics of the product will be met. For example, in a welding process, in order to verify the welding, you must enact pressure on the welding spot and see if it breaks (not so practical as a test). Process validation refers to the assurance that certain process parameters were attended when the process took place during the realization. In the example of the welding parameters, those parameters will be arc voltage, welding speed, and heat input rate.

Planning process validation involves a series of activities taking place over the realization of the product that controls the parameters. The assurance should be obtained from objective information and data gathered before and during the realization processes. An effective validation depends upon the information and knowledge that resulted from design and development. Design and development outputs indicate, determine, and define validation activities necessary to ensure that the processes, products, and services meet their requirements.

For the planning of process validation, the organization shall

- Identify the parameters that affect the process and the product characteristics
- Define tolerances or limits for those parameters
- Identify the impact of the variation of those parameters on the output of the process
- Control the variation of those parameters in a manner appropriate to the risk and impact it represents to the process and product
- Identify which changes in the process may affect those parameters and control those changes (changes in materials, equipment, production environment, personnel, and manufacturing procedures)

The organization shall develop a method for collecting and analyzing product and process data with the objective of identifying quality issues in the process—situations where the requirements will not be met. These data and information may be used for process improvements. The organization may apply sampling and testing of in-process materials that will provide evidences that processes are performing as expected and that may serve as a process validation.

Evaluating and Identifying Processes for Validation

Before we plan the validation of a process, we must decide whether we need to validate it or verify it. What is the difference between a process that needs to be verified and a process that needs to be validated? The rule of thumb is to ask whether a process can be verified with monitoring and measurement activities or needs to be validated. I prepared here a scheme that describes the process for such evaluation (Figure 8.2).

The outputs of this evaluation shall provide a list of processes that must be validated.

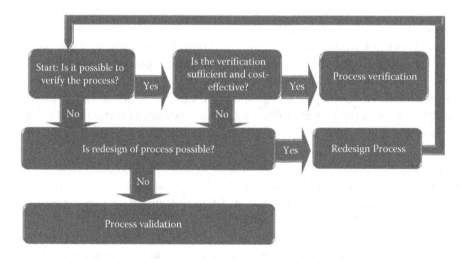

Figure 8.2   Evaluation of the necessity of validating a process.

Process Parameters for Validation

Parameters of the process affect the performance of the process; thus, the parameters are related directly to the quality, functionality, performance, safety, and intended use requirements. After defining the processes to be validated, it is required that you define for each the parameters that will be controlled. In order to implement effective validations, you must understand what the key parameters affecting the quality of the product or the performance of a process are, for example, temperature, pressure, compound of raw material, concentrations, tension, time, machine cycle, and setup conditions. Identifying key parameters will help you indicate what is to be measured.

To begin with, review and analyze the effect of each process on the specifications and characteristics of the product. For example, in the food industry, the sealing process of the package may affect the quality of the product. Values and levels for these characteristics should have already been determined (e.g., outputs of design and development). In order to reach these specifications of characteristics, the sealing process must maintain certain conditions; during the sealing process, the temperature and the pressure of the clamps are critical in order to reach the required level of strength.

When determining the parameters for control, it is important to correlate the inputs with the outputs. For example, in a plastic molding, the injection formation of the part is affected by the holding time of the part on the production tool before opening and releasing the part (a fraction of a second can be critical). The input (holding time) is directly related to the output (form of the part). These parameters will indicate the stability of a process; thus, these parameters will need to be controlled.

After identifying these parameters, it is necessary to relate them to the tools and the elements (in the process) that are influenced by those parameters, namely, machines, software, production tools, human resources, and infrastructures. These elements will supply us with the data and results regarding the process parameters.

Failure in equipment, tools, systems, and components may affect the process and the quality of the product. The organization must assess the impact of such failures on the product and its characteristics. The impact may be direct or indirect:

- Direct impact results in the failure of the system used in the realization of the product, for example, machines, production tools, raw material, and components.
- Indirect impact results in the failure of systems that support the realization systems, such as utilities and infrastructures, human resources, and work environments.

The evaluation of the impact may assist in planning the control and validation tests more effectively. Risk analysis may serve as an input for such assessment.

Acceptance Criteria (Desired Outputs of Processes)

After identifying the parameters for validation, you need to determine, for each process, the target values and their acceptance criteria. Target values are the optimal values for the process parameter. These values will demonstrate the stability and capability of a process and ensure that requirements are met and accepted. Maintaining process parameters within the criteria ensures that it will deliver expected product characteristics and specification.

The criteria are used for the evaluation of the results compared to the requirements. The limits of the criteria (target values, upper and lower control limits) are based on product or process specifications. The objectives of the criteria are to demonstrate the effectiveness of processes and to support decisions for judging, evaluating, and determining by facts, values, and data the compliance of the processes. The criteria will support the validation by indicating the status of the parameters, that is, either accepted or rejected. The criteria will be planned according to the types of data and statistical analysis methods. Deciding upon limits depends on the analysis method used, for example, action levels, control levels, acceptance levels, specification levels, and worst case conditions. The objective of these limits is to instruct the organization when process adjustments are required; for example, when results show that the process is deviating, then an action needs to be taken in order to restore compliance. The criteria may be objective or subjective:

- Objective criteria mean distinctiveness, invariance, and controllability; the results are being compared to a set of determined values, and a decision is made based on the comparison: either accepted or rejected.
- Subjective criteria are when the results are being evaluated and submitted for assimilation by an individual. The criteria are subjected using knowledge and experience. Subjective criteria need to be supported with justifications for process conformity.

The criteria will be documented on the test protocols and are required to be approved before submission for use. The following issues are to be considered:

- For each process that requires validation, the organization shall develop criteria for the evaluation of the performance of the process and the acceptance of its outputs.
- The organization will develop a policy for establishing and determining the criteria: scope, relation to the processes it serves, objectives, activities, method for documentation, and methods for reporting. For more information about this policy, I refer you back to chapter 8.1, paragraph Establishing criteria for acceptance of products and services, of this book.
- The criteria will be designed in order to provide alarms regarding the status of the output and to define when the results are regarded as nonconforming.
- The criteria shall refer to documented product requirements or characteristics that they control—materials, functionality and technical specifications, product characteristics, reference to drawings, and quality requirements.

- The criteria will be integrated into the quality requirements (e.g., can be documented into the quality plan).
- When external requirements for criteria for a certain type of process or product are applicable (standard requirements, regulatory requirements, or technical specifications), it is necessary that these criteria be implemented in the validations.
- When activities for addressing risks in the realization (outputs of risk-based thinking) specify criteria, they will be incorporated in the validation.

The next issue regarding the criteria is worst case conditions. This principle is required, for example, in the automotive or the medical industry. The worst case conditions are a set of conditions covering the upper and lower limits of process parameters. These conditions simulate possible circumstances during the realization process, which indicate a clear unwanted trend in the process (regarding its performances). The conditions will be defined according to the risks related to the process. Reaching these levels does not indicate that the process is nonconforming yet, but they warn and notify the manufacturer that the situation might pose a risk to the outputs of the process and may impact the integrity of the product. For example, there are products that must be realized in an environment that maintains a certain level of cleanliness. The parameter level that ensures the level of the cleanliness during the realization is already defined. For these levels, one determined the values for lower and upper limits and the target results. Now one must determine the lower and upper limits in the level of cleanliness in a worst case condition.

Selecting a Method for Analyzing the Data

Methods and techniques for analysis and demonstrating parameter results for validation must be determined. The purposes of the techniques are

- To demonstrate objectively, through data, that processes meet their specifications
- To replicate actual use conditions

The method shall be selected and developed according to the type of data that are required to be collected and analyzed. Using the method will provide you with an actual and accurate status report regarding the processes:

- The method will allow comparison to the objectives and the goals of parameters.
- The method will provide the organization with alarms regarding the processes.
- The method will enable the organization to compare new data with old data.
- When international, national, or local standards for analysis techniques are demanded, they will be implemented.
- The use and instruction for implementing the method will be documented or referred to (e.g., on the quality plan).

I suggest here several methods for statistical analysis:

Method	Objective	Statistical Technique
Acceptance sampling plan	Determining through the product the acceptance level of the process	Sampling of a product and through analyzing particular characteristics of this sample deciding whether to accept or reject a process
Capability study	Evaluation of the ability of a process to consistently stay in its limit specifications by sampling its outputs	Examining the processes over a period of time by sampling outputs of the process and placing them on a control chart, comparing them to control limits and criteria
Control charts	Identifying and analyzing changes in the process and mapping their trends. The identification and examination will detect the root cause of a change	Sampling process units according to a plan and identifying and analyzing changes in the process. After the change is identified, trends are analyzed. Such a technique assists in identifying process parameters with high potential to induce a change
Dual response approach to robust design Robust design methods Robust tolerance analysis	Robust processes by detecting their optimal values for process parameter	Reducing variations of process outputs by selecting the optimal process inputs and introducing them to a statistical calculation
Response surface study Tolerance analysis	Planning the optimal process output	Calculating the statistical relation between process inputs and process outputs. Using such methods can assist in identifying the optimal combination between different inputs and augment the process
Process simulation	Identifying exceptions and providing important insights on required interrelations between tasks, operations, and systems	Mathematical simulation of a process under various scenarios, conditions, or loads

Revalidation

Revalidation is required in situations where you detected that process parameters may not provide the necessary outputs, for various reasons, and you need to apply an action to restore conformity. The objective of the revalidation is determining how much a process is stable and capable after a change or correction was applied.

The following events may require the initiation of revalidation:

- When corrective action was implemented in a process or a product
- When changes or improvements in the product design, and thus the process, were initiated and implemented
- When the personnel operating the process were changed, and it affected the quality of the product
- When work instructions or work methods were changed

- When inputs to the process such as material or components were changed and may affect the quality of the product
- When a new supplier is introduced to the process
- When a new tool or equipment was introduced to the process
- When deviations or negative trends were detected and measures such as corrective actions or changes were applied
- When periodic maintenance to equipment or infrastructure is not performed as scheduled and there is a risk that they will fail
- When there are changes in the packaging processes of the product
- When a certain process produces excessive defects or indicates negative trends
- When evaluation of troubleshooting of a process is needed
- When controls to the process were changed or replaced

The revalidation of tools, machines, and equipment is referred to as requalification. Requalification is needed

- When a service, maintenance, or major repair was carried out to the machines or equipment
- When production tools or production molds were amended
- When adverse trends in equipment performance occurred
- When the process was relocated to another environment, machine, or facility
- When the replacement of critical spare parts was initiated

The need for revalidation shall be evaluated and documented on the validation plan (the validation plan will be discussed later on). For each revalidation activity, criteria will be defined according to the type of revalidation. When revalidation is required, I suggest to include it in a validation plan.

Periodic Revalidation

The QMS elements that participate in the realization of the product or the service are submitted to constant changes or situations that may alter them. In order to regularly control capability of a process, a periodic revalidation is required. The following are reasons for applying periodic revalidation:

- The process wears out equipment or the infrastructure that supports the process.
- Historical data of the process suggest that the process may not be stable in the long run.
- When a manufacturer of equipment or tool suggests or even implies performing periodic revalidation.
- When a regulatory requirement demands the performance of periodic revalidation.
- When risks associated with the process are known and cannot be reduced.

When processes need a periodic revalidation, you must plan them in advance at planned schedules. This plan will be retained as documented information. For each periodic revalidation activity, criteria will be defined according to the type of revalidation. When periodic revalidation is required, I suggest to include it in a validation plan.

Validation Circle

In order to initiate and implement effective validations of your processes, the matter must be considered by the interested parties related to the realization of the product. This is why I suggest initiating a validation circle in the organization. In the validation circle, you assemble a group of people who are involved in the design of the product and the process realization of the product and discuss the process by having each person in the group describe each activity and its process outputs. Each one will share his or her opinions and aspects on the matter, and together they will plan an effective validation activity. Who is to contribute to the circle?

- Sales will be responsible for communicating the important customer requirements (reflected as product characteristics).
- Research and development will deliver the relevant product characteristics that need to be validated.
- Production team will share its opinion on how to implement the validations effectively throughout the realization process.
- Regulatory affairs will provide all the relevant inputs for the regulatory requirements that must be accounted for during the realization processes and their effect on the process validation.
- Quality will be responsible for the following two issues:
  - Implementation of the validation throughout the QMS, for example, through documentation such as procedures and work instructions, forms and records, training, and by implementing the verification in the internal audit plan
  - Planning and implementing of the monitoring system and processes, the analysis of the data, and the reporting and alert system

Validation Plan as Documented Information

Validation plan, also known as a validation protocol, is one way to implement effectively validation processes in the organization. It defines a set of rules and events describing how validation activities will be performed. The objective of the validation plan is to present a documented approach, strategy, and methodology in conducting validation processes with reference to the relevant elements that will support the validation. The plan will identify the processes to be validated and determine the activities, timelines, criteria, and schedules for the validation, the interrelationships between processes requiring validation, references to documentation, and the timing for revalidations. In practice, one needs to develop a document that describes how the validation of products or processes in the organization is conducted. The plan will include

- The scope of the plan, that is, which processes and products are included under its scope.
- Identification of equipment, tools, systems, and components that may affect the process.
- Description of activities and their sequences of the validation processes, such as activities, timelines, schedules, and locations.
- Reference to relevant documentation, for example, procedures, process flow diagrams, test protocols, customer requirements, instructions, technical specification, and technical details.

- Reference to the criteria for acceptance or rejection of a process.
- Description of roles and functions that are involved in the validation activities. When defining the role, it is necessary to relate each role to its duties and specific responsibilities in the validation processes.
- Instructions on how to detect, evaluate, and react to deviations.
- The requirement for revalidation and the events where revalidation activities are needed will be identified.
- The outputs of the validation plan; the format of the results of the validation and the expected records.

The results of the test and validations shall be retained as documented information:

- A designated format or system to collect the data shall be planned.
- The format or the system shall be suitable to the type of the data it is collecting.
- The appropriate analytical method shall be applied for the analysis of the data.
- The results and the information shall be accessible to the appropriate persons in the organization or interested parties to ensure a quick reaction to trends and events in the process.

8.5.2 Identification and Traceability

Implementing identification and traceability allows the organization to effectively trace back the activities, operations, processes, and process outputs from both sides of the supply chain (customer and supplier) and enables the identification of all elements that participated in the realization of the product or provision of the service. Management of status enables the organization to distinguish the different elements that combine the product: materials, parts, and components, with different characteristics from the different stages of the realization process. The ISO 9001 Standard requirements include the following:

- The organization shall use methods and means suitable to its processes and process outputs for the identification of process outputs when it is necessary to ensure the conformity of products and services.
- The status of process outputs shall be established and identified with respect to monitoring and measurement requirements throughout production and service provision.
- When traceability is a requirement, the organization shall control the unique identification of process outputs and shall retain documented information necessary to enable traceability.

Goals of Identification and Traceability

Traceability refers to the development of a method for collecting data related to the use, history, and location of an object and following it based on a recorded identifier. The traceability will be guaranteed through the linking of characteristics of processed materials with products and its deliverables, for example, applying a serial number to a product or batch number to a produced batch. Managing a unique batch or a charge enables the traceability when a problem arises, the flow of material will be assured, and the disposition of the materials to the realization will be much more accurate. Traceability refers to the

- Origin of materials and parts
- The processing history
- The distribution and location of the product or service after delivery

Identification refers to the establishment of a method or an activity with means for identifying an object and providing evidence to its identification and status.

Managing identification and traceability before, during, or after the realization activities has the following clear goals for supporting the QMS:

- Describing the areas, scopes, materials, products, parts, components, or finished goods that are under the scope of identification and traceability
- Defining the inputs for the identification requirements: customer requirements, regulatory requirements, risk management outputs, supplier instructions, and design and development requirements
- Determining the activities with which the identification and traceability shall be accomplished
- Defining which records are supposed to be generated to enable the identification and traceability
- Enabling forward and backward tracking of the product and its constituents.
- Instructing employees on how to handle the product, parts, or materials before, during, or after realization
- Specifying the information and details that the identification and traceability shall include: type of product or material, quantities, units of measure, reference to documentation or process certifications, batch numbers, serial numbers, details of production, manufacturer details (including addresses), and communication details
- Indicating responsibilities and authorities required to perform the identification activities
- Avoiding incorrect identification of nonconformed products, submission to realization, false delivery to customer, or contamination of conformed products
- Enabling the application of corrective actions when defects are detected during or after the realization
- Allowing follow-up of product characteristics like validity date or management of FIFO
- Allowing the identification of personnel during the realization processes where qualifications and training are critical

Planning the Identification and Traceability

The identification and traceability of the product are regarded as the collection of related information throughout the supply chain management that may indicate or affect the quality of the product. Thus, it is necessary to identify which information and data are needed, which activities generate the data, at which points of time, and by what means or methods can the data be collected. The final stage is linking the product or process outputs to these data.

When planning identification and traceability methods and activities, bear in mind that some questions need to be answered when a person samples a product from a process or a location:

- What is the origin of the item?
- When was it manufactured?
- From what is it constructed, and are the materials or components that construct it approved?
- What is the usability status of this item?
- Where should it be stored?
- Where is this item heading? What is its next stop? To whom does it belong?

Planning identification and traceability will be done according to the nature and characteristics of the product or the service and when the organization finds it necessary.

A correct definition and implementation of identification and traceability will provide full transparency regarding the product or the service:

- It will allow the identification and tracking of all of the elements mentioned earlier throughout the realization processes until the delivery or disposal of the product, that is, storage of raw materials, components, and other materials used for production, tracking inventory in process, storage of finished goods, delivery to the customer, distribution, installation, and service.
- In case quality problems occur, the identification shall support the traceability of the product or service, facilitate fault diagnosis, and assist the organization in detecting defective products.
- The identification shall provide information regarding the product's usability status and, where applicable, shall relate to the expiry date or shelf life of materials.
- Identification will ensure the use of appropriate materials, components, or items during the realization of the product.
- The identification shall eliminate the risk of mixing products of different status.
- The identification shall indicate to the user when the product was manufactured or assembled, how it was done, and by whom.
- The identification will support action for addressing risks, for example, the ability to identify product characteristics related to the risks or hazardous situations during the realization of the product.
- The identification and traceability shall assist in minimizing damages and implications of nonconformities, when such are detected, through enablement of tracking of associated processes, operations, and process outputs.

Through the assignment of activities, means, methods, and identifiers throughout the material flow such as codes, tags, or product serial numbers, products will be correctly integrated in the realization processes. The production elements that are to be identified and tracked include

- Raw materials, components, and parts
- Purchased goods or services
- Materials used for the production
- Lots or batches
- Finished goods

- Installations
- Defective or returned goods
- Operations (in case of service)

Implementing Traceability and Identification

The implementation of traceability and identification must be suitable to the type of the process outputs, in other words, processes must support this method. Each process output has its own identification needs and possibilities. And the identification and traceability provides its relevant information and details. For example, the identification of a raw material must refer to its expiry date, MSDS, date of manufacture and of arrival, storage conditions, and COC. Another example is that a manufactured part shall be identified with the catalog number, marking date of production, edition, model, lot or batch numbers, and relevant quality protocols. Data that may serve for identification and traceability include the following:

- Product numbers and catalog number
- Batch/lot number
- Serial number
- Dates like validity dates or manufacturing date

When implementing traceability and identification, you shall refer to the following issues:

- The identification details and references shall be planned according to the properties, attributes, and characters of the identified process outputs.
- For each product, the needs for identification and traceability management will be implemented with a unique identification. This method shall be applied in those process stages where the identification and traceability affect the quality of the product.
- The organization shall plan techniques or methods for maintaining identification and traceability. Each techniques or method for indicating the identification shall refer to a type of data: a catalog number identifies the product and its edition, batch number indicates information about its manufacturing, and quality status indicates the usability of the product. A quality plan, a work instruction, or a procedure may indicate which methods and means will be used for the identification at a certain process stage.
- The identification shall be implemented on the product according to its nature but where it is accessible to all. When the status is not explicit, users or persons will be informed where to look for the status. The type of identification shall be planned in proportion to the effect of the product on the process or the process output. The means of identification shall be planned in a way that will not harm the product or its materials, its properties, its intended use, or its packaging.
- When components, parts, or materials that construct the product or activities that constitute the service may cause it to not meet its requirements, the organization shall define special traceability activities. The traceability will be implemented in a way that will enable the link between the end product and its materials or components or the service between its relevant activities.
- The association or relation to specific processes, realization operations, or activities will be possible and comprehensible. It will be possible to use the identification in order to receive details, information, and history about the object sampled. A job order number is one example.

Status of Products

Status of products shall be integrated in the management of materials, parts, components, and finished goods. When it comes to services, the status shall indicate whether the service was completed and the target object of the service is usable. The status refers to the relative position or a state of product in a particular point in the process stage and the certain characteristics it bears at this point. Applying status management ensures that the product or the service, components, parts, and materials are being controlled and introduced to the necessary activities, tests, inspections, or concessions before they are released to the next stage (i.e., introduced to a process, packed, delivered, or installed). The syllogism is that a status of the product indicates its history. Allow me to explain. When I pick a product with a certain status, it means that the product went through different realization stages through which it earned its status. Objectives of the status management include

- Eliminating the possibility of mixing materials, parts, components, or products with different status
- Indicating in which process stage the product is situated
- Situating a product in its life cycle
- Providing information regarding the fulfillment of the product's requirements

The status management shall be integrated with the identification and traceability methods and techniques used in the organization. The following are a few process factors that may influence or determine the status of a product:

- The situation of a product regarding its realization
- Activities that were applied on the product
- Results of inspections and tests that the product has undergone

Indicating the status of goods and products is critical. One should be able to sample any product, material, or component and declare its status. Status of products will be determined with respect to the relevant monitoring and measurements requirements. In other words, the results of the monitoring and measurements activities shall determine the status of the product. The identification of the product shall present the user of the QMS or the customer the status of the product in terms of the following realization:

- Tested/not tested
- Released/not released
- Useable/not useable

The status shall be identified using auxiliary tools such as routing cards, labels, usability tags, signposting, or serial numbers. The important thing is that the status must be clear to all users and personnel participating in the realization processes or using the product. When the status cannot be indicated physically on the product (due to the product's nature and characteristics), it is necessary to define other methods to identify its status, for example, the location or segregation of products or logical management.

Types of Status

Different types of product status shall be determined in order to clarify the significance and consequences of each status. The definition will determine how products receive their status classification (e.g., their quality status, undergoing a certain process, comprising certain materials, being defective, being given quality status after assembly, or before packaging). In case it is applicable, criteria shall be assigned to the transition from one status to the next.

Management of status must refer also to product under service and to post-delivery activities. During the service or maintenance activities, an indication regarding the usability of the product will be determined, that is, usable, in repair, or defective. The status will be clearly shown on the product in a location where all users may view it. It is possible to deactivate or disable only a part or function of the product. Service and maintenance activities may create defective parts or components. Therefore, it is required that one initiates and defines the status of those nonconforming parts or components.

Reference to the Quality Plan

The determination of the traceability, identification, and status may be integrated with the quality plan and may cover the entire life cycle of the product: material, development, production, storage, installation, and servicing. The status can indicate where the product is situated regarding the plan. Another way to put it is that the progress of a product in the life cycle determines and indicates its status.

Reference to Monitoring, Measurement, and Validation Activities

Monitoring and measurement activities, as well as process validations, determine the status of process outputs. The status of a product indicates which inspections the product underwent and their results: whether they were successful or not and whether the process output meets its requirements or not. These tests, inspections, and controls carried out during the realization perform a comparison between the product's specifications and the actual results of processes and determine the status of the product: either accepted or rejected. Releasing a product or a component to the next realization stage or changing its status and updating its usability is done through a set of predefined verifications and validations.

Putting it into practice, you will need to prove the link between monitoring and measurement activities, as well as the verifications of process outputs and validations of processes, and the status of the product. For example, each time a product, a batch, or a lot passes a quality test, the status is modified. This will ensure that only products with the appropriate status reach the next stage.

Usability of the Product

Status of the product indicates its usability and determines the allowed applicable activities, for example, it can be used in certain conditions, it is allowed to be assembled, it needs to be reworked, it is rejected, or it needs to be disposed of. Types of usability status include the following:

- The product is approved for use or further realization.
- The product is approved for use or further realization but with restrictions.
- The product is approved for use or further realization under concession.

- The product is suspended or put on hold.
- The product is rejected from use or further realization.

This combination of the identification, traceability, and status shall indicate when the product was manufactured or when the service was provided, what is the status in terms of usability, and to whom it belongs. The identification shall also eliminate the risk of mixing products, parts, or materials from different origins or with different status.

The identification shall provide the required information regarding the status of a product referring to a process stage and the meaning and significance of this status to the product: controlled material before processing, sterilized products, products or packages provided by the customer, batch numbers before release, devices for delivery, defective parts, and so on.

Integration of Risk-Based Thinking

When the risk analysis indicated that there are some critical characteristics that must be traced in the long term, identification, traceability, and management of status will assist you in controlling this characteristic. The necessary controls of the detected risks may be identifiers like validity date or labeling that indicate how the product may be used. Let us review the following examples:

- When conditions of the work environment may affect materials, parts, or components of the product, and cause it to fail to meet the requirements, the organization shall include these under traceability. And so traceability shall enable the tracking of the conditions in the work surrounding and the parameters that created these conditions during the realization of the product.
- When activities and actions of human resources during the realization processes may affect certain materials, parts, and components in the product, and cause it not to meet the requirements, the organization shall include these under traceability. And so when actions that were indicated as critical to the integrity of the product, identification of personnel that performed these activities will be included under traceability.
- Identification shall be applied to process inputs when they may affect the conformity of the process outputs. A good example is cleaning operations where the cleaning of products may affect its quality and intended use. In some cases, identification of the cleaning material will have to be managed. When a quality problem occurs, the manufacturer may verify that the input to the operation of cleaning was conformed to.
- When a recall of products in case of safety problems is required, the organization may use the identification and traceability as means to track down products and maintain operations for limitation and restriction and minimizing damages of nonconformed products. This activity shall span the entire life cycle of the product: from the storage of materials until the delivery of the end product to the end user.
- For preservation and maintenance of property belonging to the customer or external provider, the organization may use traceability and identification as a means for control.
- When segregation and separation of products is necessary, the organization shall develop appropriate methods and will use identification to support this method.

Identification and Traceability of Monitoring and Measuring Devices

Identification and traceability refer also to the use of monitoring and measuring devices; when monitoring and measuring devices were found to be inadequate or not suitable, one might need to revalidate the series of batches of products that were controlled and released based upon the results of those monitoring and measuring devices. Through identification and traceability, you will locate those batches and product relatively easily.

Customer Requirements

When required, the method shall integrate customer requirements for the identification of traceability methods: requested labels, barcodes, identifiers, protocols, or markings. In some cases, the identification will be part of the packaging instructions. When the customers deliver materials, parts, or components for assembly or use in the realization process, they will be identified in order to enable the requirements of clause 8.5.3—Property belonging to customers or external providers.

Regulatory or Statutory Requirements

International standards, directives, or regulations related to the realization of the product or provision of the service may demand the implementation of identification methods of the materials, products, process entities, or components used in the realization. These shall be taken into account when applying traceability and identification. Requested documentation such as specific forms and tests or work instructions shall be planned and implemented. Records that are needed to prove the performance of the requirements will be maintained.

Identification and Traceability of Externally Provided Products or Services

Requirements for identification and traceability shall cover also products and services that are provided externally. By introducing the purchasing process to traceability control, the organization can trace back all the products or services that were realized using externally provided products or services. The main purpose is to provide the ability to trace back all purchased products. When the externally provided products or services are considered as critical to the quality of the product, and it is necessary to control those components, materials, or purchased processes, the traceability of the purchased products would begin from the purchasing order and continue till the final product itself (after delivery).

Configuration Management

Configuration management is a method that can be used to meet the product identification and traceability requirements specified in the ISO 9001 Standard. Configuration management is a means through which the organization may trace its products and maintain information regarding them. Implementing the configuration management shall provide full transparency of status and the components of which a product is comprised. Anyone who inspects the product could extract the information that was documented and trace its history. Another objective is to eliminate the possibility that wrong components could be applied to the product. The activities for configuration management must be planned in accordance to the nature of the product and its constituents. Here is a short review of the configuration management principles:

- A unique identification of each item under configuration management.
- A unique identification of the items that construct the unique product is to be planned (e.g., material, components, and parts) where the items are distinguished by unique, durable identifiers or markings, where appropriate.
- A unique identification of the version or edition of the product and its attribution to a group or category of products.
- Identification of the status of the product in its life cycle: design and development, realization, delivery, service, disposal, out of date, and so on.
- Identification of the changes or updates that were introduced to the product.
- Tracking of points of time where changes or updates were introduced.
- Definition of activities, responsibilities, and necessary tools and their use for updates and changes in the product.
- Introducing controls and tests to the product after changes or updates.
- Definition of release activities after update or changes.
- Definitions of the necessary documented information to support the configuration management.

Documented Information for Enabling Traceability and Identification

Where traceability is a requirement (thus affects the quality of the product), it is necessary to retain the documented information for enabling traceability. The objective of the records is allowing the organization the ability to relate a unique product to its historical information. In plain words, when you sample a product with a serial number, for example, it will be possible to identify information about the product, its realization, and its version and track its records and data effectively. So, by reviewing a single product, one can trace back (through a logical link with identifiers and numbers) all the records related to the realization of the product. Examples for such records are as follows:

- Tags
- Tickets
- Stickers
- Packages
- Signs
- Forms
- Labels
- Shipping documents
- Marks
- Barcodes
- Stamps
- Etiquettes
- Quality protocols

The type of record and its application on the product shall be defined in accordance with the nature of the product, and the application will not harm the product or its intended use. This documented information shall maintain several types of data and information:

- General information about the product, for example:
  - Identification number
  - Title
- Information regarding the realization of the product, for example:
  - Batch/lot number
  - Serial number
  - Dates like validity date
  - Regulatory or statutory requirements
- Information about changes, updates, or configurations, for example:
  - Version of the product
  - Revision status
  - Records of change history

Which situations may require documented information?

- When identification and traceability of parts or materials that are used to realize the product or activities that constitute the service are critical to the quality of the product, define the necessary documented information to trace and identify them.
- When the process environment may affect the quality of the product and must be traceable, the organization shall define which documented information will support this traceability.
- When traceability is a requirement and the organization distributes the products through distributers, they will maintain the appropriate documented information to support this traceability.
- When localization of the product is required for its realization, traceability may include data and information such as addresses and contact details, for example, in services where it is required to know where the product is situated.
- When it is required to identify process outputs before they are submitted to another process, the appropriate records shall accompany the process outputs.

These records are to be protected from unauthorized modifications and be submitted to the control of documented information as suggested in clause 7.5—Documented information.

8.5.3 Property Belonging to Customers or External Providers

Property belonging to customers or external providers refers to objects that were provided to the organization without charges and must be used in the realization of the end product. Property belonging to customers or external providers must be handled with care, cautioun, and responsibility. Such property may play a vital role in the realization of the product and may gravely influence the later production stages and quality of the product. The organization must take the necessary precautions in order to preserve this property. The ISO 9001 Standard requirements include the following:

- The organization shall handle customer or external provider property while it is under the organization's control or being used by the organization, with care, caution, and consideration of predefined requirements.

- The organization shall define a method for the identification and verification of customer or external provider property.
- The organization shall define activities and controls for the protection and safety of customer or external provider property.
- The organization must define communication channels with the customer or the external provider to notify them when its property has been damaged, lost, or found unsuitable for use.
- The organization shall retain documented information regarding the communication with the customer or external provider.
- Note—customer's or external provider's property can include materials, components, tools and equipment, premises, intellectual property, and personal data.

Property Belonging to Customers or External Providers

Property belonging to customers or external providers refers to property, for example, goods, authorization to use premises, or data, that were delivered cost-free to the organization for further processing: for use during the realization processes, for incorporating into the product, or for the support of realization processes. The owner of this property is the customer or the external provider. The requirement applies for anything that they provide:

- Raw materials or components supplied for inclusion in the product
- Packaging materials
- Products supplied for service, repair, maintenance, or upgradation
- Products supplied for further processing (e.g., coating or painting)
- Design and development tools and software that the customer lent for a defined period of time
- Production tools, molds, and equipment delivered for work
- Monitoring and measuring tools and equipment for the measurements of particular products
- Knowledge in any form or kind and on any media or carrier
- Goods that are stored for consignation
- Software provided to the organization (e.g., terminal to the supplier's system)
- Diagrams, product plans, and technical specifications
- Customer's intellectual property
- Customer's or supplier's premises when a service or installation is provided on the premises of the customer, including the use of customer's facilities and information
- Customer's property that has been delivered to the organization for service activities
- Personal data or confidential information
- Trade secret information
- Intellectual property such as texts, patents, or professional contents

Agreement

The most effective way to define the scope, conditions, restrictions, and requirements regarding property belonging to customers or external providers is to limit the issue within a contract or an agreement signed by the organization and its partner. This will ensure that both sides have reviewed the matter and the requirements; have defined the needs, controls, and activities; and have reached an understanding regarding the care and exercise of property. This agreement may also appear on the purchase order or customer order as defined text blocks.

Verification of Property

The organization shall define acceptance and verification activities of property belonging to customers or external providers. Once the property enters the organization, these activities will be carried out, and a documented approval will be produced. The objective is to create control of incoming property before it is submitted for the realization processes. If problems arise, they could be detected earlier. For each type of property, the organization shall define a specific incoming test. The definition shall relate to issues such as

- Responsibility: for each type of property, a responsible party shall be assigned for the control and verification. Naturally, the responsibilities shall be assigned according to the type and nature of the property.
- Compatibility of the delivery: The first test would be to examine that the expected or agreed goods are really what was delivered.
- Quantity: An examination of quantity will be defined.
- Intactness after transportation: An inspection is required for the verification that nothing was damaged during transportation.

This test shall be documented on designated forms or checklists. Fulfilling this requirement would save you a lot of controversies later on.

Identifying the Property

All properties belonging to customers or external providers must be identified throughout the organization premises, and for each type of property, it is required to define the type of identification. The method for identification shall be determined according to the nature and characteristics of the goods. How you shall identify the property may vary according to the following cases:

- If you are managing a customer's mold for plastic injection, a tag or a sign that will be visible on the mold can be applied.
- If you are saving a technical diagram on the company's server, it is required that you identify the folder where it is filed.
- When this diagram is printed out, it shall carry identification details enough to assign it to a specific customer.
- If the customer delivered the diagram without details, you may add a staple or write it by hand.
- When you manage for each customer a different product with different item number, the product number on the diagram may serve as an identifier.

Protection and Preservation of Property

The organization shall plan and implement protection measures for property belonging to customers or external providers in order to safeguard it from any harm, danger, or potential danger. Dangers for property may come in various manners and cases. In order to fulfill these requirements, a few measures will be initiated:

- Capturing specific requirements for handling the property, such as instructions for use, treatment, storage, or maintenance
- Defining specific requirements for internal or external transportation
- Identifying risks that the property may be exposed to while being stored, realized, or used

These requirements could be critical, for example, when service is provided at the customer's premises. The requirements are to be documented and available to the relevant personnel in order to implement and control them effectively. When it is required by the customer, the external provider, or a relevant regulation, the organization shall maintain the proper documented information as evidence.

Storing of Property Belonging to Customers or External Providers

For each type of property, you shall define the storage activities, conditions, and controls. In particular, the organization shall review the environmental conditions in which the property is stored. It may be that one will need to evaluate the type of goods and conduct a small-scale risk analysis in order to examine which environmental conditions may affect the quality and integrity of the products or equipment. The customer or the external provider may deliver goods of a new type and nature to you that require special conditions. Where maintenance activities are required, they will be identified, and I recommend that they should also be documented.

It is most applicable when the partner delivers equipment or tools for realizing the product, which needs periodical maintenance. In such cases, you need to clear the matter with the partner and identify all the maintenance needs. The appropriate personnel shall be identified, and the requirements will be made available.

Service Operations

Handling customer property refers also to service and maintenance activities that are performed, whether in the organization's facilities or at the customer's premises. Service is a complex issue with critical implications. It is advised that one conducts a review of one's service activities in order to examine how one may deal with customer property and to define the appropriate controls. If products are delivered for service and repair, the requirements apply to them too. The property of one's customers must be properly identified, safeguarded, and preserved. The organization must define how it shall maintain the requirements:

- If you are operating a service team that visits the customer's premises, it shall exercise care and caution when handling the product and the surroundings where operations are being conducted.
- If you are operating a service center that receives products for service or maintenance, the personnel that operate the center will be introduced to the controls of the customer's property including identification and traceability.

Notification to the Customer or the External Provider

The standards demand that one notifies the customer or the external provider in cases where their property has been damaged, lost, or found unsuitable for use. The matter requires that one develops a communication channel with one's partner concerning the issue:

- The organization shall maintain the partner's details needed for the establishment of communication: contact person and communication details.
- The organization shall appoint an authorized person for the notification.
- Documented information of this notification and approval of acceptance must be retained.

Notification may be performed with any of the traditional documented communication tools: fax, mail, and e-mail. Telephone notification alone may not be sufficient since one is required to maintain documented information. In which cases are you required to notify the partner? The general principle is very clear: where property is damaged, lost, or found unsuitable for use. However, the question is as follows: Which actions or activities may create such situations? These include

- Damages that occurred as a result of transportation activities
- Delivery from the partner of wrong goods or wrong quantity
- Delivery from the partner of defective property
- Quality problems that occurred vis-à-vis the goods or goods that do not meet the requirements
- Goods or property with expired validity date
- Goods that require rework
- Damages that occurred during the realization processes or storage
- Safety issues (property was harmed or stolen)

The ISO 9001 Standard requires retaining documented information that describes the occurrence and provides information that judged the property to be unsuitable for use. An effective way will be to deal with such cases like a nonconformity and to submit the occurrence to the process of handling nonconformities. Sufficient documentation will follow naturally.

8.5.4 Preservation

Preservation refers to the conformity and integrity of the product. The requirements for preservation of products are valid to all realization processes and span the time the materials or parts are received in the organization and continue to when the product is delivered to the customer or maintained at the customer's premises, and in some cases the control may extend to service and maintenance activities. The ISO 9001 Standard requirements include the following:

- The organization shall act to preserve process outputs during the realization or the service provision.
- The preservation of process outputs shall ensure their conformity to the requirements.
- Note: Preservation can include identification, handling, contamination control, packaging, storage, transmission or transportation, and protection.

Preservation of Product or Service

Controlling and maintaining preservation of the product deals with the identification and management of all factors that may adversely affect the product, its constituents, or the operations of service provided: personnel, infrastructures, or work environment. Damages may occur due to realization processes, transportations, storage, time deterioration, service activities, or other harmful factors, such as

- Mechanical factors: movement, vibration, shock, or abrasion
- Cleanliness factors: segregation from other products, dust, and particles
- Environmental factors: corrosion, temperature variation, electrostatic discharge, or radiation
- Human factors: handling the products without care or improperly

The organization is to provide instructional requirements for product preservation. The requirements for the preservation shall be planned in accordance with the nature of the product, its components, or the work processes and will:

- Identify products that are required to be preserved or areas that are required to be controlled
- Identify responsibilities and authorities
- Determine necessary trainings and qualifications
- Describe activities for preservation
- Create necessary documentation to support those requirements
- Define the requirements for records and evidence that preservation activities are being performed when the preservation is critical to the quality of the product

Reference to Risk-Based Thinking

When planning the activities for product preservation, the organization must refer the outputs of risk-based thinking and implement the suggested controls and preventive measures in order to reduce risks related to product preservation.

External Interested Parties

All of the requirements mentioned in this clause are applicable to external parties that may take part in the product realization. These parties are obligated to implement the required activities, use the documentation, and provide the appropriate records:

- Suppliers of services, processes, storage, or transportation
- Distributors that store, provide services, or transport and deliver the product
- Customers that use the product temporarily (e.g., renting)

Storage Activities

The storage of the product will prevent mix-ups, contamination, or any other adverse effects that may harm it or its characteristics. Storage requirements shall relate to all materials, parts, components, packages, and accessories related to the realization of the product and that may affect its quality.

Storage may be implemented using segregation areas, special areas, and the identification of products and products' usability status. Where required, environmental control systems shall be installed. When storage is outsourced and the organization uses suppliers, their services and activities must be included under this control and the measure shall be implemented at their premises too. The suppliers must provide evidence that they have followed all the requirements. The storage conditions must be defined, and the following issues shall be referred to:

- Work environment and storage parameters such as temperature, humidity, cleanliness, and dust (particles count) shall be defined and, when necessary, documented.
- The storage facilities will not harm the product and will provide it with sufficient protection to preserve its characteristics (i.e., intended use, functionality, performance, and safety).
- Special tools or equipment required for the storage of different types of parts, components, or materials shall be defined and allocated.

When necessary, documentation of those requirements shall be defined and distributed to the right persons at the right process stages, for example, printing on a receiving slip the storage instructions of incoming goods. When possible you may demand such instructions from the supplier.

Packaging

Packaging activities, operations, and conditions must be verified in order to ensure that they do not harm the product or influence its quality. For example, when you are packaging the product in a container, you are required to prove compatibility between the container and the product; that is, perform a test that the container does not harm it or alter its characteristics. In some cases the matter must be validated and supported with evidences. The organization is required to implement controls and verify that the product is packed according to specifications. These tests are normally done and verified during the design and development stages. But still, it may be that the customer requires a new form of packaging that has not been verified.

Security, Safety, and Protection

Product or their constituents may be exposed to external factors, which are not part of the realization processes like theft or damage from unauthorized persons. Security and protection of the product must be defined, and appropriate measures and definitions shall ensure the safety of the product. Activities such as handling of product, components, materials, and all other items related to the product that may affect its quality including controlled access to products and safety measures shall be implemented:

- Identification of devices, components, parts, or materials is needed in order to indicate their usability status and allow the implementation of suitable preservation measures.
- Authorization and access to the products, components, parts, or materials shall be defined and implemented according to organizational rules.

- Safety measures shall be taken regarding storage facilities.
- Protection measures to preserve the product during realization processes are required.
- When required, documented information shall be maintained: definitions of the measures and evidence for practicing.

Transportation

Transportation conditions must be defined and documented. Preservation of the product during transportation means undertaking all necessary measures in order to protect the product and its package and preserve their characteristics:

- The product, its characteristics, quality, and package shall be protected from external factors during transportation.
- Special tools, equipment, or accessories required for transportation activities shall be defined and controlled.
- Only certified employees may perform certain transportation activities using designated tools and equipment (cranes, forklift, etc.).

Transportation refers to external transportation, such as by air, sea, and land, as well as internal, that is, performed within the premises of the organization (as by forklift, or between processes). Each transportation type shall be analyzed, and suitable actions and activities shall be planned. Requirements for certification may appear on work instructions. The definitions shall be appropriate to the nature of the product.

Products with a Limited Shelf Life

Special attention shall be given to products with limited shelf life. The objective is to eliminate the use and distribution of products whose characteristics have been damaged or altered. The characteristics and quality of products with an expiry date deteriorate over time. Such products have a defined period of life with an exact date indicating when the status of the product changes and the product can no longer be used, in other words the exact date on which the product is declared as nonconforming. From this date on, the product shall be treated, handled, and disposed of according to one's specifications of control of nonconforming products, as required in clause 10.2—Nonconformity and corrective action.

This requirement refers to products that the organization produces and realizes as well as purchased goods stored in the organization facilities (that have a limited lifetime). These products will be identified, indicated, and treated with different measures. In practice the organization shall make sure that

- Those products are identified
- Effective controls, activities, or operations over these products are implemented
- The usability of those products is evaluated before submission to realization
- When the product is no longer valid, it is notified to the person who needs it, an activity is taken to alter the status of the product, and it will be disqualified for use or distribution (as a nonconforming product)

When necessary documented information shall be maintained. The documentation may appear on a procedure or a work instruction and shall include and describe specific control activities. When necessary evidences and records will be defined to assist and maintain this control. Types of possible evidences include

- Labeling on the products
- A designated log for the product
- An inventory or stock management system with management of characteristics

When the product is distributed or delivered by an agent of the organization or distributors, they must follow these requirements:

- The appropriate requirements and documentation shall be available to them, or they will maintain similar requirements.
- They shall provide the organization with evidence that these requirements have been applied.

Products with Special Storage Conditions

Special reference shall be given to products requiring special storage conditions. These are products whose characteristics or qualities might deteriorate due to a failure to provide them with specific and defined storage conditions; thus, the status of these products changes and their use is forbidden. In other words, failing to comply with these specifications may result in a situation where the product will be nonconforming; and since the organization cannot allow such a situation, it must ensure the implementation of all necessary measures in order to create the optimal conditions and, of course, provide suitable evidences. The requirements refer to products that the organization produces and realizes as well as purchased goods stored in the organization facilities. The requirements refer to materials, parts, components, or packages. These products will be identified, indicated, and treated with different measures.

Special conditions may be required during the storage, realization processes, or transportation activities. These conditions and their controls are normally defined during the design and development stages. When planning activities for handling special storage condition, the organization shall refer to the following issues:

- Such products shall be identified.
- The special storage conditions shall be defined.
- Measures in order to provide these conditions shall be planned and implemented.
- Tools, equipment, or facilities necessary for maintaining these conditions shall be identified and allocated.
- Effective control activities for ensuring these conditions shall be planned and implemented.

When necessary, documented information shall be maintained and may appear on a procedure or work instruction. The documentation is to include and describe specific control activities. Records and evidence of maintaining the requirements shall be kept:

- Environmental conditions
- Performance of activities

When deviations or departures from these conditions are identified, the product shall be reevaluated and the usability status reexamined. The examination may be performed according to specified criteria or designated quality tests (e.g., worst case scenarios). The reevaluation shall be documented and records shall be maintained.

When the product is distributed or delivered by agents of the organization or distributors, they must also follow these requirements:

- The appropriate documentation shall be available to them or they will maintain similar documented requirements.
- They shall provide the manufacturer with evidence that these requirements were applied.

8.5.5 Post-Delivery Activities

The realization of the product may not be completed with the delivery of the goods or the services to the customer and may require further actions. These actions are referred by the standard as post-delivery activities and include operations like maintenance plan, support, or service contracts. The ISO 9001 Standard requirements include the following:

- The organization shall meet requirements for post-delivery activities associated with the products and the services.
- While identifying the requirements and planning the activities for post-delivery activities, the organization shall refer to the following issues:
  - Statutory and regulatory requirements
  - The nature, use, and expected lifetime of the product or the service
  - The potential undesired consequences associated with its products and services
  - Customer requirements
  - Customer feedback
- Note—post-delivery activities may refer to activities that are considered under warranty, contractual obligations, maintenance engagements, and additional activities such as disposal and recycling.

Definition and Planning of Post-Delivery Activities

The first operation performed on the product after its delivery to the customer is regarded as a post-delivery activity. While managing and controlling those activities, the organization shall plan and provide an overall description of the actions to be carried out including the sequence of events. The objective is to provide representatives of the organization, for example, a service technician, with detailed specifications regarding what has to be done in certain events, conditions, and cases concerning a product or one of its constituents. The organization shall plan and implement a process for managing the post-delivery activities. This process has the following goals:

- Clear identification of the product that comes under post-delivery activities
- Identification and analysis of the need for post-delivery activity (e.g., periodic maintenance, warranty claim, service, malfunction, or disorder)
- Planning of the activity
- Identifying the events, conditions, or business cases in which these activities are required
- Verification or validation of the activity
- Informing the customer of requirements for post-delivery activities when such are not known to the customer but are necessary for the specified intended use
- Documenting and reporting

The planning of the activities shall refer to the following issues:

- Identification of the product that is under the post-delivery activity.
- The definition of responsible personnel to perform the activities, including required qualifications.
- Planning and specifying the activities according to events (e.g., periodic service or required maintenance).
- The definition of accessibility to the product, with locations or the need for authorizations.
- Indicating the necessary documentation, such as technical specifications, drawings, plans, or user manuals.
- Identifying relevant regulatory requirements that may dictate certain activities.
- The specifications of tools and equipment that are required to be used during the activities.
- The acceptance of validations and verification of the activities.
- Ensuring product preservation; the product features will not be harmed during the activities.
- Ensuring preservation of property belonging to the customers or external providers.
- Defining the need to provide training associated with the specific post-delivery activity.
- Reference to relevant documented information needed to support these post-delivery activities.

Types of post-delivery activities include the following:

- Service
- Maintenance
- Actions under warranty provisions
- Supplementary services such as recycling or final disposal
- Product modifications or updates
- Help desk
- Support of any kind
- Trainings

Responsible Parties

Since the post-delivery activities such as service or maintenance can be complex and may have implications, for each activity the personnel that will perform it must be defined with the required qualifications. Naturally, the suitable trainings will be defined in the QMS, as required in clause 7.2—Competence—and incorporated into the training program.

When the service or maintenance is performed by a third party (the customer or supplier), the organization must define in detail which post-delivery activities are expected and provide them with the necessary documentations answering the requirements mentioned earlier. When the organization is outsourcing the service activities, it is the organization's responsibility to control and verify it. Thus, it is required that the organization train the supplier appropriately in order to qualify and control its services. The performance of the supplier during the service activities will be evaluated and included in the general evaluation of the suppliers, as required in clause 8.4.2—Type and extent of control. The supplier is obligated to perform all the service requirements as specified by the organization and to provide the necessary evidence and process validations.

Time Intervals for Service and Post-Delivery Activities

The intervals for carrying out post-delivery activities such as service or maintenance are to be defined for each product, part, component, or service. The definition for the period may be

- An output of the design and development stages
- Instructions of the supplier from whom the organization purchased the product
- Instruction of a third party for which the organization provides services as a subcontractor

The definition shall refer to the identification of the product; required service or maintenance activities; documentation of the date when the activity was performed, for example, when the service was delivered; and schedule for the next activity. Another way to define it is to maintain a service or maintenance plan. The performance of the periodic service activity shall be visible on the product, for example, like a sticker with the date of the service, identity of the person that performed it, and the date of the next service.

Status Identification

During post-delivery activities (service or maintenance), an indication regarding the usability of the product will be determined, that is, whether it is usable, in repair, or defective. The status will be clear to all users, that is, on a location or with means where all users may view it. It is also possible to deactivate or disable only a part, or a function, of the product as long as the intended use, functionality, performance, and safety are not affected; but do not forget to identify the status. When necessary the action of deactivating only a part of the product shall be planned, reviewed, approved, and documented by an authorized person before allowing a service employee to perform the deactivation.

Service and maintenance activities may create defective parts or devices. Therefore, it is required that you initiate the interface between the service and control of non-conforming products.

The Nature, Use, and Expected Lifetime of the Product or the Service

The post-delivery activities are decided according to the nature, use, and expected lifetime of the product or the service. Here the organization must identify in which situations or conditions the post-delivery activities are critical or necessary for maintaining the expected lifetime of the product and initiate those activities. What types of cases may we encounter?

- Cases where the organization must complete the realization activities after the product was delivered to the customer, for example, configuring the product after installation
- Cases where a periodic service or maintenance is necessary in order to ensure the functionality and intended use of the product, for example, a periodic service for a lift that ensures its safety

Those activities are to be

- Planned, for example, on a quality or maintenance plan
- Communicated to the customer, for example, on a contract or an order (see clause 8.2.3—Review of the requirements for products and services)
- Scheduled
- Recorded or documented as documented information

Reference to Regulatory and Statutory Requirements

When national or local regulations require a specific activity to be carried out or specific data to be recorded during the post-delivery activities, this requirement will be implemented and the coverage of required data will be incorporated into the activities. Regulatory and statutory requirements, in the context of post-delivery, aim to mitigate ill effects that may result from post-delivery activities. For example, in the medical branch, such requirements refer to special reporting in case of malfunctions or disorders in the medical device, or when its performance might pose a hazard to users or patients when such are detected during service or maintenance activities.

During the planning of the activities you will need to identify which requirements are dictated by national, regional or international regulations and apply to your product, and to implement the demands into one's post-delivery processes: what information is to be gathered, how the information will be forwarded, and the subjects of the report.

Potential Undesired Consequences

Post-delivery activities may be associated with risks concerning the use of the product or its handling after delivery: a service, repair, or maintenance may affect the product directly or their components and thus affect its functionality. Therefore, the risks during the post-delivery activities are to be reviewed and addressed. In other words, it is necessary to analyze and understand how post-delivery activities including service and maintenance may cause nonconformities. The results of this review are the appropriate controls to be implemented, for example, user manuals, trainings, or web page support online. These controls will be reviewed and approved by authorized personnel.

In practice, I recommend that each post-delivery activity will be reviewed by an appropriate employee in order to validate and verify it. Thus you may ensure that certain activities will not affect the product or other components, parts, or processes in the product. The evaluation of the activity will be cross-referenced and will include all relevant aspects of the product.

What quality tools do we have in order to control potential undesired consequences?

- Verification—verification is necessary to ensure that all the required post-delivery activities were performed by the appropriate person, on the appointed time, and on the designated product. In case of service or maintenance activities, for example, a sticker will be placed on the product indicating that maintenance was performed along with the next schedule for treatment. The sticker is a classic quality record. A checklist specifying all the necessary actions is also acceptable for verification. The verification will be documented.
- Validation—validation, on the other hand, is to ensure that after post-delivery activities have been carried out, the product still functions according to its requirements, that is, intended use, functionality, performance, and safety. This is more complex. In order to ensure its effectivity, plan a validation test for every service activity. Measurement procedures and material references will help. The functionality of the product must be validated after the service activities. The validation will be documented. I refer to chapter 8.5.1 paragraph Validation and periodic validation for more details on how to implement validation effectively.
- Monitoring and measuring devices—in some cases monitoring and measuring devices are used to measure outputs of processes during the post-delivery activities such as service and maintenance activities. These devices must be calibrated and controlled. Their use will be documented in order to prove their compatibility. Documentation may indicate their traceability, for example, an internal number and serial number. In this manner, you prove that the monitoring and measuring devices were qualified during use.

Updates and Product Modifications

Modification of the product or a part of it may influence its functionality, its intended use, and its safety or may cause negative implications on the quality of the product. When post-delivery activities create a change or modification in the product, you will need to evaluate the implication on the product and control, verify, and validate that the functionality, intended use, and safety were not affected by the change. For example, an update of software may bring new functionalities or improve others and must be appropriately communicated to the customer. Management of product configuration is an effective tool for controlling such changes or updates and was discussed in detail in chapter 8.5.2—Identification and traceability. This topic was also discussed in chapter 8.3.6—Design and development changes—where the evaluation of changes in the product and their implications must be identified and controlled. Those outputs of the requirements mentioned earlier (8.3.6 and 8.5.2) should be incorporated into the post-delivery activities.

Reference to Customer Requirements

Some post-delivery activities shall be planned according to customer demands for operations after the product was already delivered. Those requirements were referred to while reviewing the requirements for products and services (clause 8.2.3) where the organization is required to confirm that it has the ability to perform those requirements before forwarding the request for realization. The bottom line—before you are realizing post-delivery activities that are required by the customer, you must ensure the following issues:

- When post-delivery activities are part of the product or service or the customer demands them, they shall be reviewed.
- The organization shall review the requirements for post-delivery activities prior to the engagement with customer.
- The organization shall ensure that differences between the specifications conveyed by the customer and the actual requirements submitted to the customer are settled.
- The organization shall ensure that the requirements for post-delivery activities are defined, clear, and understood.
- When the customer does not provide documented requirements or specifications, the organization must obtain an approval before acceptance (performing the activities).
- The results of the review shall be maintained as documented information (according to clause 8.2.3).
- Where and when new requirements for post-delivery arise from the review, they shall be maintained as documented information (according to clause 8.2.3).
- Where the customer is responsible for certain activities, they will be clearly defined, for example, allowing access to certain locations or information or preparing for the post-delivery activities.

These requirements are supposed to be available in QMS on tenders, contracts, or order conformations. The most effective way to handle this requirement is to transfer information from those tenders, contracts, orders, or changes to contracts or orders to the quality plan or post-delivery plan (maintenance or service plan).

Reference to Customer Feedback

Post-delivery activities such as service maintenance or warranty claims are to be used as a means of identifying quality problems in the product related to the postproduction phases. The issue relates indirectly to clause 8.2.1—Customer communication—where it is required to obtain customer feedback relating to products and services and collecting data and information regarding the use and performance of the product through the planning of communication channels. Data and records related to the post-delivery activities are one way to reach this goal. During the post-delivery activities, one is required to document the problems or events, the causes or factors, and the corrective or preventive measures that were carried out. One may manage the maintenance history of a product. The following details for records are just a suggestion and may be incorporated on a format that will gather the following feedback information:

- The unique identification of the product (product number and characteristics such as serial number where it is applicable)
- The identification of post-delivery activity (e.g., service call number, service ticket number) and date of service
- Identity of the personnel who performed the activities
- The identity of the contact person at the customer's premises who received the service
- A description of the activities that were performed
- Identification of designated tools or equipment used to perform the activity, such as measuring and monitoring devices (for traceability needs)
- A list of parts or components that were repaired, maintained, or replaced, including identifiers such as model, catalog numbers, serial numbers, or batch numbers
- The tests and inspections that confirmed that the post-delivery activity achieved its objective (verification or validation)
- Description of how and where the product or the service did not meet its intended use or specifications and the events that brought about this situation

Communication with the Customer

The organization must develop and maintain a process, an interface, and a communication channel between the different interested parties of the organization (internal such as service teams as well as external such as subcontractors) and the organization itself. The goal of this process is to enable the reporting of problems and issues that were identified during post-delivery activities or were reported by the customers. The methods for the communication shall be planned according to the nature of work of those interested parties:

- Identifying the internal and external parties that are involved in post-delivery activities and their particular needs
- Identifying resources available to address communication
- Identifying and selecting possible communication methods

Here I will refer to clause 8.2.1—Customer communication—where one is required to define and establish communication channels between customers and the organization regarding post-delivery, among other issues. With respect to service and maintenance, the organization needs to define communication channels, interfaces, and processes covering both transmission of information and execution of activities:

- Transmission of information: definition of how a customer may contact the organization or service provider in case of need, for example, via e-mail, fax, telephone call, and company website, and which information must be gathered or shared in order to provide efficient service or maintenance.
- Execution of activities: definition of how the provider can access the customers in order to perform the activities, for example, in-person visits, service center where the product is delivered for repair, and telephone support.
- It may be that the organization will have to initiate contact with the customer when maintenance for the product is periodically required.

When a contract or an agreement for customer support services exists, the organization shall ensure the following issues:

- Identify which organizational units are involved; it may be that not only the customer support is taking part in the processes or the services but also units that are responsible for the realization of the product.
- Identify all the resources that are necessary and available for managing customer service applications.
- Plan and arrange training for the involved persons of how the processes, operations, and activities of the service are to be performed.

Issues to be addressed are as follows:

- Planning of service routes or courses
- Requirement for training users of the product
- Management of supply of spare parts
- Technical support
- Verifying qualification of employees

Documented Information for Post-Delivery Activities

There is no specific requirement for maintaining and retaining documented information for the post-delivery activities, but when such documented information may influence the quality of the product and the achievement of customer requirements, one may need to manage it. In other words, it is up to one to decide whether documented information is required. If one is managing the documented information, it will be made available to the relevant parties in the appropriate places and at the right times or events.

The objectives of the documented information are as follows:

- To provide the ability and means to perform post-delivery activities according to a plan
- To prove that post-delivery activities were performed as planned and the results are satisfying

Relevant parties may be members of the organization like service technicians as well as the customers. An appropriate place can be the company's server, help module on the user's application, technical files, web pages, accompanying documentation for technicians, and service manuals for customers. The right time or event is the definition of the use case in which the documentation is to be used. The documentation may include

- Procedures
- Work instructions
- Reference material
- Troubleshooting documents
- Reference measurement procedures
- Quality plan

Evidences of post-delivery activities may be required when it is needed to prove that the activities have been done according to a predefined plan and were verified, controlled, and accepted and the product is approved for use.

The history or log of post-delivery activities of the product plays an important role when servicing or maintaining it. The history describes the functionality and performance of the product concerning its intended use. Thus, the history of servicing and maintenance is to be documented and managed.

8.5.6 Control of Changes

Clause 8.5.6 refers to changes that may occur in the controlled conditions of the realization or service provision and which the organization is committed to provide. The objective is ensuring quality requirements after changes have been submitted to the realization or service provision. It is important to understand how the change may affect or impact other aspects or conditions of the realization and to react accordingly. Such monitoring should enable the organization to identify, assess, and manage the changes in the conditions of the realization. The ISO 9001 Standard requirements include the following:

- The organization shall review and control changes for production or service provision.
- The control and review shall ensure continuing conformity with requirements.
- The organization shall retain documented information describing
  - The results of the review on changes
  - The person or persons authorizing the change
  - Any necessary actions arising from the review

Understanding the Changes in Conditions for Realization

Clause 8.5.6—Control of changes—deals specifically with changes that relate to the conditions of the production and service provision, in other words those conditions that affect the operations and activities that consist of the production or the provision of service of the product. Such changes may influence the interaction between processes or activities and the workflow between the different related areas in the organization and affect the specifications of the product. Thus, such a sequence of events may affect the quality of the product. When planning and executing the realization of the product, one provides certain inputs to a process:

- Materials or components
- Technology
- Knowledge
- DI
- Resources
- Work methods
- Human resources
- Infrastructures
- Process environment

When the realization begins, respectively, when you process the inputs, they are going through the process stages. Each process stage uses its necessary inputs (mentioned earlier). The end result is process outputs according to a specification. When one of those conditions is unexpectedly changed, for example, a breakdown of infrastructures, absence of employees, or delivery failure from one of the suppliers, it will have implications on the final result of the process, and the goal of clause 8.5.6 is to control and manage those changes.

Identifying What Was Changed

Change is regarded as any modification to an element or to a combination of several elements that may affect the realization and thus the quality of the product or the service or the ability of the organization to realize the product or the service according to the specifications. There are two types of changes:

- Proactive changes—changes that are planned. In such cases the organization shall control the changes that are implemented as planned.
- Reactive changes—changes that are being under taken in reaction to a change of situation or of conditions. For such changes, the organization shall review whether the consequences of the unintended changes impact the process or the product.

The proactive changes to the QMS and for the realization of the product will be addressed according to the requirements in clause 6.3—Planning of changes. The requirement of clause 8.5.6 deals with the reaction to unplanned changes—changes that were forced upon the organization—and the organization did not have the time to plan them and assess their impact. While controlling these unplanned changes, one reviews and assesses consequences that may occur after the change has been enforced. Possible reasons for changes are as follows:

- Changes requested by customer or dictated by the business environment.
- Update of regulatory or statutory requirements may mandate changes to the product or its realization.
- Technological changes may bring changes to the realization of the product.
- A key process indicator may indicate the need for changes in the realization.
- Corrective actions as a response to nonconformities.
- Changes originating from risk assessment, for example, risks related to changes in technology.
- Changes referring to specification or expectations of external providers or relationship with external provider.
- Changes in external documented information such as standards, technical specifications, or procedures.

Which process elements might be changed?

- New specification for the product due to customer request
- New materials or components that are submitted to a process
- Qualification of personnel that are realizing the product or providing the service
- New technology that is being used during the realization

- Lack of knowledge (due to changes in personnel)
- Update of documented information
- New resources or replacement of resources due to a breakdown
- Applying new work methods
- Changes in human resources due to absence
- Changes in infrastructures due to breakdown or malfunction
- Changes in the process environment

Reviewing of a Change

Changes in the conditions of the realization may affect the process and the process outputs and may have implications on the end result of the process and the quality of the final product. The intensity or effect of a change depends upon the nature and complexity of the related element, process, or its output. The organization is required to review and control the changes to the extent necessary to ensure continuing conformity with requirements. In order to determine the extent necessary to ensure continuing conformity with requirements, one must understand how the change may affect the product and its conformity. What does it mean to ensure continuing conformity with requirements?

- The change will not affect the quality and requirements of the product—process outputs, after the change was applied, still meet the predefined requirements.
- The change will not affect the integrity (the state of being complete or whole) of the process or the realization.
- The change will still enable the organization to meet the needs and expectations of interested parties.

Reviewing the change shall address the following process aspects:

- Inputs—does the change have any influence on process inputs? Are there any new inputs? Which documented information supporting the process is needed to be amended (where procedures, diagrams, work instructions, or forms are influenced) or amended (where records are affected)? How will these new inputs be transmitted?
- Does the change have any influence on tools and instruments and resources that are already being used in the process? Do these process elements have the capacity to support the new situation in the first place?
- What is the influence on activities? Do current activities need to be changed?
- Do any verifications or validations need to be changed as well?
- Were the risks relevant to the change and were the impacts of the change on other product or product realization elements addressed? It is important to understand the impact of the change on other elements involved in the realization.
- What are the new process outputs that resulted from this change?
- Are the actions required to implement the change carried out? Planned processes and activities, their sequences, and the interactions between them necessary to carry out the change must be reviewed.
- Were all the elements involved in the process where the change takes place reviewed? Is it necessary to identify all the elements that are involved and their interrelations in order to understand the influences of the change?
- Was the change communicated to all relevant participants of the process such as employees, external providers, and, when necessary, customers?

In order to evaluate the effect of change, one must have access to supporting data and information before and after the changes were applied. When a change is known to be implemented in the realization, I would use my existing quality tools to help in the review:

- Results of monitoring, measurement, analysis, and evaluation that will indicate whether process outputs still meet their requirements, for example, verifications of process results or validation of process outputs will indicate whether a change has affected the process or its ability to provide conformed outputs.
- I would initiate a process audit and evaluate whether the change still receives its needed conditions and practices its requirements and whether the ability of the changed process to achieve its goals is still maintained.
- Feedback from interested parties may provide indications whether their expectations are still being met after the change was implemented.

Retaining Documented Information

The organization must retain documented information regarding the review of the change. The documented information shall include

- Results of reviewing the change—here one will need to document whether the change achieves its objectives or not.
- Identity of the person that authorized the change—the goal here is to ensure that persons with the appropriate competence reviewed and authorized the change, for example, the process owner and head of the department, but bear in mind that the authority should be available to react and authorize the change quickly without the need to wait too long for its authorization.
- Documentation of actions that were derived from the change and its review.

To make the documented information practical and not complicated, you will need to integrate such documented information in your daily operations, which means allowing and training the persons who are operating the process to initiate or at least review the change. The authorization will be done by a responsible party. As a matter of fact, any documentation that will collect the information mentioned earlier may be suitable:

- The review can be designed as a checklist or questionnaire where all the issues mentioned earlier are listed. But allow space for references or remarks.
- If one is practicing a process audit in the QMS (see chapter 9.1—Monitoring, measurement, analysis and evaluation—paragraph Process audit), one may use this method for evaluating the results of the change.

Documented information of any necessary actions arising from the review refers to the following cases:

- The objective is achieved and the change can be fully integrated in the QMS—then one will need to update documentations, schedule trainings, and so on.
- The objective was not achieved and an improvement is required—then one will need to initiate the process once again or submit it for improvement.

- The nonconformity was not removed and one must address the issues once again—try to implement the change once again with modification or try it another way.
- Initiate contingency action (an output of the risk-based thinking), actions that must be undertaken in order to respond to unplanned changes or events.

One may incorporate it directly on the same documentation, but I suggest that one should refer to the relevant documentation. Each of the three (nonconformity, improvement, or change management) is an input to another quality activity.

Changes in the Quality Plan

The quality plan is an ever-changing document and might be impacted by unplanned changes. Therefore, it will be submitted to control that will check its accuracy and relevancy and will update it when needed. Where processes, operations, or activities may be changed over time or where conditions are changed, their relevant documentations such as procedures or instructions as well as relevant resources may need to be modified and updated.

These changes must be approved by all relevant parties taking part in the realization of the product and who influence the quality of the product. If required, changes will be submitted to a confirmation by the customers. In case the quality plan is controlled, it shall be submitted to the control of the documented information including as specified in clause 7.5.3—Control of documented Information.

Tracking Changes in the Realization of the Product

I recommend managing traceability of changes. The traceability will allow you to take one step back when a change does not achieve its goals or creates a quality problem. To deploy a change effectively, the organization should

- Identify the relationships between its processes and activities and the sequence and interactions of the processes
- Identify the risks that are associated with the change and plan the appropriate controls
- Keep track of what was changed or modified in those elements

8.6 Release of Products and Services

Release of the product or service refers to the required activities and their documented approval that products or services that are delivered to the customer fully meet their requirements. The objective of release activities is to ensure that the realization of the product or the service has been successfully completed and that the product or the service meets all its requirements and is ready for delivery to the customer. This will be done with the definition of final acceptance activities or final release tests, that is, activities that will ensure that the product or the service was realized according to the requirements and that all controls were implemented and the product meets the acceptance criteria. These activities will be supported with records. The ISO 9001 Standard requirements include the following:

- The organization shall implement arrangements and activities to verify that product and service requirements have been met.
- These activities and arrangements shall take place at the appropriate process stages.
- Release of products and services to the customer shall proceed when the planned arrangements have been satisfactorily completed.
- When the planned arrangements have not been completed, the release shall be authorized by a relevant authority or by the customer (when applicable).
- The organization shall retain documented information on the release of products and services. The documented information shall include
  - Evidence of conformity with acceptance criteria
  - Traceability to the person that authorized the release

Release of Products and Services to the Customer

Release activities refer to making products or services available for the use of the customer. The release activities will examine that the product was assembled or manufactured from conformed materials or components, all the necessary operations were performed (for goods as well as service provision), the required controls were applied, and their results are satisfactory.

The general idea claims that the realization of a product is combined from various elements. And at some point of the overall process, before delivery to the customer, the organization must stop and review the realization and decide whether the product meets its requirements. Thus, releasing a product or service to delivery to the customers means

- The realization of the product or provision of the service is completed; the product or service went through all the required realization stages and activities.
- Each process was controlled: tested, verified, validated, and approved according to specified criteria.
- All the required validation activities and quality tests (for products, parts, and processes), during and after the processes, have been completed, and the results are satisfactory.
- Actions derived from risk-based thinking have been appropriately implemented.
- The required realization records are completed and available.
- The product is identified, labeled, and packed as specified.
- The product is stored under appropriate conditions.
- The product was transported to the customer as specified.
- All the required records needed to prove that realization is completed.
- Traceability is maintained as specified.

When product or a service operation is updated or modified, their release to the customer will be controlled. All the interested parties shall be informed of how to release an updated product or service.

Actions and Arrangements for Verification of Product
and Service Requirements

Fulfilling the requirements of clause 8.6 is the definition, planning, and implementation of arrangements needed to verify that the product and service requirements have been met. As mentioned earlier, the organization must, at a certain point of the realization and before the final delivery to the customer, stop and review the realization and decide whether the product meets all requirements. This is why the review before release must cover all the elements that are involved in the realization of the product and may affect its quality. Results of the release activities may differ:

- Release of the product or service for delivery to the customer
- Release under concession for delivery to the customer
- Suspension or quarantine or declaration as scrap of a product
- Transfer to rework

These arrangements shall be integrated in the realization processes at the appropriate process stages. That means that although the product was delivered to the customer, in some cases the release activities will need to be performed after the delivery:

- A common example is installation by the customer—the goods were delivered but the total of the requirements include an installation. After that the product may be released.
- When the organization provides services, maintenance, or warranty operations, the release activities shall be integrated into them as well. For example, after a service was performed, the employee runs a list of tests before redelivering it to the customer.
- In some cases the environment, location, or conditions for the release will be exceptional, for example, the release must be conducted in the premises of the customer.
- Release activities may be externally required, for example, customer requirements or regulatory requirements. In that case those release activities must be planned and integrated into the realization of the product.

In order to maintain effective release activities, one must ensure that all required inputs and data are available to the relevant person at the time of the release. This information shall assist the person in deciding whether the product meets its requirements or not:

- Evidences that the product went through all the necessary process stages.
- Characteristic evidences that the product meets its requirements: quality tests, verifications, and validations are available and clear.
- All the required resources were available at the time of the realization.

I would integrate the release activities into the quality plan where it will be clear to the employees which release activities must be performed, which inputs are needed, in which process stage must it be taken, and which records are expected. The organization may use processes that already exist, for example:

- If one decides that after realization and the storage of the product on the shelves the product is released for use, one may define that the last operation shall be considered as release activity. In that case, one shall need to provide all evidences that the product meet its requirements.
- When the organization installs equipment or devices at the customer's premises and the customer approves and signs the delivery of the installation—this may be considered as a release activity and the signature and approval may be used as documented information for the release.

Acceptance Criteria or Standard Measures for Release

The release will be approved against defined acceptance criteria.

The objective of the criteria is to assist the person releasing the product to decide whether the product or service is ready for release.

The release activities must verify compliance to the product requirements, and therefore an acceptance criterion is needed. When releasing a product, certain tests and reviews of the functionality, characteristics, or intended use of the product must be conducted using standard measures, such as criteria. It does not mean that one needs to perform all the tests and verifications of the product all over again. One needs to provide the service person with a list of the tests or checks and the expected results for the release. The release will be authorized when the results are satisfying:

- For each type of product or service, criteria for the release will be established. The criteria will be planned to ensure that the product meets its requirements or specifications.
- The criteria for the release will be available to the appropriate persons at the appropriate process stage.
- The use of the criteria will be clear to the person who carries out the release activities.
- The stages in which the measurements shall be conducted shall be defined.
- The manner in which the results shall be recorded will be defined.

Developing an effective release test will cover the following issues:

- The release tests will ensure that the product does achieve its objectives and intended use.
- The release tests prove that all the specified characteristics are available.
- The release ensures that additional requirements are delivered like special delivery requirements or schedules.

In order to define release criteria and conduct the release effectively, I suggest to conduct a small-scale audit and develop a tool (a format, a checklist, or a procedure) that will measure and determine the extent to which the requirements of the product are fulfilled. What issues shall this audit cover?

- Identification of the product or the service that is to be delivered to the customer.
- Verification that all process inputs such as materials, components, or information were available for the realization.
- Verification that the employees realizing the product were qualified and trained.
- Verification that the equipment, tools, and infrastructures that were used to realize the product were controlled and whether there were any disturbances during the process.
- Verification that processes were verified and approved by authorized parties.
- Review of product and process controls and verification indicating that the results of the tests are within limits or tolerances.
- Verification that all the validation activities were performed and the results are satisfactory.
- Verification that the monitoring and measuring devices that were used to realize the product are maintained and controlled.
- Verification that the required documentation was available.
- Verification that packaging activities were controlled and that the product was packed in the appropriate way.
- Verification that the package was stored under appropriate conditions before delivery.
- Verification that the storage conditions are appropriate.
- Review of all records and verifications to ensure that they are complete and accurate.
- When quality problems are detected during the release activities, they shall be documented and managed.

Authorizing the Release

The release of the product or service shall be performed by an authorized person of the organization. This person shall authorize the release of the product for further use after a defined verification (the release activities) and with satisfactory results. The authorization of the release shall be retained as documented information. Documenting the person who authorizes the release provides traceability to qualifications and then accountability of the person. This traceability will assist the organization in case quality problems arise later.

According to the nature, type, or operation of the realization, a defined role will be designated for reviewing and authorizing the release. The definition of the authorizer may appear in a job description or a procedure. But in any case, the identification will be documented in a clear manner indicating that he or she authorized the release of the product: a manual signature on a form or allowing an authorized user to perform a function in the ERP system. This person may be an employee but also an external party like an external provider or a regulator. A good example is the construction industry where an external supervisor approves the results of the organization.

Concession for Releasing Products or Services

There are cases where not all planned arrangements have been satisfactorily completed, and as a result the product may not fulfill 100% its requirements but will nonetheless be released. This is also known as a release under concession. Concession refers to a documented approval for the release of a product or service that did not go through all the planned arrangements, at appropriate stages, and that all its requirements have been met. The release of products under such conditions is a state where although the product may not fully meet its requirements, under the appropriate approval, the release of the product for delivery was authorized.

The allowed deviation or tolerance must be defined whether by the customer or through a regulatory requirement (when the release is done according to regulations). A good example of the concession of a product is when a process was performed, but with deviations. These deviations were reviewed, and it was decided that they were not crucial, they do not affect the quality of the product, and therefore the product was approved and released.

The release under concession shall be authorized by a designated person in the organization. It may be that the person authorizing the release may not have sufficient qualifications to approve under concession.

When planning how concession will be managed in the QMS, one must refer to the following issues:

- It will be defined whose responsibility it is to review and approve a concession.
- Products that were released under concession must be identified. The identification shall link to the documented information of approval. The goal is to be able to locate approval of a certain product.
- A reference to the suitable records shall be available, for example, the results of the measurements and the justification for the concession.
- The approval will be accompanied with the identity of the approver.
- The concession and the release under concession shall be documented.

If a reported problem has a temporary solution or a work-around that does not impair the quality of the product and its intended use or pose any risks, this solution may be considered as a condition for the release.

Retaining Documented Information of the Release

The review, approval, and release will be documented and maintained as documented information. The goals of the documented information are to identify the release activities and to prove that the product requirements are met. The documented information will contain evidences of the release activities or at least the critical activities that ensure that the requirements of the product are achieved. The documented information will

- Define to the user of the QMS what the release activities are and which acceptance criteria he or she must use
- Document the execution of those release activities
- Provide evidences of conformity with the release criteria: the product meets its requirements

The documented information shall enable traceability of the release activities to the product or service operations, for example, by reference to a serial number, a batch number, or a service ticket number. The traceability will allow the users of the QMS to link the activities, the criteria, and the released product. Which information and data may be recorded?

- Unique identification of the product or service operation.
- The person who authorized the release.
- Data and information regarding traceability of the product such as batch number or serial number.
- Date of release.
- Relevant acceptance criteria (or reference to it) that were used to approve the release.
- Results of monitoring, measurement, analysis, and evaluation whose results were used to compare with the criteria.
- The results of the release—documentation of the actions that provide the user with understanding that the product and service requirements have been met.
- When monitoring and measuring devices were used in the release activities, it will be mentioned.
- When conditions of the environment for the operation of processes are critical for the release of the product, they will be documented.
- When necessary, according to the nature of the product, as per customer request or regulatory requirements, the documented information of the release will be retained for the lifetime of the product or for a predefined period.

8.7 Control of Nonconforming Outputs

As soon as nonconforming outputs are detected within the organization, they must be identified, recorded, and controlled; and most importantly, they must be treated. The objective is to prevent the release, delivery, or use of nonconformed products or services by the customers. Treating and managing nonconformities is one of the basic objectives of the ISO 9001 Standard. Nonconforming outputs refer to the outputs of processes that do not fulfill a requirement that is a need or an expectation of one of the interested parties of the organization that is stated, implied, or obligated. Controlling the nonconforming outputs is planning activities for detecting nonconforming outputs and deciding what is to be done with them. And the ISO 9001 Standard specifically mentions what options one has. Let us review the ISO 9001 Standard requirements.

8.7.1

- The organization shall ensure that outputs that do not conform to requirements or expectations of the interested parties are identified and controlled to prevent their unintended use or delivery.
- The organization shall take the appropriate action based on the nature of the nonconformity and its effect on the product or the service.

- Those actions shall refer and be applied to products or services that are being realized or were already delivered or provided to the customer.
- The organization shall manage nonconforming outputs in one or more of the following ways:
  - Correction
  - Segregation, containment, return, or suspension of provision of products and services
  - Informing the customer
  - Obtaining authorization for acceptance under concession
- When nonconforming outputs are corrected, they will be verified once again in order to ensure their conformance.

8.7.2

- The organization shall retain documented information for
  - Describing the nonconforming outputs
  - Describing the action or actions taken to manage the nonconforming outputs
  - Documenting accepted concessions
  - Identifying the authority that approved the actions for managing the nonconforming outputs

Detecting the Nonconforming Outputs

Controlling the nonconforming outputs will allow the organization to minimize their effect. Nonconformance can occur in raw materials, components, in goods that are in process or finished goods, or service operations. When nonconforming outputs are detected, the first step is to identify them. In order to effectively identify nonconforming outputs, one must understand which quality tools will indicate that a process output is nonconforming. The ISO 9001 Standard requires quite a few controls that may be of use:

- Verification activities—when verifications deliver unsatisfying results, it may mean a nonconformity. The verifications may refer also to conditions that were necessary during the realization but were not achieved, for example, cleanliness conditions— the product itself may be conforming, but the required conditions of cleanliness in which it should have been realized were not. Therefore, the product may not meet its requirements.
- Monitoring, measurement, analysis, and evaluation activities—results of monitoring, measurement, analysis, and evaluation activities may indicate that process outputs do not meet their specifications. If the quality test shows that a certain measure of the product is unsatisfactory, it may indicate that the output is nonconforming.
- Controls of external providers—when controlling products or services delivered from external providers, products or services may be found as nonconforming.
- Release activities—the release activities are a list of verifications that the product went through the necessary operations before it is delivered for the use of the customer. When the release activities reveal that certain activities were not performed or were performed but their results are not satisfactory, it may indicate that the

output is nonconforming. If a service summary report describes a problem in one operation, it may be indicating a nonconformity.

- Audits and process audit—during audits, product or process outputs may be sampled and found to be nonconforming.
- Risk-based thinking—during the deployment of activities of risk-based thinking, the organization may judge that certain risks may produce nonconforming outputs.
- Customer feedback—feedback from customers may point to cases in which nonconforming outputs were delivered to the customers.

The ISO 9001 Standard expects that this information will flow to the system of managing nonconforming outputs at the right time before the output is submitted to the next process stage.

Reaction to Nonconformity

As soon as nonconforming outputs are detected, the organization must react to them. The ISO 9001 Standard expects several basic actions for managing nonconformity. Let us review them and how they may be implemented in the QMS:

- Identification
- Segregation
- Correction
- Containment
- Informing the customer
- Suspension of provision of products and services

For more details and information of each action, please see the designated paragraph in this chapter.

Let us take, for example, a case where one, as a productive company, delivers parts as a supplier to one's customers. A customer called and reported that one delivery contains nonconformed product. You will (as suggestion for reaction)

- Ask that the whole delivery be removed from the environment and be marked as nonconforming
- Organize a new delivery
- Get access to the nonconformed parts
- Examine the nonconformity and gather all the required information
- Begin the process of analyzing the nonconformity

Identifying the Nonconforming Outputs

After spotting the nonconforming outputs, one must identify them as a means of classifying them. The standards specifically demand the identification of nonconforming outputs in an injective manner. This means that each nonconforming product or service will be identified independently (physically as well as logically). The objective of the classification is to prevent the use of those outputs before it is decided what is to be done with them.

One objective is to determine which products are involved in the nonconformity. The identification will be done according to realization process identifiers—production time intervals, batches, production machines, production areas, or products. Bear in mind that when one defective product is detected, the whole batch may be suspect.

A second objective is differentiation, segregation, and distinction of the nonconforming output from the conformed outputs in an explicit manner. This ensures that the nonconforming product would not contaminate the conformed ones. Defective product may contaminate, pollute, infect, or trigger any other defects to others. Therefore, segregation from the conformed products will be applied; that is, the nonconforming products will be removed from the realization area to a specific controlled area. In case a component used in the assembly of the product, or a raw material used to realize it, has been detected as nonconforming, you will have to consider identifying the finished product. Processes of traceability and identification as planned according to clause 8.5.2 (Identification and traceability) will support the distinction of such products. The segregation may be physical or logical, for example, in the case of software.

A third objective is ensuring, beyond any doubt or possibility, that the nonconforming product will not be used, applied, submitted to realization processes, or delivered to the customer, before it is decided what to do with it. The status of the product must be changed in order to ensure this. The product's status shall indicate its conformance or nonconformance. In other words, the status shall indicate to a user or an employee whether the goods (finished product, service that was provided, or materials) are usable or not. The status shall be clearly identified, and any employee should recognize the status at any given moment at any location—within the organization or at the customer's premises. The means of identification of nonconforming outputs shall be maintained as documented information. How are these identified?

- When it is a tangible output like a product or a component, it will be clear to the user of the QMS that this is nonconforming: tags, signs, stickers, ribbons, changing serial numbers, certain boxes, areas, or locations in the organization where the outputs are stored, for example, a red strip that announces the segregation. Consider also access authorization to such areas.
- If the output is not tangible, one may need to identify it in an appropriate way. That is the case with services. If the organization is providing a cleaning service and the customer is not satisfied, one cannot suspend the entire house of the customer as a nonconforming output but one may create some kind of a format that will indicate that this service provided on this date by this employee is nonconforming. A ticket system is the classic example.
- If one revealed a deviation in a process, one may then track down the time when the deviation occurred and trace potential nonconformed products. Areas impacted by the nonconformity should be identified and tested as well.

The issue of identifying nonconforming outputs is relevant to products that have already been delivered for distribution or are in use (and one has the information regarding their status). Since one cannot physically identify products that are outside one's premises, one needs to identify them logically; that is, create a list with all the products according to production identifiers that indicate these as nonconforming.

Obtaining Authorization for Acceptance under Concession

Concession refers to the approval of a product with restrictions. This is a state where the product does not fully meet its requirements or specifications, but under the appropriate concessions, the release or use of the product is approved. Concession may be regarded as the release of a nonconforming product. The concession may minimize the costs of repairs, rework, or recall of the product. The concession is generally approved to the delivery of products and services within specified limits or quantities.

The risk lies in cases where the organization may not correctly assess the severity of the nonconformity or may prefer to release a nonconforming product instead of investing resources in correcting the failure. Furthermore, when initiating a concession, one must identify regulatory requirements relevant to the matter and, when applicable, find out how they refer to release under concession.

When determining the process or procedure for obtaining a concession, a few conditions must be kept in mind:

- The concession must be evaluated carefully for its effects and potential effects on the product, on the end user, on the environment, and so on.
- The concession is to be limited in time, parts, products, models, components, and so on. These limitations shall be documented. The concession should be limited precisely to a batch, a product with serial number, production dates, a certain delivery, or a customer. The objective is to avoid "flowing" of the concession to other products.
- The concession is to be approved by an authorized person, and it will be defined whose responsibility it is to review and approve a concession.
- The concession and the decision are to be documented, and a reference to the suitable records shall be available, for example, the justification of concessions.
- The acceptance criteria that assisted in the decision-making are to be documented or referred to.
- The approval will be accompanied with the identity of the approver.

A good example of the concession of a product is when a process was performed, but with deviations. These were reviewed, and it was decided that they were not critical, and therefore the product was approved. Such a concession may create a special status for the product.

The review is to be registered as documented information. It is important to indicate that the documentation will include the identities of all relevant individuals and authorities that participated in the decision-making and approval of the concession. You may involve the customers in the review and accept their opinion on the matter. However, the approval is to be conducted solely by the organization. The main goal is to explain why the nonconformity is being approved. I suggest planning a format that guides the responsible parties through all the required stages, investigations, and questions.

Responsibilities and Authorities

The organization is required to appoint and certify authorities and responsible parties for identifying and controlling nonconforming products throughout the processes of managing them. Responsibility and authority are separated into two main roles regarding the identification of the nonconforming outputs: the responsibility to detect them and the authority to decide what to do with it.

Each member of the organization that takes part in or operates a process or a part of it shall have the competence, skills, knowledge, and means to identify nonconforming outputs throughout the processes. The authority for reporting and documenting nonconformity to a supervisor shall be granted to personnel in the organizational structure:

- A production employee will have the qualifications to assess process outputs during the realization and decide whether they are conforming or not. For example, before each batch, I used to conduct a small-scale training (right in the production area) regarding the quality of the produced part and known quality problems and to place a sample of a conformed product as a reference.
- A quality person responsible (e.g., quality assurance) will perform the needed examination in order to control the process.
- A storage employee will have the qualifications to determine whether goods, materials, components, or finished goods were stored appropriately before submitting them to realization or delivering them to the customer.

For such a person, it will be clear

- When he or she shall look for nonconformities—when arranging incoming goods on the shelves, when preparing goods for realization, when removing a product from the machine, or when packing a product
- Which product characteristics the person must control
- In case a nonconformity is detected, which actions must he or she perform: put the outputs aside, stop the machine, report it, and so on
- To whom, how, and what information must he or she report

The issue will be implanted in each training program. This will ensure the timely detection and treatment of nonconformities.

The second level responsibility is the authority to make a decision or initiate an action based on the discovery. This person shall be identified:

- The authority shall be granted: on a job description or a procedure describing the treatment of nonconforming outputs.
- This authority shall determine when and how a decision and an action regarding the nonconformity will be initiated. This identification shall appear on the designated documented information.

It may be an individual or a number of persons. The objective, however, is clear: at any given moment throughout the process, a responsible party will be appointed and qualified to react to nonconformity.

Correction of Nonconforming Products

Corrections of nonconforming products may be initiated as reworking, removing, or reprocessing. The correction includes the repair, rework, reprocessing, or adjustment of the product in order to eliminate the nonconformity and ensure that the product meets its specifications. Correction refers to parts, component, materials, or the entire product or the reexecution of service operations. Like everything else, parameters of the correction that may affect the quality of the product will be evaluated:

- Scope of correction and identification of the product—each component or part of the product that requires correction will be identified.
- Objective of the correction—define which problem or nonconformity the correction will solve. One may include a reference to the documentation of the nonconformity that initiated the process.
- Planning of correction activities—define which activities and actions are necessary in order to achieve the objective.
- Risks
  - Evaluate whether the correction may pose new risks to product or service.
  - If there are controls that need to be implemented during the realization of the product due to risks, they will be applied during correction as well.
- Responsibilities—identify the roles and responsibilities of personnel that are to perform and approve the correction (certain roles, employees, and authorities). One may include this matter in the training plan.
- Tool and equipment—determine which tools and equipment are needed to perform the correction.
- Materials, parts, and components—determine which materials are needed to perform the correction.
- Status—define the status of corrected product (before correction, after correction, after revalidation).
- Necessary documented information—identify the necessary documentation needed for the correction (work instructions, diagrams, charts, drawings, etc.), for example, you may document the correction in a designated work instruction.
- Tests, validations, and revalidations—the required controls are to be implemented in order to verify that the results are achieved.
- Revalidation—define revalidation tests to the product after the correction is done.
- Training—submit correction to training program in order for it to reach the appropriate roles.
- Statistical analysis—data related to the correction will be recorded and will serve later for product analysis.

Correction may require a change in the product. Therefore, before submitting the product to correction, it is necessary to evaluate the feasibility and the effects (or potential effects) of the correction on the product specifications, that is, the functionality, performance, safety, and intended use, as well as on other components in the product:

- Each person involved in design and realization of the product will evaluate and give their professional opinion regarding the effect of the correction on the product, for example, engineering, design and development, quality department, manufacturing, and the customer.
- The review of the authorized roles will include the development of corrective activities. Each activity of correction must be reviewed and approved by an authorized person. If one is replacing an electrical component in a device, the

appropriate engineers must evaluate the replacement and its effects and give their approval.

- This review will include the design of revalidation; each correction process and activity must be revalidated to ensure that the product will meet its specifications after the correction. The revalidation will include the regular product validations and tests that prove that the product meets its specifications, but also verification of the correction activities.
- Correction may lead to the creation of a new version or configuration of the product. In such cases when the organization finds it suitable, there is a need to implement controls as required in clause 8.5.2 (Identification and traceability).
- This review will be documented and retained as documented information.

After the correction is applied to the product, the conformity to the requirements shall be verified. This may be achieved by

- Applying the product verifications and validations once again before submitting it for use
- Applying the revalidation or reverifications for the corrected product

This verification shall be documented and retained as documented information:

- The description of the tests that shall be performed including the characteristics that must be proved and the test actions that must be applied
- The criteria that support the test
- The results of the test and the evaluation whether the results are satisfactory or not

Return of Nonconforming Products

Return of a nonconforming product is an action taken to address a problem that occurred in the product after it was already delivered to the customer that exceeded the requirements, specifications, and accepted tolerances. Delivery to distributers counts here as well. Return of the product is also known as a recall. When referring to the return of nonconforming outputs, the organization shall define a few ground principles:

- The organization shall define in which cases or under what severity of nonconforming outputs the product must be returned.
- The organization shall define how it will communicate to the customer or how the customers may communicate with it in order to coordinate the return. These communication channels shall be planned according to the requirements of clause 8.2.1—Customer communication.
- The organization may plan actions of how the product will be returned to the organization:
  - Repair center
  - By post
  - By special delivery
- The organization should define what actions shall follow the return of the product:
  - Replacement of the product
  - Correction
  - Delivery of another product

- Local or regional regulation or statutory requirements may set rules regarding the return. Such requirements must be identified and implemented, for example,
  - Requirements regarding the severity of the nonconformity that requires return
  - Requirements regarding the communication with the customers regarding the nonconformity
- The return process generates data and information regarding the use of the product. I definitely suggest that you develop a systematic collection of data and the analysis and submission for improvement.

Of course, each case, product, or service type demands a separate reaction. I certainly suggest implementing return activities with a documented plan or a procedure with objectives, implementation measures, simulations, analysis, and the submission for improvements. In order to implement such a plan effectively, there are a few issues that must be referred to:

- The identification of relevant regulations in accordance with the region where products are being distributed.
- The implementation of the requirements in your activities for returning products.
- The creation of effective communication processes and channels between the organization and the subjects of the return, for example, customers, users, distributors, and other responsible parties (such as authorities). The requirements for an effective communication channel are discussed in clause 8.2.1 (Customer communication).
- The initiation of an effective interface between the return processes and plan and the quality improvement processes (corrective action and risk-based thinking), in other words how the returned product shall be submitted to the process of handling of nonconforming products.

Documented information:

- Which products must be returned. This documented information may include reference or a link to the relevant customer.
- Documentation of the event of receiving the nonconforming output in the organization:
  - Date
  - Receiver
  - Product

Segregation of Nonconforming Products

Segregation is the act of separating the nonconforming process outputs from the conformed ones by transferring the nonconformed items out of a production or testing environment into a separated environment. The objectives are differentiation and distinction by transferring the nonconforming process outputs into an environment different from the realization environment and distinguishing the nonconforming products from the conformed products in an explicit manner. This ensures that the nonconforming outputs would not affect the conformed ones and will not be used. The nonconformed products will be removed from the realization area to a specifically controlled area. One may define designated area.

When segregating, one may prepare a list of the products under segregation. The classic way is managing a stock for segregated products. Pay attention that within the segregation area one may need to separate each product from another because of risk of contamination.

Containment of Nonconforming Outputs

Containment refers to the initiation of actions to prevent nonconformities from spreading and extending to other processes or other process outputs. When a nonconformity is detected during the realization process, the person who identified it or is responsible for the process shall take suitable action to prevent the nonconformity from spreading further into other outputs of the same process, other products, or other processes. The difference from segregation is that it may be necessary not only to separate the nonconforming outputs from the conformed ones but also to apply further measures that shall involve elements of the QMS to prevent the nonconformity from spreading as mentioned:

- Replacing human resources in the process due to incompetence
- Replacing inputs to the process such as materials or components due to nonconformity
- Replacing resources that support the process such as infrastructures or process environment due to failure or disorder

Let us review an example: during a batch production, one noticed that realized parts (process outputs) are nonconforming. A short investigation reveals that one of the purchased material used in the process is noncompliant to its specifications. This material is also being used for other parts for other products. As a containment, one should

- Detect all the material that is in the stock and suspend it from being released for further realization of other products
- Inform the supplier of the problem
- Request the supplier to replace with new and conformed material as soon as possible

The actions that one is initiating shall be documented:

- Description of the QMS element that was found to be relevant
- Analysis of its impact on the process outputs
- Measures undertaken to prevent the nonconformity from spreading

Suspension of Provision of Nonconforming Products and Services

Suspension of provision of products and services refers to the action of preventing the delivery, provision, or use of nonconforming products and services:

- Identifying or detecting other products or services that were delivered to the customers that are or may be nonconforming
- Contacting distributers and informing them about the nonconforming outputs or products

Let us review an example from the service sector; let us assume that one detected a nonconforming service operation due to lack of knowledge—a service technician performed an operation wrongly because he or she did not have the correct know-how (regarding this operation). The result is a nonconforming service operation. One has identified the required knowledge and provided it to the technician. In order to prevent other technicians from repeating the same error, one must inform them of the nonconformity and provide them with the required knowledge to avoid this mistake.

Informing the Customer

In case the defective product was delivered or the nonconformed service was provided to the customer, the customer must be informed. In clause 8.2.1—Customer communication—one is to determine the communication channels between the organization and the customer regarding several issues. The subject is widely dealt with in chapter 8.2.1 of this book. It is just as important to add that the ISO 9001 Standard expects you to maintain a communication channel and method between the organization and the customer when a nonconforming output is detected. The goal is to inform the customer of

- Which product is affected
- Who is the contact person in the organization
- What is the nature and cause of the nonconformity
- Whether the customer can still use the product
- What is going to happen with the product
- Whether it has to be returned
- How is it planned to handle or correct the nonconformity
- What are the next stages of the handling of the nonconforming:
  - Return
  - Replacement
  - Submission for correction

This subject is very important, and during the years I have noticed that organizations do not give enough importance to the fact that their customers are not sufficiently informed when nonconformity occurs. This leads to dissatisfaction among customers. I urge you to refer to chapter 8.2.1—Customer communication—where I discuss the principles, objectives, and methods of communicating with the customer.

Another issue to be informed to the customer is the reference to contingency actions (which is also discussed in chapter 8.2.1). Contingency actions are executed as a response to unplanned events and require the attention of the customer. Those events may be the detection of nonconforming outputs and require the attention of the customer.

Retaining Documented Information

Documenting a description of the nonconforming outputs is required by the ISO 9001 Standard as a documented information. I regard this documented information as critical because it includes the primary information gathered about the nonconforming output and any information that would help you to investigate the nonconformity later on. The objective of the information is to assist you in mapping the problem and tracking down the root cause. Bear in mind that this documentation is the first step in a process that will later lead to a corrective action. Details, information, and data regarding the nonconforming output that will be documented include the following:

- Statement of the starting date.
- Identification and detail of the product (e.g., product name, model, catalog number, serial number, batch number, date of manufacture).
- Identification of the individual that detected or identified the nonconforming output (employee, customer, supplier, regulatory authority, etc.).
- Type of nonconformity (production error, customer feedback, supplier error, etc.).
- Time and date of receiving or detecting the nonconformity.
- Quantities that were identified as nonconforming.
- Description of how the nonconforming outputs were labeled.
- Specific areas where the nonconforming output was detected or related to the product (department, production hall, machine, etc.).
- Description of the nonconformity (e.g., why it fails to meet the requirements—one can write here a literal description, or even attach other documents such as statistical analysis or photos).
- Reference to documented evidence.
- Immediate remedial action that was applied to the matter (correction, segregation, etc.).
- Relevant procedures or work instructions.
- Categorization of the nonconformity—this is not required by the ISO 9001 Standard but is highly recommended and will assist you later with a statistical analysis of the nonconformities.

You may design a format that includes all the details that are relevant to your organization and type of products.

The second requirement for documented information is the reaction to the nonconformity. The standard suggests a few types of reactions: correction, segregation, containment, return, or suspension of provision of products and services. They were all discussed earlier in this chapter. Each action that was undertaken must be described and documented. The documented information shall include the identification of the authority deciding on the action in respect of the nonconformity.

In order to be practical, I would include these two requirements of documentation in the format or technique that you use to describe the nonconformity. So you maintain the information in one place.

The next requirement for documented information refers to the documenting of the concession obtained. Please refer to the paragraph Obtaining Authorization for Acceptance under Concession for details and information about the expected documented information.

9 Performance Evaluation

9.1 Monitoring, Measurement, Analysis, and Evaluation

9.1.1 General

Measurement, monitoring, analysis, and evaluation are critical for the assessment of the performance of the quality management system (QMS). The goal is to reflect the quantitative and qualitative performance of the QMS and to report the degree to which processes meet their stated objectives. This activity is regarded by the ISO 9001 Standard as one of the tools that promotes improvement of the QMS. Clause 9.1.1— General—lays the required principles to be considered while determining and creating methods for monitoring measurement analysis and evaluation of processes or process results. In other words, when you plan, determine, and define how you will monitor, measure, analyze, and then evaluate a process or its outputs, make sure that you are applying and practicing the following principles (the ISO 9001 Standard requirements):

- The organization shall determine which quality elements shall be monitored and measured.
- Methods for monitoring, measurement, analysis, and evaluation shall be determined. The methods shall define which activities are needed to ensure valid results of monitoring and measurements.
- The methods shall define the stages and the intervals in the process when activities of monitoring and measurements must be performed.
- The methods shall define when results of monitoring and measuring shall be analyzed and evaluated.
- The results shall enable the evaluation of the performance and the effectiveness of the QMS.
- Documented information on the results of the monitoring, measurements, analysis, and evaluation shall be retained.

Terms and Definitions

Before we begin, let us understand the difference between monitoring, measuring, analyzing, and evaluating a process:

- Monitoring—a continuous inspection or observation of process performance or process output for a special purpose through a defined scope (e.g., with a sample size or over a period of time) and maintaining records of those observations
- Measurement—the activity of delivering data to a method in order to define objectively a quantitative or qualitative measure and capturing a situation without any references to the significance
- Analysis—a set of techniques for examining trends and tendencies of measurements of an output (process or product)
- Evaluation—the action of comparing a process or process output measurements against given criteria to determine the performance of the process or conformity of a process output

Goal of Monitoring, Measurement, Analysis, and Evaluation

The goal of monitoring, measurement, analysis, and evaluation is to provide the decision makers an understanding through a situation report concerning the performance of processes. The data that the monitoring, measurement, analysis, and evaluation provide shall relate directly to the controls suggested by the standard such as supplier evaluations or control of nonconform products. Those activities shall indicate the effectiveness of the QMS and the extent to which the QMS achieves its quality objectives. This type of reporting shall regularly provide the different organizational units data and information regarding quality issues.

These needs should be known and available through the outputs of the process approach—which processes must be measured in order to control their performance and verify their conformity. In clause 4.4—Quality management system and its processes—you were required to

- Determine the processes needed for the QMS
- Evaluate these processes and implement any changes needed to ensure that these processes achieve their intended results
- Improve the processes and the QMS

In other words, for each process, with outputs having an effect on the quality of the product (or service), a suitable method for measuring and monitoring must be planned and applied. The method shall be planned in accordance to the nature of the process and the nature of its outputs. Regarding the quality objectives, for each area in the QMS, the relevant quality objectives are defined. The definition of the quality objectives should relate to specific realization activities. Now it is required to define how those quality objectives shall be monitored and measured or in other words to break down the quality objectives into operational measurements:

- Which processes and activities must be monitored?
- Which process outputs must be monitored and measured?
- Which methods shall be used for monitoring the activities and measuring the data?

What are the advantages of implementing monitoring, measurement, analysis, and evaluation?

- Implementing a systematic control of the realization processes
- Identifying deviations in time and submitting them to a controlled process for treatment
- Allowing users of the QMS to make decisions regarding results of processes
- Allowing the users of the QMS to prevent nonconformities by identifying gaps in a process and preventing the transition of nonconform outputs to the next process
- Determining the effectiveness and efficiency of processes
- Using the monitoring, measurement, analysis, and evaluation as means for continual improvement
- Promoting the achievement of quality objectives through key process indicators related to the objectives of the organization

Defining the Methods for Monitoring, Measurement, Analysis, and Evaluation

Each monitoring and measuring activity must be conducted according to a defined method. The goal is to identify, for each process, the parameters of outputs that affect its quality and determine the activities necessary to ensure valid results, activities that will ensure that the monitoring and measurement deliver results that can be analyzed and evaluated. The methods shall enable a comparison of the results with the objectives, for example, measuring the performance of a machine where the results are ranked in within upper and lower limits. Such ranking provides the status of the process—either within tolerances or not. The new revision of the ISO 9001 Standard (2015) does not separate the requirements for monitoring, measurement, analysis, and evaluation of products or processes. But there are some distinctive requirements that I would like to discuss.

When monitoring and measuring a process, the following issues shall be referred:

- Identification of the processes or process outputs that must be monitored and measured
- Identification of those parameters or outputs that may indicate how the process behaves
- Identification of the function or the role responsible for conducting the activities of measurement (designated employees, a certain department, organizational unit, etc.)
- Definition of parameters for monitoring such as process conditions, interval, sampling rates, batch loads, and quantities
- Description of the stages during the process where measurement activities shall be undertaken
- Tools, equipment, or software required to perform or assist in the measurement
- The activities and the use of techniques of measuring the process
- Where and how the results must be documented
- The criteria necessary for the evaluation of the results (the objectives)
- Action required in case nonconformities are detected

When monitoring and measuring a process output or a product, the following issues shall be referred:

- Definition of the objectives of the measurement, the characteristics of the product or the service that is to be measured—a specific material, product, part, or component for the monitoring or a service stage or a specific activity

- Identification of the characteristics of those outputs that may indicate how the product was realized
- The definition including the extent of the test, that is, the time interval and the sampling rate
- Tools, equipment, or software required to perform the test
- Identification of the function or the role responsible for conducting the activities of measurement (designated employees, a certain department, organizational unit, etc.)
- The exact realization stage where the product or service is to be sampled and controlled
- The sequence of events in order to perform the monitoring and measurement activities and the conditions for the test that are to be defined, such as room conditions, material conditions, and specific processes
- Where and how the results must be documented
- The availability of the criteria needed for the comparison of results and evaluations
- Action required in case nonconformities are detected

Defining the Process Stage, the Interval and Timing for Monitoring and Measurement

Each activity is measured from the time the input flows into a process and initiates the activity until the desired output is accepted (this activity might include subsequent activities). The measurements, collecting the data regarding a process or a process output, must be conducted in a defined stage, period of time, points and events in a process, under defined conditions, and according to a sample rate that will support decisions. The goal here is to ensure that the monitoring and measurement shall deliver valid results:

- Process stage—it will be clear to the users which are the points or events in an operation that must be measured and which process stage, phase, after which activity, or during which operation the measurements will be conducted.
- Product sample rate—the user shall know exactly how much process outputs he or she must sample.
- Conditions—the conditions under which the measurement will be conducted will be defined.
- Interval—the scope and interval of the measurement will be defined.
- Area and extent of the process—for each process the definition shall indicate which area shall be monitored and measured.

I suggest here a way to detect the critical process stages for monitoring and measurements:

1. Define the critical measures of performance related to the relevant quality objectives.
2. Define the critical triggering events, inputs, key steps, and results for the process.
3. Assess the process's current performance.
4. Determine the level of desired performance for the process expressed by the relevant quality objectives.
5. Assess the size of the performance gap between the current performance and the objective.

6. Initiate an improvement plan, which refers to the relevant process scope, that may influence the performance of a process. This improvement will be naturally submitted to the requirements as presented in clause 10—Improvement.

Analyzing and Evaluating Results of Monitoring and Measuring

The activities of monitoring and measuring generate data that must be analyzed and evaluated. The methods shall indicate in which point of the process the results will be analyzed and evaluated. The goal is to support the decision makers of a process and to provide them with the understanding of the current state of the process and its alignment with the objectives at the appropriate point of time, in other words to deliver them the data they need in time to make the right decision. According to the results, it will be necessary to derive actions for corrections or improvement of the processes. This issue is discussed in detail in chapter 9.1.3—Analysis and evaluation.

Defining Roles and Responsibilities for Monitoring, Measuring, Analyzing, and Evaluation of Processes

Each monitoring, measurement, analysis, and evaluation activity shall be conducted by qualified personnel. Thus, it is required to define which skills and competencies will enable effective monitoring and measurement of processes and products. The definition can be on

- Work instruction or any kind of documentation that describes the activities of the monitoring, measurement, analysis, and evaluation
- Job description

Training and qualifying personnel for monitoring, measurement, analysis, and evaluation shall be mentioned to on the organizational trainings program.

Evaluating the Effectiveness and Performance of the QMS

Effectiveness of the QMS refers to the extent to which objectives of processes were achieved and may indicate the degree of improvement or the potential for improvement. We know already that the monitoring, measurement, analysis, and evaluation activities enable an effective interrelation between processes by ensuring submission of appropriate outputs between processes. The approval of those outputs (the feedback from the customer of the process) allows a smooth workflow or demands changes and improvements. This issue will be discussed in detail in clause 9.1.3—Analysis and evaluation.

Maintaining Documented Information Related to Monitoring, Measurement, Analysis, and Evaluation

Retaining documented information regarding the monitoring, measurement, analysis, and evaluation is required when the organization finds it appropriate and when the organization decides that maintaining this documented information will affect the quality of the product and is necessary for the effectiveness of the QMS. Now, do not allow yourself to underestimate the requirements; it will be harder in an audit to persuade the auditor why you do not need a documented information than to present the documentation. I tend to divide the documented information into two levels:

- First level—the definitions and instructions of the monitoring, measurement, analysis, and evaluation activities: test instructions, quality instructions, etc. The effectiveness of this documented information means that the relevant personnel will know how to perform the monitoring, measurement, analysis, and evaluation activities. This type of documentation shall be affected by the complexity and the necessary knowledge for performing the monitoring, measurement, analysis, and evaluation activities. This type of documented information shall cover the following issues:
  - Describing how the practice of the monitoring, measurement, analysis, and evaluation will be applied (which activities)
  - Responsibility (who collects the data)
  - Frequency of the activities (when and how much from the process must be measured)
  - Methods and techniques for the evaluation of the data (which tools)
  - Interpretation of trends (how to analyze the results)
  - Distribution (who views the data)

  If you decide that the activities of the monitoring, measurement, analysis, and evaluation are considered as knowledge and are necessary for the operation of its processes and for obtaining and assuring conformity of products and services, it will be needed to document them. This level includes also regulatory requirements: when national or regional regulations might require the implementation and documentation of statistical methods or techniques, they must be maintained.
- Second level—this level refers to how the results are reported: the required documented information that provides data, information, insight, and feedback about processes and process outputs (the records). The main objective of this type of documentation is to enable decision support for process owners and managers. Effectiveness of this documented information means that the documentation shall indicate clearly the trends in the processes and their outputs, for example, electronic charts or SPS software and forms. I find that this level is more critical than the first one because this type actually holds the important information regarding the processes and their outputs.

The documented information shall be planned according to the nature of the processes that are measured and the nature of the interrelated processes:

- The instructional level will be planned and adjusted to the processes—if the activities are taking place in a production hall, then protected hardcopies are required. But if the monitoring, measurement, analysis, and evaluation take place in digital environment such as a computer software, a digital copy may be suitable.
- Tools for collecting results and data and information shall be adjusted to their type of data.

In any case, the documented information will be submitted to the requirements of clause 7.5—Documented information.

Key Performance Indicators

Although the standard does not specifically refer to the issue of key performance indicators (KPIs), I would like to review it shortly because it has a direct relation to monitoring, measurement, analysis, and evaluation of the business activities and processes and may be used as an effective tool for implementing the requirements of clause 9.1—Monitoring, measurement, analysis, and evaluation.

KPIs are performance measurements of processes or activities in the organization used to evaluate the effectiveness of a system. Monitoring and measuring the KPIs present the performance of processes, operations, and activities related to those KPIs. Effectiveness is determined through measurement of factors that influence those processes or activities. Influencing those factors may initiate changes in the QMS. When the data and information do not conform to their objectives, the organization may apply changes to the QMS, allocate resources better, and initiate corrective actions, preventive actions, or improvements. When choosing the KPI, the organization shall consider the following issues:

- The KPIs shall be related to the needs and expectations of interested parties, for example, KPIs may be part of the customer satisfaction or product performance objective.
- The KPIs shall be aligned with organizational strategies and drive the organization to achieve its quality objectives, for example, KPIs shall relate to quantity of nonconformities or level of customer satisfaction.
- KPIs should focus on a few high-value activities that reflect effectiveness of the process and that will deliver the most adequate data.
- KPIs shall provide critical information and data. In other words, changes that will be based on results of monitoring and measuring the KPI will induce changes and reactions on the organization.
- KPIs may be based on standard definitions, rules, and calculations, in order to enable a better analysis of the data they provide.
- For each KPI, an individual, a group, or a specific person shall be held responsible, for example, a process owner.
- KPIs shall be practical, which means it will be possible to draw conclusions from the data that are delivered.
- KPIs should relate to the activities and operations of the QMS, and the relevant personnel may understand how they may influence them.
- The KPIs shall indicate points and events in a process that must be analyzed and evaluated.
- KPIs shall relate to the business activities and will make it possible for employees to know and identify when their intervention is needed when nonconformities are detected.
- Different KPIs should reinforce each other and not compete, contradict, or arouse conflicts with other KPIs.

The KPIs are to be defined according to the nature of your organization, its processes, and its products or services. So it would not be practical to suggest KPIs in this book. But what I do suggest is to define KPIs that cover the entire life cycle of the product:

- Design and development activities
- Realization of the product
- After-sales activities

Process Audit

Process audit is another effective method that you may apply in order to measure and evaluate performance of processes, their ability to create a product according to its specifications, and the extent to which process output meets their requirements. The process audit serves to examine whether the product quality requirements are met through the assessment of processes and to determine whether a process is controlled and capable. Through reviewing the operations of the process and auditing their parameters, you measure the controls that are applied to the process and decide whether they are effective or not.

The objective of the process audit is to detect opportunities for improving a process. The process audit may be applied when the organization realizes that it has quality problems, there is an increase in the number of customer complaints, changes are incorporated in the realization, or the organization has the desire to reduce costs. Which process parameters may be examined?

- Process audit examines the conformity of predefined process parameters such as time to process and machine parameters such as accuracy, temperature, or electrical current supplied to the process. The examination provides evidences that indicate how well the process performs.
- Process audit examines resources and process environment conditions that support the process in processing inputs into outputs. In practice you audit whether the defined requirements for a process and conditions for process environment, performance of infrastructures, and deployment of competent personnel were supplied.

In order to implement a process audit, you need to define, for each critical process, a process that may affect the quality of the product—an audit plan. The audit itself may be planned in advance or be unscheduled. The plan will include

- Unique identification of the process and the process outputs that will be audited—it must be clear which process is audited.
- The structure and flow of the process, its operations, and their sequence must be defined and clear to the auditor. It is very important so he or she can follow the process effectively.
- The objective of the audit.
- Documented information for reference: flowcharts, work instructions, description of process parameters, etc.

- The actions that are required to audit the process and their sequence.
- Responsibilities that take part in the process and operate it must be defined—those persons will answer the questions regarding the process.
- How data and results will be documented.
- Tolerances and limits in order to evaluate the performance of the process and to determine whether a nonconformity has occurred.

The end result of the planning is a questionnaire or a catalogue of questions. And through answering these questions, a process and its ability can be evaluated. Let us review a few examples of how the questions shall be formulated:

- Are the requirements of the process customer and the specifications of the process outputs defined and available to the operators of the process?
  - Drawings
  - Orders
  - Contracts
- Are the flow of the process, its operations, and their sequence defined to the operators of the process?
  - Description of the process and its operations
  - Trainings
  - Documented information
  - Schedules
  - Special cases
- Are the outputs of the process defined?
  - Process outputs
  - Expectations of preceding processes
- Are the capacity and resources of the process planned?
  - Production planning
  - Allocation of resources
  - Costs of operations
  - Planning of scrap
- Were the resources and support of the process available at the time of the processing?
  - Infrastructures
  - Process environment
- Were the required conditions for the resources provided at the time of the processing?
  - Environmental conditions
  - Required knowledge and competence
  - Condition necessary to operate infrastructures
- Are the goals or objectives of the process achieved?
  - Process objectives
  - Needs and expectations of interested parties, for example, customer and regulatory bodies

It must be very clear to the auditor of the process which process or part of the process he or she must audit. The unique identification of a process should refer to a customer order, a production order, or a batch. Flowcharts, process maps, or process flow diagrams may help you orient in the process and understand which parameters you must audit and at which process stage. The output is a report that indicates the performance of the process through the results of the different proven elements that were audited. Where the process audit finds that a process is incapable, it may be forwarded to a corrective action or improvement.

9.1.2 Customer Satisfaction

One of the declared goals of the ISO 9001 Standard is to allow the organization to enhance customer satisfaction (see clause 1—Scope). This is achieved by continually and systematically evaluating whether the product or service that the organization provides meets the requirements and expectations of its customers throughout its life cycle. The strategy is to create a systematical method that evaluates customers' perceptions of the degree to which their needs and expectations have been fulfilled. The ISO 9001 Standard requirements include the following:

- The organization shall monitor customers' perceptions of the degree to which their needs and expectations have been fulfilled.
- The organization shall determine and develop methods to collect data, process information, evaluate that information, and present results regarding customer's satisfaction.
- Remark—among the methods used for monitoring and evaluating customer satisfaction or perception, the organization may use customer surveys, customer feedback on delivered products and services, interviews with customers, market share analysis, compliments, warranty claims, and dealer reports.

Needs and Expectations of the Customer

Monitoring customer satisfaction enables the early identification of problems concerning quality, performance, and functionality of the product and unfulfilled expectations of customers. The organization's success depends on its customers, and therefore, understanding current and future customer needs will allow meeting customer requirements and exceeding customer expectations. Customer satisfaction is associated with the level of expectation on behalf of the customer: What are the expectations of the customer? What had the organization delivered? Had the expectations of the customer been fulfilled? Understanding what the expectations of the customers are is the first step when wanting to evaluate their satisfaction. I tend to divide the expectations into two types:

- Objective expectations—objective expectations refer to the characteristics of a delivered product or service that were communicated to the customer and upon which the customer submitted an order. This type of evaluation shall indicate whether agreements with the customer were fulfilled or not. The evaluation of those expectations considers specific elements or aspects of the delivered product or the realization processes.

- Subjective expectations—subjective expectations refer to customers' perceptions of the degree to which their needs and expectations have been fulfilled. This type of evaluation shall indicate
  - Which customer expectations were not fulfilled
  - Which customer expectations were not identified

Each product has its characteristics that reflect customer expectations and the quality of the product. Those characteristics are results of product realization activities. And these are the activities that must be measured in terms of customers' expectations.

After understanding what the relevant expectations of the customer are, you must measure them using the method for monitoring the customer satisfaction. Let us take, for example, a software product:

- Evaluating the objective expectations will examine whether all the functionalities or characteristics of the software that had been promised to the customer were actually delivered.
- Evaluating the subjective expectations will examine whether
  - The functionalities or characteristics of the software that were delivered do fulfill the expectations of the customers.
  - Other functionalities or other characteristics of the software are expected by the customer.

Determining the Method for Monitoring Customers' Satisfaction

After understanding what the expectations of the customers are, you need to develop a method for monitoring customer satisfaction. The objective of the method is to gather data and information regarding the extent to which customer expectations are fulfilled, submit it for analysis, and initiate actions upon its conclusions. When determining the method, please refer to the following issues: The method shall

- Indicate who the target group of the analysis is
- Indicate how customer satisfaction information is to be obtained or with which tools the data and information will be collected
- Define responsibilities for obtaining the data
- Indicate which business activities are to provide that data
- Indicate which methods or tools will be used to analyze the data
- Indicate the communication channels to allow data to flow between interested parties
- Indicate the interface between the conclusion of the analysis and the improvement processes
- Indicate to whom the data and its analysis is designated

The following table lays out the method in terms of process elements:

Process Element	Method Requirements
Relevant business activity for obtaining data: • Survey • Complaint handling • Interviews with customers	• Creating proactive operations (initiated operations) that will deliver data and information regarding customer satisfaction (like customer surveys). • Identifying business activities that may provide data and information regarding customer satisfaction (like complaint handling). • Determining who the responsible person for this activity is.
Data	• For each type of activity, determine which type of data will be collected. • Determining which tools or techniques will be used to collect the data.
Resources • Employees • Information system • Additional software for reporting	• For each type of monitoring, the required resources shall be identified and allocated.
System for gathering the data	• Which system in the organization shall gather the collected data.
Analysis • Statistical methods • Quantitative/qualitative analysis	• For each type of data, it is needed to define which method or technique shall be used to analyze the data.
Reporting	• For each type of data, how the conclusion will be reported needs to be defined.
Actions that are required upon the result	The responsible parties for taking actions upon the results must be defined. • In order to ensure that actions upon the results are performed, an interface between the system and the responsible person must be defined.

Defining Time Frames

It is necessary to define time frames or periods for each of the customer satisfaction activities (e.g., those mentioned in this table). This definition of time frames or intervals has two objectives:

- Scheduling or planning the activities for measurement of customer satisfaction
- Allowing the corrective actions or improvements to achieve their objectives

However, a customer satisfaction survey may be consecutive, which means that the collection, gathering, and analysis of the data may be an everyday task. But please allow time between one measurement of customer satisfaction to the next in order to evaluate the effectiveness of corrective actions and improvements.

Defining Who Is the Customer

When measuring and analyzing the satisfaction of customers, one must first define who the customers are. I divide customers into two types:

- External customers—External customers are the entities that receive the end products or services. One of the main goals of the ISO 9001 Standard is to enhance their satisfaction by meeting their requirements. Through their purchase and the revenue that they provide, the organization can survive. Not meeting the expectations of the external customers may result in the risk that they will not purchase the goods or services again. If we refer to the sales cycle, the end customer that receives the goods is the external customer.
- Internal customers—Internal customers include those members of the organization who operate the activities and rely on assistance from other internal entities to fulfill their duties. A good example is the sales cycle—the logistics (the department that processes deliveries) is the customer of the sales department (sales personnel)—the sales delivers process outputs that the logistics must use in order to process an order. If the organization feels that the satisfaction of the internal customers may affect the quality of the end product, the organization may monitor it as well.

Methods for Monitoring Customer Satisfaction

There are many methods for monitoring and evaluating customer satisfaction (and the standard counts some of them): customer surveys, customer feedback on delivered products and services, interviews with customers, market share analysis, compliments, warranty claims, and dealer reports. The sheet here is too short to review them all, and you must choose the ones that are most suitable for the nature of your organization and your customers. But nevertheless I decided to focus on the most common and, to my opinion, the most simple ones.

Customer Surveys

Customer surveys are an effective tool for collecting data from customers regarding the use, performance, quality, delivery, after-sales service, or other issues concerning the product or the service. The survey will measure characteristics related to the product, its delivery, and the performances of the organization that are relevant to the satisfaction of the customer. The objective of a survey is to reflect the opinion of the customers regarding the quality, functionality, and performance of the product. Data from the survey may serve later on for analysis purposes. There are two types of surveys:

- Qualitative survey—a qualitative survey is designed to measure the satisfaction of customers that will assist the organization to understand the perception of the degree to which their needs and expectations have been fulfilled, for example, interview with the customer where the customer provides his/her personal opinion on the product or the performance of the organization.

- Quantitative survey—a quantitative survey is designed to measure the degree of customer satisfaction. The quantitative survey will measure the satisfaction of customers in different areas or operations of the organization in calculable or quantifiable terms, for example, with notes from 0 to 10.

While planning the survey, be sure to refer to the following issues:

- The questions—it is very important to identify the topics and issues that reflect the expectations of the customers. Those may be related to the realization processes, quality, functionality, and performance of the product.
- The target group of the survey is to be defined—that is, direct customers, end users, or distributers.
- The method for the survey—the method of how the survey will be conducted, that is, how the questionnaire will be submitted to the customers and how they submit it back must be defined. The questionnaire can be printed and filled out manually, or it can be a digital form. Today, most management systems provide the possibility of electronic questionnaires. When your management system does not provide such a solution, you may find many online solutions.
- Interval or frequency of the survey—the interval of the survey must be defined. In any case you must allow time to handle risks, improvements, and nonconformities that were detected during the last survey before you submit a new one.
- Sample size—the sample size for the survey must be determined. The size will be determined in order to obtain sufficient reliable data for the analysis. If you are managing 1000 service calls per year that involve 10 service representatives and 20 technicians, you must set a sample size that will ensure that activities of all of the involved persons are measured.

Customer Complaints

A complaint is an effective way to receive a customer's opinion regarding the use of the product. A complaint is an application communicated to the organization from a customer regarding events or the disturbances of performance, functionality, intended use, or safety of the product during use. It serves as an indication that the product has not reached its purpose or met its requirements. Thus, a complaint is a reliable source of feedback regarding the product, its design and development, and its realization processes. Using the complaint for measuring customer satisfaction, the organization shall focus on complaints that relate to the functionality, characteristics, and performances of the product. When determining that complaints are an instrument for measuring customer satisfaction, please make sure that the following issues are addressed:

- Communication: The method shall describe how customer complaints are received in the organization—a specific telephone number, an online form, a designated e-mail address, and an interaction center—such options for filing a complaint must be made known to the customer, and the contact details must be published (printed on a brochure, the firm's website, or on the product itself). The method could be incorporated with other types of inquiries (such as ordering of products or visits to customers).
- Details: The method shall define the required information and details from the complaint that are required in measuring customer satisfaction. The details and

information should be sufficient in order to ensure effective analysis. If the complaint refers to a product, then the following information must be specified—the model, identification or traceability (batch number or serial number), production details, contact person, delivery date, and any other information that would identify the product and support the analysis of (eventually) the nonconformity. If the complaint refers to a service, then specify when the service was performed, by whom, to which product (with identification details), and the service identifier. The best way to ensure that such details are provided is to design a form (manually or electronically) and verify that all fields are completed.

- Availability of data: The method for handling a complaint shall consider how the data and information necessary for the analysis will be available.
- Awareness: The function that is in charge of the process of accepting the complaints must be informed that the data that are gathered during handling of complaints are used for the purposes of measuring customer satisfaction.

Customer Feedback

One way of measuring customer satisfaction is customer feedback. While customer satisfaction may be submitted to subjective opinion, customer feedback provides an objective point of view on whether the organization supplied a product according to the specifications and expectations. Customer feedback provides an early warning about potential quality problems, effectively identifies risks or opportunities, generates the inputs for improvement, and initiates the necessary corrective actions. By measuring customer feedback, importance is given to the fulfillment of the product's requirements rather than the perception of the customer as to whether the organization has met its requirements. Customer satisfaction is a subjective matter, whereas achieving product's specifications is an objective issue that can be assessed and measured. Feedback may be measured through the following business activities:

- Customer complaints—the organization monitors customer complaints with regard to the performance and functionality of the product.
- User surveys—the organization conducts surveys regarding the functionality, characteristics, and performances of the product.
- Reviews—the organization initiates a review regarding the product and its functionality.
- Journal reviews—the organization researches sector and industry tendencies.

The feedback shall indicate the status of the delivered product or service compared to its requirements. The data and information from the customer feedback are gathered in the postproduction phase, for example, transportation activities, storage, installation, service, and use.

Bear in mind that the data collected via feedback methods are to be used at later stages as inputs for processes of analysis of data as required in clause 9.1.3—Analysis and evaluation. Thus, it is recommended that when you plan the activities for collecting and characterizing the data, you should consider future analyses. Once quality problems are detected, they are either handled as nonconformity or submitted to improvement processes. The objective of the analyses is to demonstrate the suitability and effectiveness of the QMS and to evaluate whether improvements of the effectiveness of the QMS are needed.

Communicating and Reviewing the Results

The method for monitoring customer satisfaction shall indicate how the results of the evaluation shall be communicated to the relevant interested parties in the organization. Each interested party of the organization has other interests and therefore needs other type of reviews:

- Top management will require a general analysis of the overall customer satisfaction.
- Second level management will need to view data and results concerning their area of business.
- Operational level will need to have access to detailed information in order to be able to deal with specific cases of customer dissatisfaction.

Here I recommend you to develop an interface between the results and the analysis and the persons involved. By interface I mean the interaction between the provider of the results and its client. A simple way will be to allow certain roles in the organization access to the data and its results and to plan for each level its designated reports.

Submitting the Results for Improvements and Corrective Actions

After collecting the data and understanding how customers perceive the degree to which their needs and expectations have been fulfilled, one must examine the need for further actions and when required to submit the conclusions for

- Addressing risks or opportunities
- Initiating corrective actions
- Initiating improvements

For example, if the organization detected through customer feedback that risks during the realization processes were not effectively addressed, and in certain cases led to nonconform products, the organization will be required to submit such risks to the process of handling risks. Which tools allow us the standard for submitting results for improvements?

- Management review where the results of customer satisfaction will serve as an input
- Complaint handling
- Investigating and handling exceptional cases where satisfaction is proved to be very low (Dissatisfaction may indicate nonconformity)

Retaining Documented Information

The standard requires determining a method of obtaining, monitoring, and reviewing this information. In other words, you are not required to document the method, but you are required to

- Prove that a method was planned and is implemented
- Show evidences of actions for measuring and monitoring customer satisfaction
- Show how those results were analyzed and concluded
- Provide evidences that nonconformities were submitted for correction when such were detected

During an audit, try to present the processes as a whole and not only the forms of customer survey. One way to prove determination of methods for measuring and monitoring customer satisfaction is to relate records of the QMS to its operation or activity. Take a look at the following table:

Activity	Expected Records
Customers surveys	Filled-out questionnaires
Customer complaints	Records of customer complaints and records of the investigations
Service calls	Service reports for analysis

Bear in mind that the results of the analysis are required as an input to the management review: customer satisfaction and feedback from relevant interested parties.

9.1.3 Analysis and Evaluation

Clause 9.1.3 lays out the requirements for analysis and evaluation of the performance of the QMS. The organization is to assess the progress in achieving planned results. These are to be reflected in the mission, vision, policies strategies, and objectives at all levels and in all relevant processes and functions in the organization. The performance of the QMS will be measured through QMS aspects mentioned in this clause. Let us review them. The ISO 9001 requirements include the following:

- The organization shall analyze and evaluate appropriate data and information arising from the monitoring and measurements activities.
- The results of the analysis shall serve for the evaluation of
  - Conformity of products or services
  - The extent of customer satisfaction
  - The performance and effectiveness of the QMS
  - The effectiveness of planning and implementing of the QMS
  - The effectiveness of actions taken to address risks and opportunities
  - The performance of external providers
  - The need for improvements to the QMS
- Remark—Methods for analyzing data can include statistical techniques.

Terms and Definitions

Before we begin unfolding the requirements, I would like to toe the line regarding what is analysis and evaluation in terms of quality management.

- Analysis in the eyes of the ISO 9001 Standard refers to a detailed examination or an investigation of the elements of the QMS and reaching their goals, using a set of techniques for detecting and identifying trends.
- Evaluation is the estimation of the nature, value, quality, ability, extent, or significance of those elements or structures through the results of the analysis.

Identifying the Appropriate Performance Indicators

Analysis and evaluation have the objective of interpreting and understanding the current situations of quality elements, resources, processes, products, and services, through different quality aspects using different methods. The goal is to decide whether those elements reach their desired objectives. The aspects of the QMS that are to be analyzed and evaluated are specifically defined by the standard (see the ISO 9001 Standard requirements clause 9.1.3).

Analyzing and evaluating a process indicates to which degree activities were performed and resources were used. The end result will provide a situation report regarding each of the quality aspects mentioned in this clause and will enable the organization to initiate the appropriate corrections or improvements.

Measuring and monitoring those elements generates data. The indicators should have already been determined, and you should already have applied monitoring and measurement activities that generate the data (see clause 9.1.1—General). Now it is the time to analyze the data and evaluate the condition and status of those specific quality elements with reference to their objectives. In order to be precise, you must indicate for each of the quality elements:

- Which indicators shall be analyzed and evaluated?
- How will they be analyzed or with which methods?
- What are the relevant objectives? Are the objectives relevant?
- Who are responsible for performing those analysis and evaluation activities?
- How will it be reported?
- Who is the target group that will use the results of the analysis and evaluation?

Sources for data and information for analyzing and evaluating include

- Process planning outputs
- Results of process performance measurements (quality tests)
- Analysis of process variations
- Analysis of occurrences (like nonconformities)
- Results of audits
- Results of validation activities
- Results of verification activities
- Results of supplier evaluation

Evaluating the Extent of Customer Satisfaction

When evaluating customer satisfaction, one must consider appropriate data and information arising from the monitoring, measurement, analysis, and evaluation activities that will serve the analysis and evaluation. Customer satisfaction is measured through the customers' perceptions of the degree to which their needs and expectations have been fulfilled by the delivered product. The objectives and the methods are well discussed in chapter 9.1.2—Customer satisfaction. Please refer to this chapter for more details and techniques. After defining how customer satisfaction will be measured, you are required to decide how or with which methods the data will be analyzed and evaluated. Types of data that will be analyzed may include

- Results of customer surveys regarding satisfaction
- Data of customer returns
- Data of customer complaints, customer dissatisfaction, or loss of customers
- Customer feedback

Evaluating Conformity of Products or Services

When evaluating conformity of products and services, one must consider the appropriate data and information arising from the monitoring and measurement activities that will serve the analysis and evaluation. It is required to provide data and figures that will represent the degree of conformity of products and services. The data shall refer to two aspects:

- Analyzing data of nonconformities encountered—from these data and information, it will be possible to identify the root causes of quality problems or nonconformities and to derive corrective actions and improvements of
  - Processes and realization activities of products and services
  - Products, their characteristics, or intended use
- Analyzing data of actions taken for the elimination of those nonconformities—from these data and information, it will be possible to
  - Update activities for addressing risks and opportunities
  - Apply changes to the QMS

For each type of data, you must assign the appropriate method for analysis and evaluation. Types of data that will be analyzed may include

- Data regarding the waste and scrap of process outputs
- Added value of corrective actions
- Data regarding costs of quality
- Data of customer returns
- Data of rework or corrections of process outputs
- Data from customer complaints, customer dissatisfaction, or loss of customers
- Fault of extern delivered goods or services
- Data from warranty claims

When planning the reports for the analysis of conformity of goods and services, please refer to the following issues:

- The reports must be presented in a realistic and clear format that will enable conclusions, further actions, and improvements.
- The reports will indicate trends of processes, activities, or characteristics of products or services.
- It will be possible to compare results of reports between periods in order to prove effectiveness.

Evaluating the Performance and Effectiveness of the QMS

Effectiveness of the QMS refers to the extent to which objectives of processes were achieved and may indicate how much the QMS has improved. Performance of the QMS is measured by certain indicators of processes and their relevant criteria, which suggest how much processes achieve their objectives (or not). The standard demands that the organization shall use its methods of monitoring, measurement, analysis, and evaluation to prove the ability of the processes to attain desired results. According to the results of the analysis, resources and activities shall be adjusted in order to meet business objectives.

Process performance is measured with comparison to the quality objectives. In other words, a process or an activity shall be evaluated and its performance shall be determined according to the extent its relevant quality objectives were achieved. Those activities related to the objectives of the organization as well the expectations of the interested parties are to be analyzed and evaluated:

- Processes—the objective is to review the effectiveness and efficiency of processes. For example, evaluating a process will review its progress made on achievement of the process objectives.
- Resources—the objective is to ensure that resources that are used in the realization processes, such as equipment, monitoring and measuring devices, facilities, knowledge, and personnel, are used effectively and efficiently.
- Process outputs—the objective is to ensure that process outputs such as products or services meet their specifications (customer or regulatory).

Performance indicators may include

- Achieving process objectives
- Allowable process variations
- Achieving expectations, specifications, and needs of process customers
- Obtaining intended results
- Measuring process parameters that indicate the performance of a process: duration of activities, costs, scrap, quality, and capacity
- Improvement of processes in the long term

In practice, each method shall evaluate and indicate whether processes operated in their full potential and whether they deliver the expected outputs. They shall refer to a basis for comparison or a reference point against which process outputs can be evaluated, for example, criteria that will enable decision making regarding the performance of a process—whether it has achieved its objectives or not. A simple example is the instruction for quality test, a document that specifies what the expected product's characteristics are. This is the base for the evaluation. The monitoring, measurement, analysis, and evaluation in this case should represent trends and changes in the process or the product.

Evaluating the Effectiveness of the Planning of the QMS

The ISO 9001 Standard requires that the effectiveness of the implementation of the planning of the QMS be regularly analyzed and evaluated. Evaluating the effectiveness in this case means measuring whether the processes, operations, and activities of the QMS are implemented as planned. The purpose of this evaluation is to detect necessities for the allocation and optimizations of the QMS activities. Evaluating the planning of the QMS refers to the different elements of the QMS that are used to operate it. This type of evaluation considers qualitative activities such as

- Planning—the evaluation shall assess whether planning activities are performed as required by the ISO 9001 Standard:
  - Processes are planned with process approach.
  - Actions to address risks and opportunities are applied.
  - Quality objectives are defined and measured.
  - Changes are identified and submitted to control.
- Resources—the evaluation shall assess whether the planned resources that are supposed to support the processes are applied:
  - The appropriate human, knowledge, and competence resources were allocated to the operations.
  - The appropriate infrastructures are available and maintained.
  - The appropriate process environment is provided and maintained.
- Product and service requirements—the evaluation shall assess whether activities necessary to identify product requirements are implemented:
  - Review of requirements for product or service.
  - Customer communication activities.
- Processes—the evaluation shall assess whether all process activities are performed as planned:
  - Inputs are flowing as required.
  - Controls are applied as planned.
  - Results of process are controlled against criteria.
  - Outputs are released as planned.
  - New processes or modified processes achieve their objectives.
- Design and development is being managed as required:
  - A design and development plan is in place.
  - The design and development is verified, validated, and reviewed.
- External providers:
  - The controls over external providers are defined and implemented.
  - The external providers are controlled and evaluated.
- Production and service provision:
  - Identification and traceability are controlled and activities to manage them are defined.
  - Postdelivery activities are defined and controlled.

Tools for obtaining data regarding the effectiveness of the planning include the following:

- Internal and external audit—audits provide you with data regarding the degree of implementation of processes against their planning.
- Inputs to the management review—those inputs include information regarding the implementation of the QMS.
- The degree of implementation of a quality plan—which defines the planning of processes.

When evaluating the planning of the QMS, the results shall indicate whether the processes are implemented, partly implemented, or require a corrective action. The results may be qualitative and quantitative.

Evaluating the Effectiveness of Actions for Addressing Risks

Effectiveness of actions for addressing risks refers to the achievement of their goals: mitigate, eliminate, or reduce the risk to an acceptable level. The objective is to create an iterative process of ever-improving the controls. Evaluating the effectiveness is done through two aspects:

- Verification that the required controls were implemented as planned
- A verification that the controls reduce the risks as planned

The results of the evaluation shall indicate

- Whether further controls are required
- Whether to accept the risk or to plan the reviewed QMS element again

This issue is discussed in detail in chapter 6.1—Actions to address risks and opportunities under paragraph Evaluating effectiveness of actions to address risks and opportunities. Please review this paragraph for more detail.

Evaluating the Performance of External Providers

Evaluating performance of external providers refers to their abilities to deliver products or services in accordance with predefined requirements. An effective and objective evaluation considers the significant parameters regarding the purchased products or services. The objective is to establish an ongoing control process over the external provider in order to foresee events that might become nonconformities or quality problems. The evaluation shall assess the risks and effects of the purchased product on the end product.

This issue is discussed in detail in chapter 8.4—Control of external provision of products and services under paragraph Evaluation of external providers and the following paragraphs. Please review this paragraph for more details.

Evaluating the Need for Improvements to the QMS

Last but not least is the requirement to provide data and information that will suggest the need for improvements to the QMS. Analysis of data and information related to the quality goals mentioned earlier will provide information regarding situations of areas in the organization or scopes of the QMS and will indicate their need for improvement. The evaluation of the data can lead to improvements, redesign, or reengineering of business activities, and the results of the analysis and evaluation will allow the organization to determine whether certain opportunities shall be forwarded to the process of continual improvement. The data and information that will be analyzed shall refer to

- The degree to which products or services answer the expectations of customers or regulatory requirements
- The influence of nonconformities on product or service characteristics
- The performance of the processes and operations of the QMS

The evaluation of the information shall suggest

- Improvements of elements or scopes of the QMS
- Improvements of realization processes
- Improvements of products or services

How these data and information shall be used and how improvement shall be developed will be discussed in chapter 10—Improvement.

Selecting Methods for Analysis and Evaluation

In a note the standard indicates that methods to analyze data can include statistical techniques. And I would like to add some inputs regarding selection of a method for the analysis and evaluation:

- A method should be selected upon its ability to provide data that will accurately indicate process performance.
- The relevant personnel shall have the qualifications to use those methods, collect the data, and analyze it. When required you may include the issue in requirements for competence of persons who operate the relevant processes.
- The methods shall support and provide relevant personnel with the correct data needed to evaluate the progress of a process with relation to its desired results or objectives.

9.2 Internal Audit

Internal audit is an effective tool that is used for self-assessment of the organization and to determine the extent to which the QMS requirements are fulfilled. The results of the audit (the audit findings) shall demonstrate the effectiveness of the QMS and identify nonconformities and opportunities for improvement. Let us review the ISO 9001 requirements.

9.2.1

- The organization shall conduct internal audits at planned intervals.
- The internal audit shall provide information on whether the QMS conforms to the organization's own requirements for the QMS.

- The internal audit shall provide information on whether the QMS conforms to the requirements of the ISO 9001 Standard.
- The internal audit shall provide information on whether the QMS is effectively implemented and maintained.

9.2.2

- The organization shall define, implement, and maintain an internal audit program. The program shall refer to
  - The frequency and intervals of the audit
  - The methods for conducting the audit
  - Roles and responsibilities that take part in an internal audit
  - Planning requirements
  - Reporting the results
- The organization shall define the scope of the audit.
- The results shall include reference to
  - The importance of the audited processes, activities, or operations
  - Changes that may affect or might have affected the organization
  - The results of previous audits
- The organization shall define the criteria for the audit.
- The organization shall select auditors and conduct the audit to ensure objectivity and the impartiality of the audit process.
- The organization shall ensure that the results of the audit are communicated and distributed to the appropriate relevant managerial levels.
- The organization shall ensure that nonconformities are addressed and corrective actions are applied without unnecessary delays.
- The organization shall retain documented information referring to the findings and results of the audit as evidence of the implementation of the audit program and the audit results.
- Remark—the ISO 19011 Standard may serve as guidelines for establishing and implementing QMS auditing.

Audit Program

The organization must maintain a documented program for conducting audits (internal and external). An audit program is a series of steps or specifications required by the organization in order to be able to conduct the audit. The goal of the program is to identify the required organizational elements that will be audited and determine when they will be audited. The program has several objectives:

- The audit program will ensure that the audits are conducted as planned.
- Through publication of the program, employees and personnel will understand that the internal audit is a continuous measure of the QMS and not a capricious decision made by the top management.
- The audit program shall introduce the auditor with the scope and objectives of the audit (fields, subjects, departments, locations, sites, products, areas, roles, processes, or the specific status of processes).

- The program shall specify the authorities and responsible parties that will participate in the audit (the auditor or audit team, employees, specific roles, management representatives, technical experts, etc.).
- The program shall detail the resources required for the audit (meeting rooms, personnel, records, products, production lines, etc.).
- The program shall give a description of the agenda or topics and issues that will be audited and discussed.
- The program shall indicate scheduled time frames for the different audit stages.

It is recommended that you publish and communicate the audit program. The program can appear as a list or a procedure. The following is a suggestion of an audit program:

Example of an audit program
Time and date for part 1 of the audit:
- 8:00–12:00 12/12
Deportment:
- Assembly
Topics to be reviewed:
- Accomplishment of work instructions
- Accomplishment of cleanliness instructions
- Performance of trainings
- Nonconformities
Resources:
- All work instructions are to be available
- All required records are to be available
References:
- Audit plan
Time and date for part 2 of the audit:
- 13:00–17:00 12/12
Deportment:
- Warehouse
Topics to be reviewed:
- Accomplishment of work instructions
- Accomplishment of cleanliness instructions
- Performance of trainings
Nonconformities
Resources:
- All warehouse employees are to attend the audit
References:
- Audit plan

Some organizations also include detailed tests, investigations, or examinations that are conducted during the audit. I will refer to it as the "plan" of the audit. The program must be documented and will be submitted to the controls suggested in clause 7.5—Documented information.

Audit Scope

The first step in implanting effectively the internal audit is to define the scope of the audit. The audit scope defines the extent and the boundaries of the internal audit: the areas and limits to which the audit is applicable and which the audit must control. The scope shall cover the following issues:

- The purpose of the audits (e.g., compatibility to the ISO 9001 Standard requirements).
- Physical locations (like branches or affiliations).
- Different organizational units like divisions or departments.
- The relevant or related activities and processes to be audited.
- The time period to which the audit results are valid and after this period a new audit shall be conducted (a year, a quarter, a month).
- Additional standards or regulatory and contractual requirements that may serve as audit criteria.
- The expected records.

This scope will be managed and implemented through the audit program and plan (which will be discussed later on in this chapter).

Audit Plan

The audit shall include and specify the activities necessary for organizing and conducting the audit and describe the required resources for the audit. For this purpose you may define and use an audit plan. By audit plan I mean a specification (may appear on a document) that describes all the tests and examinations that must be performed during an audit and needed to be done in order to evaluate the extent of meeting requirements. The plan will direct the auditor during the audit. The objective of the plan is to ensure that all aspects of the scope are covered and to allow auditees to prepare and organize for the audit. While auditing the QMS, you are required to evaluate it through three distinct aspects (each of them will be discussed in detail separately):

- Implementing all the relevant ISO 9001 Standard requirements
- Achieving the organization's requirements for its QMS: quality plans, processes and procedures, and maintenance of records
- Extent of effectiveness of implementing the QMS: achieving quality objectives, customer satisfaction, and improving the QMS

The audit plan should include reference to the following issues:

- The audit scope
  - Locations where the audit activities are to be conducted
  - The audit topics
- Time schedules
- Information for the opening of the audit
- Changes in the QMS that must be mentioned before the audit begins
- Results of the last audit
- Open nonconformities

- Required resources
- The audit criteria and reference to relevant documents
- The auditor or the audit team
- Roles and responsibilities of the auditee
- Other required accompanying persons
- The audit tests, inspections, and examinations

You may manage a general plan that will refer to all the organizational units and will ask to evaluate performance of procedures and work instructions, evaluate quality procedures, and sample evidences of executing those processes. Such a plan will be applicable to the entire organization. But I recommend adopting a more specific plan designed to audit one specific organizational unit and is therefore not applicable to other units. Such a plan will refer to specific processes related to this unit, will consider the interrelations of this unit with other organizational units, will present with the appropriate criteria for evaluation, will examine specifically the quality requirements of this unit, and will ask to review documented information related to its processes. This plan is more effective; for example, if the auditor is auditing a warehouse, a specific plan will direct them to the appropriate processes and activities, support them with the right criteria, and describe to them which records must they sample.

Defining Planned Intervals for Internal Audits

The audits will be planned and conducted at planned intervals. The intervals shall be documented on the audit program. The goals of planning ahead are

- To allow the different units of the organization time to prepare for the audit
- To maintain a periodical plan for internal audits (usually an annual plan)
- To ensure the continuity of the audit

In practice, you may define appointments for the audit on the organizational calendar. The ISO 9001 Standard does not require specific dates but it is more practical. If you would like to perform "unexpected" audits, then define them on the program but do not publish them.

Conformance of QMS to the Organization's Requirements

The organization has determined its own requirement of the QMS while designing and developing it. The internal audit is one instrument for self-review of whether those requirements are achieved. Which type of requirements?

- Quality plans for the product: Any requirement for product realization must be evaluated on whether it was performed as planned. The best way is to sample and evaluate a product against its relevant predefined criteria. Sample a product or an output of a process, review its quality plan, detect its specifications, and check whether the product was realized according to the plan. Document the results.
- The identification and implementation of international, national, or local regulations will be evaluated. The audit shall examine the identification of appropriate regulations, their introduction to the quality processes (such as management review,

or integration in the training program), implementation throughout the realization processes, and maintenance of the appropriate records.

- Processes and procedures: The audit must evaluate whether the realization processes are performed as required. It could be correlated with the quality plans. The evaluation would refer to required results or predefined criteria. Generally speaking, an audit must sample the processes and outputs and evaluate two things:
  - Whether the process is managed as planned: according to its definition (process chart, diagram, SOP, etc.)
  - Whether the process achieves its objectives: if the parameters of the process are valid and the outputs are as expected

Conformance of QMS to the ISO 9001 Standard Requirements

Aside from the organization's requirements, the audit shall evaluate the implementation of the ISO 9001 Standard requirements. The implementation of the ISO 9001 Standard requirements for quality processes includes, for example,

- The specific documentation requirements
- Addressing risks and opportunities
- Implementing the required purchase controls

This part of the audit must be conducted throughout the entire organizational units related to product realization, or are under the scope of the QMS. For each organizational unit, it is required to audit how much quality management tools are implemented, for example, management of resources, knowledge and competence, maintenance of quality documented information, or management of nonconformities.

Effectiveness of the Implementation and Maintenance of the QMS

The audit must provide the ability to evaluate whether the QMS is effective or not. The effectiveness of the QMS refers to the extent to which objectives of processes were achieved and may indicate how much the QMS has improved. This will be reviewed through two aspects of the QMS:

- Quality objectives: The audit must indicate whether the organization has achieved its quality objectives. During the audit the relevant quality objectives will be reviewed for their relevance to the process and how they are measured and to what extent they were achieved (the results of the measurement). When it is found that an objective has not been reached, the cause must be presented and the measure taken to handle it.
- Improvement of the QMS: The audit must indicate whether measures and actions were undertaken in order to improve the QMS and whether those actions were effective— the improvement was achieved.

The Auditor

Conducting the audit pretty much depends on the personal and professional competence of the auditor. It is equivalent for appointing an experienced investigator. The auditor shall be objectively related to the organizational unit or function that he or she is auditing and try as much as possible to conduct the audit objectively and impartially. This alone demands from the auditor the ability to

- Accept opinions that are different from his or hers
- Look at things with open- or broad-mindedness
- Understand and accept other points of view of the issues that are being audited

Beside its personal approach, an auditor must have a minimum acquaintance with the field of the organization in order to evaluate the processes and their results beyond the working procedures, work instructions, and documentation (the documented criteria). This kind of knowledge can give him or her the ability and the consideration to evaluate situations while identifying any nonconformities or faults. The ISO 9001 Standard refers us to the ISO 19011 Standard (guidelines for quality and/or environmental management system auditing), which specifies the required auditor qualities:

- Ethics: An auditor will possess personal characteristics like credibility, integrity, and honesty and provide reliable information and results regarding the unit the auditor is auditing.
- Open-minded: An auditor will be willing to listen, learn, and accept new ideas and to reflect them on the situations or requirements. Sometimes auditor may encounter new approaches or opinions. An auditor must have the ability to assess and accept different points of view as long as they achieve the requirements.
- Diplomatic: An auditor will be polite and well-mannered; an auditor serves as the representative of the top management.
- Observer: An auditor will have the ability to recognize and evaluate what he or she sees and to understand and interpret events without deep interrogation.
- Perspective: An auditor will have the ability to evaluate situations beyond the person's appearance, with a systematic view of things, and will have the ability to understand the organizational consequences of the evidence he or she finds.
- Versatile: An auditor will have the ability to mobilize from one situation to another without losing direction. One moment the auditor may audit one field; the next moment it may be another. An auditor must be able to stay focused.
- Structured: An auditor will advance and progress the audit according to a defined method or plan and within the boundaries of the scope.
- Persistence: An auditor must be persistent with their objectives, so that when they ask a question, they must receive an answer to it and not be diverted by interferences or disturbances.
- Independent: An auditor shall have their own opinions on things, will not be influenced by the environment, and will be free from conflict of interest.
- Decisive: An auditor must be ready to make decisions even when they are hard or will not satisfy the auditee.

Roles and Responsibilities

The audit program shall specify the responsible parties that will participate in the audit. The roles may induce the following:

- The auditor or audit team
- The management representatives

- Employees or specific roles such as technical experts, specific managers, or specific functions that will be audited
- External parties such as specialists or consultants

The top management has a special responsibility; it is obligated to monitor the internal audit process by

- Ensuring and verifying the conducting of the audits according to the program and the plan
- Reviewing and evaluating the results of the audit
- Verifying that all nonconformities detected during the audit are addressed and closed

Submitting regularly the results of the audit to the management review (as required in clause 9.3.2—Management review inputs c, 6) answers this requirement. In big organizations, I used to perform a regular internal audit in the different units or affiliates (according to a planned program). The results were then sent to a contact person at the top management on a regular basis.

Audit Criteria

In order to conduct effective tests or inspections, you need to assign and document criteria to each test. Criteria are set of policies, procedures, or requirements used as a reference against which audit evidence is compared. The objectives of the criteria are

- To support decisions for judging, evaluating, and determining by facts, values, and data the compliance of the outputs to the requirements; the auditor samples a process or process output, views it, turns to the criteria, and decides whether the process was effective or not
- To enable the determination of the extent of conformity of processes or process outputs

The audit plan shall refer each test to its appropriate criteria. The criteria may be quantitative such as tolerances and limits of processes, number of trainings per year, and number of accepted nonconformities or qualitative such as estimations of knowledge and skills. One important property of the criteria is the ability to evaluate audit findings against it. In other words, the criteria must be adjusted to the type of findings; if you are auditing the processes of certifying new personnel, the criteria should be a procedure or a checklist that describes which activities must be accomplished when accepting a new worker.

The criteria will provide a successful validation by indicating whether the findings are accepted or rejected. The criteria will present a method for the evaluation and will refer not only to products, parts, or components but also to realization processes and conditions for realization. Types of criteria include

- Policies
- Working instructions, test instructions, and procedures
- Drawing and specifications
- Management system requirements

- Quality plans
- Standards and technical specifications
- Laws, regulations, and directives
- Documented customer requirements like orders or contractual requirements
- Industry or business sector codes of conduct

Audit Results (Findings and Evidences)

During the audit, the auditor samples, observes, asks questions, examines processes or processes outputs like records or products, or observes locations where a service was provided. The inspections or observations shall be considered and recorded as the audit findings. The findings will be used for the audit's conclusions and will describe what the auditor saw and revealed. These shall then be compared to the predefined criteria and shall indicate the status of the processes or products: conformity, nonconformity, or opportunity for improvement. The findings include the evidence that the auditor reviewed (e.g., processes, products, records); for each, he or she may document

- What was presented or found—whether it is a product or a record
- The context: what is the relevant process or product
- The requirement or the relevant criteria
- The significance of the process to the QMS and its influence on the quality of the product
- Changes that may affect the QMS (will be discussed in the next paragraph)
- Results of the previous audits (will be discussed in the next paragraph)

Indicating the significance is important for the classification of the findings, particularly when the auditor reveals a deviation from the specifications and must decide whether to submit it as a nonconformity (when the process is significant) or just as an opportunity for improvement. That is why the auditor must have the knowledge, the ability, or at least the minimal understanding of the area he or she is auditing for deciding what is significant and what is not.

Accuracy of the documentation is important for the next step; the classification, for example, good, opportunity for improvement, or corrective action, is required (which will be reviewed later). This is the most important part of the audit report.

Changes Affecting the Organization

One of the issues that must be reviewed during the audit is the effect of occurred changes on the QMS and their consequences. If such changes are relevant, they and their consequences will be mentioned to the auditor at the beginning of the audit. A good example is changes in the product—the auditor must know about such a change in order to assess whether it was properly implemented. Changes to the QMS are referred two times on the standard:

- 6.3—Planning of changes—where it is required to develop a method for planning the implementation of changes
- 8.5.6—Control of changes—where it is required to ensure that changes do not affect the ability of the organization to continually provide goods or services that conform to the requirements

Those two aspects of changes will be reviewed during the internal audit, and the auditor will review

- Changes that occurred in the QMS since the last audit
  - Changes in the specifications or requirements of the product
  - Changes in processes
  - Change in resources
  - Changes in documentations
- Changes that were submitted to a planned method for implementation
- Changes that were reviewed for assurance that they do not affect the ability of the organization to continually provide products or services that conform to the requirements

In practice, the audit plan shall include references to changes in the QMS, shall indicate their consequences and the results of the audit, and shall prove that those changes were reviewed.

Results of Previous Audits

The results of previous audits shall be referred in the internal audit. This requirement refers mainly to nonconformities that were revealed during previous audits but may refer also to opportunities for improvement. The organization shall prove to the auditor that

- For each nonconformity, a corrective action was taken within the scheduled time frame, the treatment was effective (objectives were achieved and the nonconformity was removed), and the nonconformity did not occur again
- Where opportunities were adopted, the implementation was effective

In practice, the audit plan shall include references to the results of previous audits, and the results of the internal audit shall prove that those findings were addressed.

The Classification of Audit Findings

Any findings during the audit shall be indicated with one of the following classifications:

- Conformity—the process or product sampled was in accordance with the relevant requirements and criteria.
- Opportunity for improvement—in the auditor's opinion, an improvement can be applied to the matter, and the organization may or may not adopt this opportunity and submit it to the control of opportunities as required in clause 6.1—Actions to address risks and opportunities.
- Nonconformity—the process sampled was not according to the requirements and the audit criteria.
- If the organization feels the need to add another classification suitable to its nature or processes, it may do that.

This classification is important for the auditee to know afterward where he or she must invest resources; nonconformities must be removed while opportunities for improvements may be considered.

Nonconformities Revealed in the Audit

Nonconformities revealed during the audit must be addressed, documented, and submitted for correction. The nonconformities may be documented three times during the audit process:

1. Within the audit report along with the audit findings. We can also refer to it as the report itself.
2. Where it is suitable, as nonconformities. Any audit report should contain a summary of the nonconformities revealed during the audit.
3. As an input for a corrective action followed by the audit.

When nonconformities are revealed, they should be applied to a controlled process. The ISO 9001 Standard specifically requires that for each nonconformity, a decision and an action will be determined without unnecessary delays in order to ensure that they will be handled and removed. The goal is to verify that the nonconformities are removed or eliminated and will not be repeated. In order to ensure the submission to the correction, you need to initiate the interrelation between the internal audit process and the control of nonconformities (as required in clause 10.2—Nonconformity and corrective action), where the outputs and the nonconformities discovered in the audit shall serve as inputs to the control of nonconformities and opportunities shall serve as inputs to the continual improvement process.

Communicating the Results

Any audit must have a summary report that will communicate the results to the appropriate persons in the organization. The auditor should gather all the information, data, findings, nonconformities, and opportunities for improvement and present them together in one report. The goals are to provide the organization with a status report regarding the QMS and for follow-up during the next audit, to review the treatment, and to verify that all nonconformities are closed. Who are the target groups of this report?

- The function that is responsible for the auditee
- The auditee—workers of personnel if the organization finds it appropriate
- The top management

This report is a tool for those target groups for understanding the status of the QMS and of the organization with reference to the requirements or the criteria. Therefore, it is recommended that the report be designed in a format that would be easy to understand.

Retaining Documented Information

The documented information of the internal audit is the output of the audit process. The organization shall retain documented information as demonstration for the planning of the audit as well as evidences for implementing the program and plan and conducting the audit activities. The records of the internal audit have several goals:

- Planning the audit and the audit activities
- Ensuring that the audit was conducted according to its scope
- Proving that activities were performed according to the plan
- Proving that objectives of the audit were achieved

The documentation shall be submitted to the control of documented information as required in clause 7.5—Documented information. The ISO 9001 Standard chooses not to refer to the distinction between two types of the documentations:

- Documented information used for the planning of the audit such as the audit program and plan
- Records and evidence of the implementation of the audit program and the audit results such as audit protocol or audit report

Which documented information may we expect?

- Audit program—a document describing the program
- Audit plan—a document describing the activities of the audit
- The audit report
  - The reviewed issues such as changes or follow-ups of previous audits
  - The findings of the audit—a documentation of what was inspected and observed
  - List of detected nonconformities and opportunities
  - Summary report

ISO 19011

The ISO 9001 Standard refers us to the ISO 19011 Standard—Guidelines for auditing management systems and for more references and guidance regarding planning and conducting the internal audit. I definitely recommend reviewing the ISO 19011 Standard because it broadens the issue of the audit and provides good techniques and ideas for conducting an effective internal audit.

9.3 Management Review

The management review is a management tool in the hands of the top management for evaluating the QMS. It is one method (among others) dictated by the standard for monitoring the QMS and evaluating its performances and effectiveness. The management review is an activity usually performed as a meeting where representatives of the top management are presented with data and information regarding the performances of the QMS. The objective of the review is to give the top management a chance to periodically evaluate the QMS. Let us review the ISO 9001 Standard requirements.

9.3.1 General

- Top management shall review the QMS at defined intervals.
- The review shall be planned in advance.
- The review shall ensure the consecutive suitability, adequacy, and effectiveness of the QMS.
- The review shall ensure that the QMS is planned with alignment to the strategy of the organization.

9.3.2 Management Review Inputs

For effective review and decision making, the management review shall take into account the following issues:

- Follow-ups from previous reviews and status of decisions and actions taken.
- The review shall discuss changes in external and internal issues that are relevant to the quality management and their influence on it.
- Information regarding the performance of the QMS with reference to trends and indicators such as
  - Results of customer satisfaction and feedback measurements
  - The extent to which quality objectives have been met
  - Process performance and product conformance supported by evidences
  - Audit results: internal, external, and customer audits
  - Information about nonconformities and status of corrective actions
  - Results of monitoring and measurement activities
  - Issues related to suppliers and external providers that may affect products, services, and the QMS
- The adequacy of resources.
- The effectiveness of actions taken to address risks and opportunities.
- Opportunities for improvement.

9.3.3 Management Review Outputs

- The outputs of the management review shall include decisions related to
  - Actions for improvement or further decision regarding opportunities for improvements
  - Any need for changes to the QMS
  - Any need for resources
- The inputs and the results of the management review shall be retained and handled as documented information. Evidences for actions taken as a result of the management review shall be retained as documented information as well.

Goal of the Management Review

The ISO 9001 Standard clearly specifies the goal of the management review—to ensure its continuing suitability, adequacy, effectiveness, and alignment with the strategic direction of the organization of the QMS. Let us interpret it:

- The strategic direction of the organization was determined while discussing the context of the organization and should be documented (e.g., in the quality manual in case you are still choosing to maintain one). In practice, you may review the quality policy in order to ensure its suitability to the purpose of the organization.
- To ensure the continuing suitability of the QMS refers to the review whether characteristics of the QMS are appropriate for their objectives.
- To ensure the adequacy of the QMS refers to the review whether the QMS can act correctly in order to fulfill its requirements and achieve its objectives.
- To ensure the effectiveness of the QMS refers to extent of achievement of quality objectives.

Participants of the Management Review

Participants of the management review are clearly defined by the standard—the top management. But who is considered as the top management? According to the ISO 9000, the top management consists of persons or group of people who direct and control an organization at the highest level. To this definition I choose to add a person or a group of persons who have the authority to appoint resources for the QMS. But when the scope of the QMS refers only to a part of the organization, for example, the logistic center or the service center, the top management is not considered as the board of directors of the whole organization but the organizational role who steers and directs the logistic center or the service center. This fact makes it simpler to define the relevant top management.

Intervals for the Management Review

In the standard it is written specifically that the management review shall be held *at planned intervals*. The ISO 9001 Standard knows that the top management is busy and sometimes the QMS is not on its top priorities. Sometimes it is really hard to summon the top management and to convince them that the management review regarding the situation of the QMS is important. And when an appointment is already set, it is exposed to postponements. The standard shows understanding but insists that the QMS shall be reviewed by the management at planned intervals. Planned intervals refer to a planned period of time from the last review. The interval will be determined with accordance to the nature and complexity of the activities. There are two main goals for this requirement:

- The management review must be performed on a regular basis in order to ensure continuing involvement of the top management in the QMS.
- The periodic management review allows the top management to determine trends of the QMS and degree of achievement of its objectives.

I suggest here a practical method for obtaining this requirement. As I functioned as a representative of the top management, I recommended that the QMS will be reviewed by me (a delegate of the top management) and other delegates from the different organizational levels (who are not considered as top management) once a week as a quality circle. In addition, the entire top management will review the status of the QMS twice a year. In cases and situations where a higher decision or intervention is needed, an exceptional meeting with the rest of the top management will be summoned ad hoc. In order to institutionalize the matter, I created a procedure that defines it. But keep in mind that a procedure is compelling and you will be requested in an audit to demonstrate the implementation of this procedure.

Inputs for the Management Review

As you might have noticed, the standard is quite explicit regarding the information and data that are required to be reviewed during the management review. This list of inputs for the management review should reflect trends in the QMS and involves aspects of performance, control, consulting, changes, improvements, and risks. The information (the processed inputs) should assist in determining the achievement of planned results. The challenge here will be to identify the sources of data and the responsible parties in the organization that can provide these data, collect them in an orderly manner, and present them in an acceptable way. From my experience, I can give you the next advice—do not wait until month or two weeks before the review to summon and organize the data; you will find it hard and one cannot ensure that the data will be reliable. Instead, try to create a system of reporting according to the management review that will generate the data on itself. The objective here is to automate the gathering of the data. For example:

- If you are implementing a system that gathers and monitors nonconformities, make sure this system can provide you with a report appropriate for the management review: the nonconformity, the details, the corrective action that was taken, and the effectiveness of the corrective action.
- If you are implementing a system to monitor and control processes (SPC), try to see whether this system generates a report with annual results that show trends of the processes.
- If you implemented a system to collect and analyze customer feedback, I am almost positive that such system can offer a summary report.

The data and information may be presented and maintained as documented information that will support top management in discussing and making decisions: results of surveys, reviews, audits, summary reports, presentations, and statistical charts. What are the required inputs?

Follow-Up Actions from Previous Management Reviews

Decisions for actions determined in previous reviews will be presented and discussed: what was decided, what was carried out, and what was turned down, rejected, or postponed. The review will examine not only whether they were applied and implemented but also whether the actions were effective and achieved their objectives.

Changes Affecting the QMS

Changes regarding internal and external issues that are part of the QMS and might have affected the QMS or the product will be reviewed for their influence and impact on the QMS. It refers to all events that occurred since the last management review and were not referred by the top management. This is a chance for the top management to be informed of those changes and address them. The review shall relate to the nature of the change and its relevance to organizational issues—internal or external.

Internal issues relate to internal strategies, policies, and objectives that steer the QMS and to the resources that are required to implement and achieve them. For example, an implantation of a new ERP system is an internal issue of high importance because it refers directly to the processes and has a direct impact on the operations of the QMS. When such event occurred in the organization, the management review should assess it (as a change) and its consequences. Which internal changes may we encounter?

- Changes in product specifications or customer specifications
- Changes in the production installations
- Organizational changes
- Changes concerning human resources

External issues are environmental factors in which the organization is active: political, legal, regulatory, social and cultural, financial, technological, economic, and natural. When events regarding those issues occurred in the organization and had an impact on the QMS, the top management must review it: receive the information about it and understand the impact on the QMS. For example, when a regulation relevant to the realization of the product has been changed, the top management will review the change and the requirements for the implementation of new or revised regulatory requirement. Which external changes may we encounter?

- Political changes
- Regulatory or legal changes
- Economic changes
- Technological changes
- Changes in work regulations

The type of review of changes will ensure that changes that affect the QMS (internal and external) have been taken into account and when applicable the quality policy has been updated appropriately. In practice, when changes to policy were required, add (in the documentation of the management review) that the policy was updated. But pay attention; if you did change the policy, be sure to submit the revision to the requirements of clause 7.5—Documented information. A good example is the revision of the ISO 9001 Standard—I am convinced that your current quality policy refers to the last revision ISO 9001:2008. It will be necessary to update the policy and change the reference to the new revision of the standard. I recommend that this will be reviewed and referred in the management review.

Status of Quality Objectives

One of the most important purposes of the management review (in my opinion) is the examination of quality objectives. The management review shall review the extent to which quality objectives have been met. Quality objectives are an efficient tool for maintaining the effectiveness of the effective QMS, and the control and review of the top management play an important key role in this mechanism. The examination will focus on two areas:

- The achievement and fulfillment of quality objectives like quality plans, objectives of performances, marketing objectives, customer-related objectives (e.g., complaint or feedback related), or quality management objectives (e.g., audit or training related).
- The compliance of the objective and their level of effectiveness concerning the QMS. This examination will determine whether the objectives are appropriate to the nature of the organization (will be discussed in details in the next paragraph).

Result of Audits

Evidences of whether the organization's activities and processes conform to planned arrangements and are effectively implemented and maintained will be presented through audit results. This includes results from internal, external, supplier, and customer audits. The review shall refer to the findings, the rejections, the required actions, and the recommendations for improvements. Results of audits demonstrate the compliance of the auditee against the given criteria.

Result of Monitoring and Measurement Activities

The result of monitoring and measurement activities will be introduced as inputs to the review. The results shall indicate the achievement of process objectives and conformity of process outputs. During the review, the top management shall evaluate the results and decide whether they are sufficient or improvements are needed. The format and manner with which the results will be presented depend much on the type of data and information.

Customer Satisfaction and Feedback

Information and results of monitoring customer satisfaction and feedback will be presented and will enable the top management to evaluate whether requirements and expectation have been met. The information that will be presented

- Shall reflect the performance of the QMS, which means it shall shed light on how the activities of the organization affect customer satisfaction
- Shall allow the top management to understand how the customer perceives the organization, its processes and activities, and the products or services

This review may include passive feedback data that were received in the organization like customer complaints and feedback activities that the organization initiated such as customer feedback surveys or analysis of data gathered during service activities.

Supplier and External Provider Issues

Issues related to external providers and suppliers shall be presented in the management review. The objective here is to present the issue, review the performance of the external providers, and explain the impact on the QMS. In case a problem was discussed to which a solution was not found, a decision of the top management is required here.

Process Performance and Product Conformity

Data regarding process performance and product conformity are needed to demonstrate ability of processes and quality of products. Evidences that demonstrate the ability of realization processes to achieve planned results and the conformity of products (process outputs) to the acceptance criteria will be presented. The information will include data from statistical analyses of product characteristics and process performances. In other words, allow top management to evaluate the performance of processes and products in comparison with the criteria. Try to reflect successes and failures of certain cases onto other cases.

Nonconformities and Corrective Actions

Status of actions taken to eliminate the causes of nonconformities or potential nonconformities since the last review will be discussed. Nonconformities and the corrective actions that were initiated throughout the organization are to be discussed and analyzed. In particular, relate to the effectiveness of the actions and to the extent to which objectives were achieved. Again, try to analyze and reflect successes and failures. I bet that the top management would be highly interested in this information because this is their opportunity to be informed of those cases. In practice, I would prepare in advance a report that gathers all the nonconformities and indicates the trends in the QMS.

Adequacy of Resources

Adequacy of the resources of the QMS shall be discussed in the review. The goal here is to indicate how the resources perform and how well they meet their requirements. This can be expressed through the following perspectives:

- Relevant nonconformities
- Related risk or opportunities
- Data of performance
- Suggestion for improvements

For example, when an infrastructure cannot support the activities of the QMS anymore and must be replaced or improved, the management review is an opportunity to discuss the matter with the top management.

Opportunities for Improvement

Opportunities for improvements in the QMS will be reviewed. The top management must review each opportunity, assess its impact on the QMS, refer to internal and external issues, and decide whether the improvement is feasible or not. Improvements may occur in

- Products or services
- Achievement of customer requirements
- Realization processes and quality management processes

The source for opportunities must be identified:

- Employees—one of the most common sources for opportunities for improvement is employees. In many organizations, I encountered a form that allows an employee to suggest an improvement. It is very effective and normally the employee is rewarded when a suggestion is adopted. But the true question is as follows: Where does this form end up, and how is it reviewed?
- Audits—another effective source is recommendation for improvements from audits; an objective party had visited the organization, sampled activities or product, detected an opportunity for improvement, and suggested it.
- Customer feedback—suggestions from customers may indicate unfulfilled needs and may serve as opportunities for improvement.

Addressing Risks and Opportunities

A new requirement of the ISO 9001:2015 is addressing risks that may prevent the organization from achieving its intended results or create undesired effects in processes, products, or services and addressing opportunities that will reduce the probability of undesired effects and enhance customer satisfaction. Undesired effects include nonconformities, not achieving customer requirements, and the lack of customer satisfaction. Another aspect of risks is risks related to changes and improvements that were implanted in the organization and their potential impact on the organization and its processes. This issue is well discussed in chapter 6.1—Actions to address risks and opportunities, where the organization is required to develop and implement plans and activities for this identification and handling of related risks or opportunities. The management review shall refer to those issues, examine new risks and opportunities, discuss their relevance, and assess the effectiveness of the implemented plans for reducing or eliminating those risks. Which inputs related to risk and opportunities can we expect?

- Risks or opportunities related to strategic changes
- Risks or opportunities related to availability of the resources for future activities or potential scarcity of resources
- Risks or opportunities related to infrastructure and their use
- Risks or opportunities associated with the relationships with external interested parties such as suppliers or customer (partners)
- Risks or opportunities related to changes in knowledge and technology
- Risks or opportunities related to new developments
- Risks or opportunities that were identified in audits

Management Review Outputs

The outputs of the management review are decisions and actions that must be carried out as a reaction to the data and information that were presented and discussed during the review. In practice, the outputs should include decisions and actions related to

- Opportunities for improvement of the QMS
- Need for changes in the QMS
- Needs of recourses

Implementing Decision and Actions

The decisions and required actions of the management review shall be presented and suggested in a way that will facilitate the implementation of activities for process improvements. In other words, try to make feasible decisions instead of calls for actions lacking in meaning. Your objective is not to publish the wishes for improvements but to allow and enable their implementation:

- Be precise—which business activity must be changed or improved.
- Delegate responsibilities and indicate who will be in charge of implementing.
- Set definitive objectives for the decisions.
- Allocate resources.

Always bear in mind that in the next review, you are required to assess the success of those decisions. For example, instead of declaring "we must enhance our customers' satisfaction," try "we should enhance customers' satisfaction during installation activities." The second form indicates a precise business activity that may be improved and ease the implementation of such decision. Plus in the next review, it will be easier to show whether an improvement was achieved or not. These decisions should be communicated in the organization in order to create awareness among interested parties (intern as well as extern).

Opportunities for Improvement

The outputs regarding those opportunities for improvements shall be formulated with actions: which action shall be undertaken in order to promote those improvements. The outputs regarding those opportunities for improvements shall be formulated with actions, which action shall be undertaken in order to promote those improvements.

Benchmarking may serve as a tool for implementing improvement as output of the management review. The objective of benchmarking is to allow the organization to develop plans for improvements; the management had just reviewed the status of the QMS and recognized the need for changes or improvements. Now it is recommended to develop plans on how to plan resources, implement the tools and techniques, or adapt specific best practices, in order to achieve those improvements. But benchmarking may be quite a challenge for the organization.

There are many methods for conducting benchmarking. This is done through a research that brings data and information and eventually points out the critical topics that require reference and improvement. Which methods can we use?

- Interviews with customers, employees, or suppliers
- In-depth marketing research
- Quantitative research
- Surveys
- Questionnaires
- Reengineering analysis
- Process mapping
- Quality control variance reports

All these techniques require inputs such as

- Relevant interested parties or populations (for the interviews, in-depth marketing research, or questionnaires)
- Performance data of processes
- Methods for the research like techniques for collecting data, interviewing technique, or methods for analysis

All these techniques generate outputs that shed light on trends and status of processes, activities, and goods or services such as

- Information
- Conclusions and suggestions for improvements
- Results of measurements or analyses

Need for Changes

The inputs of the management review and the followed discussion may generate planned changes to the QMS. The outputs regarding those changes shall be formulated with actions, which activities are required in order to implement and control those changes. Changes may occur in

- Strategic direction of the organization and the quality policy
- Quality objectives
- Processes, operations, or activities
- Resources
- Products

For example, if strategic changes in the QMS objectives are required, an update of the quality policy will be considered, suggested, and eventually carried out. If qualitative or quantitative changes in the objectives are required, the relevant objective controls will be updated. The next management review will verify if those changes were implemented.

Resource Needs

The output of the management review shall assess the adequacy of available resources and review requirements for further resources necessary for

- Improvements in the QMS
- Changes in the QMS
- Addressing risks or opportunities
- Addressing changes in the QMS
- Achievement of quality objectives
- Corrective actions

It is necessary to mention for each decision which resources are required. The reference will act as a commitment on behalf of the top management for the allocation of those resources. In other words, when an auditor observes allocation of resources on the management review outputs, he or she would expect to see that those resources were invested.

Documented Information

The organization is required to retain (to secure and keep the records of the review for future uses and to eliminate from lost or discard) documented information. It refers to the inputs as well as outputs of the management review; the inputs of the review shall be documented and the decisions and required actions of the review shall be documented. The standard does not specify any particular method for documenting the review. The documentation may appear on a designated form, a meeting summary, or a special report—whatever suits the organization. But always remember that this is the beginning of a long-term review process, where information and data from last reviews will always be queried and will be needed to be presented and re-reviewed. It should be also documented in a way that allows follow-ups and comparisons at later stages: a report, a form, or a presentation. I suggest designing and maintaining a designated form for the matter that will document

- The date and location where the review took place
- The participants
- The issues that were discussed
- Evidences and records that supported the issues or referred to them
- Decisions including schedules, resources, and responsibilities for execution

Bear in mind that I mentioned a form, but in today's electronic environment, a form may be digital and not the traditional paper form. In any way the documented results of the review shall be submitted to the control of documented information as required by clause 7.5—Documented information of the ISO 9001 Standard.

10 Improvement

Improvement in the eyes of the ISO 9001 Standard is a concept that will drive the organization into meeting customer requirements and enhance the achievement of the quality objectives and customer satisfaction. Improvement means, in a way, finding the parameters that affect the attainment of goals and submitting them to change. The challenge is identifying the specific processes that have the most effect on conformity of goods to customer satisfaction. The ISO 9001 Standard requires the identification of those poor in performance and effectiveness processes or process outputs (products or services) and the implementation of controlled changes that will improve them. As a result, the quality management system (QMS) and its effectiveness will be enhanced. Let us review the ISO 9001 Standard requirements.

10.1 General

The ISO 9001 Standard requirements:

- The organization shall determine and select opportunities for improvement and implement actions necessary to meet customer requirements and enhancing customer satisfaction. This includes
  - Improvement of products and services in order to meet current as well as future expectations of interested parties
  - Corrective actions
  - Preventive actions
  - Actions for reducing undesired effects
  - Improvements of the performance of the QMS
  - Improvements of the effectiveness of the QMS
- Note—Examples of improvement can include correction, corrective action, continual improvement, breakthrough change, innovation, and reorganization.

Improving Products and Services

We plan, implement, and control improvements to meet customer expectations and to maintain their level of satisfaction. When planning the identification and implementation of improvements in the organization, the actions for improvements must achieve two objectives (clause 1—Scope):

- Meeting customer requirements—This objective refers to the improvements of products and services in order to meet current expectations of customers. These improvements relate to the objective expectations of customers from the product or service and its characteristics that were communicated to the customer upon sale or delivery. This objective refers also to regulatory requirements.
- Enhancing customer satisfaction—These improvements relate to a subjective perception of the customer to the degree to which their needs and expectations have been fulfilled:
  - Which customer expectations were not fulfilled?
  - Which customer expectations were not identified?

Correcting, Preventing, and Reducing Undesired Effect

Undesired effect refers to an impact caused by a quality problem. For example, lack of qualification in a work station may cause a high level of scrap and maybe even lead to nonconformity. Undesired effect has the risk of affecting other processes or other process outputs further on the supply chain. These undesired effects were supposed to be identified and addressed during the review of risks and opportunities. After identifying events that may cause the undesired effect, the goal of improvement is to remove the obstacles, interferences, and interruptions. The standard demands initiating improvements that will correct, prevent, or reduce them. If we look back at the example of the scrap, then the organization will need to initiate actions to correct, prevent, or reduce the undesired effect:

- Analyze the root causes for the scrap related to the performance of personnel—which activity is not performed well and why?
- Suggest ways to handle those root causes—what kind of training may help personnel to do their job better?
- Control the improvements—measure the scrap before and after the improvement.

Improving the Performance of the QMS

Performance of the QMS is measured by certain indicators of processes and their relevant criteria, which suggest how much processes achieve their objectives (or not). The performance of the QMS is monitored, measured, and evaluated through many quality tools and methods suggested in the standard. An analysis of process performance may result in improvement, redesign, or reengineering of the QMS activities.

The standard requires that the organization improve the ability of the processes to attain desired results. Based on the results accepted from the monitoring, measurement, analysis, and evaluation, activities shall be initiated in order to meet the business objectives. Improvement of process performance means improving the ability of processes to achieve intended results and thus the achievement of the quality objectives. One must initiate specific activities and promote them to achieve their objectives. Improvement may be introduced to one of the following process factors:

- Process inputs that are delivered to the process
- Activities that operate the process
- Resources used in the processes
- Process outputs accepted from the process

Improving the Effectiveness of the QMS

Effectiveness of the QMS refers to which extent objectives of processes were achieved and may indicate how much the QMS has improved. Effectiveness of the QMS is regarded from several aspects:

- Effectiveness of the planning of the QMS—Improving the effectiveness of the planning refers to the implementation of processes, operations, and activities of the QMS.
- Effectiveness of processes and activities—Improving effectiveness of process and activities refers to their performance.
- Effectiveness of performance of external providers—Improving the performance of the external providers refers to their abilities to deliver products or services in accordance with predefined requirements.

Identifying Opportunities for Improvement

An improvement in the organization will be achieved by identifying opportunities or needs for improvement, and submitting these needs to a planned implementation.

- The organization shall create the awareness and understanding that improvements concern all process stages and organizational units.
- Identification of opportunities shall span all levels or areas of the organization or in all organizational units that are under the QMS.
- The importance and significance of the matter shall be introduced to all of the relevant personnel in the different organizational units during training and certification.
- The appropriate tools shall be identified in order to provide employees with the suitable platform to promote improvements (forms, meetings, reviews, motivation plans, competitions, incentives, etc.).

- The organization shall promote improvement at external providers and distributers.
- The tools for the implementation of improvements shall be suitable to the processes of the organization as well as their outputs (products).

The ISO 9001 Standard expects the organization to define which inputs and sources of information in the QMS may indicate the need for improvement. The standard refers to some well-known and widely used quality tools and their outputs:

- Quality policy—Deviation from the general guidelines, intentions, and goals referring to quality may suggest that an improvement is required.
- Quality objectives—The purpose of quality objectives is to carry out the quality policy and implement it in the QMS. These objectives are to be aimed, planned, and implanted for achieving improvement. Lack of obtainment of these objectives may suggest that an improvement is needed.
- Results of audits (internal or external)—Audit is an effective tool in identifying opportunities for improvements, since the outputs of the audit indicate whether a nonconformity (which is, by definition, a lack of achievement of objectives) or opportunity for improvement was detected.
- Monitoring, measurement, analysis, and evaluation activities—A quality activity that indicates the status of processes or process outputs with reference to their criteria may point out opportunities for improvements.
- Results of process audits—Process audit is a method that you may apply in order to measure and evaluate performance of processes, their ability to create a product according to its specifications and the extent to which process outputs meet their requirements. Where the results indicate inability of processes, it may indicate the need for improvement.
- Nonconformities—Nonconformity is an indication that the product has not met its specifications, suggesting a need for improvement.
- Addressing risks and opportunities—Improvements may be suggested while conducting activities to reduce or eliminate risks or promote opportunities.
- KPI—Key process indicators identify performance gaps and indicate areas for potential improvement.
- Customer feedback, warranty call, and complaints—These are activities that involve direct communication with the customer regarding the performance and characteristics of the product or service. These types of outputs are very effective for identifying opportunities for improvements.
- Corrective actions—There is a correlation between the improvement and the corrective actions, since each of these may be considered an improvement. Therefore, these actions may be presented as improvements. However, there are cases where corrections are temporary, and thus a long-term action of improvement is required.
- Management review—One of the outputs of the review is the suggestion for improvements and changes to the QMS and its processes.
- Employees—Another effective source for improvement is collecting ideas for improvement from personnel that operate the QMS. It is a very effective source because employees know the processes and the activities and face the daily problems. A classic method is for employees to fill in a form, where they can describe their idea for improvement.

The next aspect to be considered is the information and data that those inputs provide. In order to initiate improvements in the organization you have to make sure that relevant and accurate data and information flow to the appropriate persons:

- Information and details about the need for the improvements
- Data must be measurable, accurate and reliable, and usable to indicate need to improve process efficiency and effectiveness
- Data that will support the decision for the improvements
- Data that will enable monitoring of the effectiveness of the improvements

After collecting the inputs for the improvements, one must determine where the improvement shall be applied. Improvements may be introduced to the following QMS elements:

- Products—improvements related to the intended use of products
- Processes
  - Improvements related to the activities and operations
  - Improvements related to their interrelations with other processes
- Organizational structures
- Resources
  - Improvements related to the use and deployment of resources
  - Improvements related to competence and knowledge of personnel
  - Improvements related to infrastructure and technology
  - Improvements related to process environment
- Relations with relevant interested parties
  - Improvements related to external providers
  - Improvements related to the expectations of customers and other interested parties (like regulators)

Initiating Actions for Improvement

The organization is required to implement any necessary actions to achieve improvements that were identified. As mentioned earlier, the goals of such actions must be

- To meet customer requirements
- To enhance customer satisfaction

An improvement envisions a new process, operation, activity, or process element and the actions taken would normally bring about a change in the QMS. Each action refers to its relevant ISO 9001 Standard requirement:

- When an action causes a change in the product or service, you must determine to which QMS requirement it shall be submitted:
  - Changes in the QMS
  - Changes in the planning of processes
  - Changes to requirements for products and services
  - Design and development changes
  - Control of changes for production or service provision

- When an action initiates a new activity or a process, it shall be submitted to the planning of processes.
- When a corrective action is initiated, it will be submitted to the controlled process of corrective actions.
- When documentation is changed, it will be submitted to the control of documented information.

In addition

- The organization shall analyze and provide the required resources needed for improvement.
- Improvements must be prioritized according to their effect or impact on the conformity of products and customer satisfaction.
- Many process improvement efforts tend to focus on one functional area without considering other consequences that may influence the QMS.
- When initiating an improvement, make sure to involve those individuals and groups in the organization that have the knowledge, information, and experience of the process and who can point out how changes may be effectively applied.
- Actions of improvement may be assigned to a process owner so they may control the progress and make decisions that affect the development, maintenance, and improvement of the process:
  - Defining goals for the improvement
  - Defining the plan for achieving the improvement
  - Scheduling the steps
  - Addressing risks

Examples for Improvements

A practical method for practicing improvement in the organization is required (and will be discussed) in clause 10.3—Continual improvement. Other improvement methodologies:

- Benchmarking
- Redesign of processes
- Six Sigma
- Lean
- TQM (total quality management)
- ABC (activity-based costing)
- Performance improvement model

Each methodology has its advantages and disadvantages. You must pick the one that suits your organization and its capabilities and operations.

Introducing Innovation for Improvement

Innovation refers to the creation of a new QMS element (a new product, resource, or process) resulting from a need for improvement. Innovation must be significant in its effect in order to justify the related investment. The objective of consistently meeting requirements and addressing future needs and expectations may be achieved when applying or introducing innovation in the organization. Furthermore, many of the necessary improvements will require the application of innovation in various fields and areas of the organization. Another trigger for innovation is changes in the business environment of the organization, which presents new technologies or new market needs. A good and relevant example is the Internet that impels many organizations to develop an online platform for their distributing their products and services. This kind of change demands from those organizations to drop their traditional activities and develop new innovative operations. The following events may involve innovation:

- Organizational changes as a result of self-assessment
- Potential changes in the organizational environment that may affect the realization processes or the product directly
- Changes as well as potential changes in the expectations of customers or regulatory requirements
- Technological changes as well as potential changes that may affect the realization processes or the product directly
- Needs for optimizing the performance of the organization
- Results of monitoring, measurement, analysis, and evaluation that indicate a need for improvement

Innovation may be applied to

- Organization and organizational units that operate the QMS
- Processes that are realizing the product
- Technology in the organization used for the realization of products
- Product characteristics and features that are expected by the customer

The approach for adapting innovations may include the planning of activities for identifying needed innovations and the applied innovations may be implemented with a plan. The plan may then consider the following:

- Need for the innovation—which improvement requires innovation and which QMS element will be innovated
- Prioritization of the innovation and deciding whether it is applicable
- Introducing the innovation to a controlled process, that is, introducing it as an improvement:
  - Planning actions
  - Identifying which knowledge is required in order to implement the innovation
  - Schedules for implementation
  - Allocation of resources
  - Planning of controls
  - Review and assessment of related risks and planning actions for minimizing the risks
  - Application of corrective actions when needed

10.2 Nonconformity and Corrective Action

Corrective action is one of the foundation elements of quality management and is essential for sustaining improvement of the QMS. The main concept of corrective action promotes a systematic analysis of quality problems that have already occurred, and the elimination of any root causes of nonconformities through the implementation of controlled measures. As per clause 0.1—Introduction the organization, it is required to develop the ability to demonstrate conformity to specified QMS requirements. In other words, your customers gain confidence in the organization's ability to consistently provide products and services conforming to their requirements and specifications. The ISO 9001 Standard requirements:

10.2.1

- The organization shall react when nonconformity occurs, including nonconformities that were reported during a complaint.
- The organization shall initiate an action to control and correct the nonconformity.
- The organization shall analyze and react to the consequences of the nonconformity.
- The organization shall evaluate the need for an action to eliminate the cause or causes of the nonconformity in order that it does not reoccur or occur elsewhere.
  - The organization shall review and analyze the nonconformity.
  - The organization shall detect the root cause for the nonconformity.
  - The organization shall initiate actions to remove the root cause.
  - The organization shall determine whether similar nonconformities exist, or could potentially occur.
- The organization shall initiate and implement the necessary actions for the correction of the nonconformity.
- The organization shall review the effectiveness of the actions taken to correct the nonconformity.
- The organization shall update the risks and opportunities related to the corrective action during their planning.
- The action for the correction shall be appropriate to the effect of the nonconformity.

10.2.2

- The organization shall retain documented information as evidence of
  - The nature of the nonconformities
  - Any subsequent actions taken
  - The results of any corrective action

Terms and Definitions

Before we start, let us review some terms and definitions in order to toe the line:

- Conformity—Demonstrating fulfillment of a requirement.
- Nonconformity—An unfulfillment of a specified requirement. Nonconformity has two aspects: the realization of defective products that cannot perform their intended use and situations where the QMS does not perform in accordance to the ISO 9001 requirements.

- Root cause—An initiating cause of either conditions, factors, or parameters that led to an undesirable outcome (nonconformity).
- Corrective action—An activity that was taken to eliminate the cause of nonconformity and to prevent its recurrence, for example, rework or repair.

Developing a Method for Handling Nonconformity and Applying Corrective Actions

For handling nonconformity, the organization is expected to develop a method with the objective of ensuring that all discovered nonconformities are identified, analyzed, managed, and controlled. Handling the nonconformity should follow a scheme for categorizing and prioritizing the nonconformity, analyzing it, and suggesting the appropriate actions. The goal is the successful implementation of the corrective action. The objectives of the method are

- Capturing and identifying the nonconformity
- Reacting to the nonconformity
- Prioritizing, analyzing, classifying, and recording the nonconformity
- Analyzing and assessing the nonconformity and detecting the root cause
- Handling and solving the nonconformity through planning and implementation of corrective actions
- Tracking the corrective actions and evaluating their effectiveness

The documentation, classification, and categorization will be useful for statistical purposes, for the identification of the root cause, for trend analysis, and problem resolution. Setting priority or severity of the nonconformity is important for allocating resources and initiating the handling of the nonconformity.

The last edition of the standard (ISO 9001:2008) demanded the maintenance of a documented procedure that described how such a method is exercised. The new revision (2015) does not require such a procedure anymore. You can use this kind of documentation if you feel it will assist you in defining and implementing the method. I personally think that a detailed form may do the job.

Sources for Identifying Nonconformity

In order to handle nonconforming products, one must first detect them. For identifying the nonconformity, you shall initiate activities or operations that will measure, investigate, compare, and assess whether an attribute, a feature, a characteristic of a process, process outputs, or elements that are part of the QMS are performing as expected. These expectations indicate the quality of the product and when not satisfied a nonconformity occurs. During the planning of the QMS the organization identifies and determines activities, operations, or process stages where controls are necessary. The ISO 9001 Standard suggests several quality management tools to control activities and operations of the QMS and evaluate the results against the requirements:

- Audit—Auditing the QMS (internally or externally) is an effective way of identifying problems and nonconformities and the results of audits contain information on compliance against given criteria, that is, nonconformity. One clear objective of the internal audit, for example, is the sampling of processes and activities of the QMS and evaluating them against predefined criteria. When the test shows that processes were not performed as planned, that may indicate a nonconformity.

- Monitoring, measurement, analysis, and evaluation activities and KPIs— Monitoring, measurement, analysis, and evaluation activities determine the extent to which a process is quantitatively or qualitatively stable, capable, and predictable within defined limits and provide the decision-makers a situation report regarding the performance of processes and activities. Evaluation of KPIs may demonstrate where performance of processes is not as expected and may indicate nonconformity.

- Need for revalidation—The need for revalidation may indicate, in certain situations, where process parameters have not met their requirements and a corrective action is therefore required.

- Quality costs—The increase of quality costs may indicate that a quality problem exists and therefore somewhere along the process flow of the QMS a nonconformity exists. Quality costs may be generated through scrap, rework, deviations of quantities, unplanned tasks or jobs, time between failures, and increase in required gestures to customers.

- Customer complaints—A complaint is an effective way of receiving customer's opinion regarding the use of the product. A complaint is an application communicated to the organization from a customer regarding events or the disturbances of performance, functionality, intended use, or safety of the product during use. It serves as an indication of lack of customer satisfaction and that the product has not met its requirements. Thus a complaint is a reliable source for indicating the existence of nonconformity.

- Risk-based thinking—Risk-based thinking outputs may assist you in evaluating the severity and need for controls over nonconformities.

- Quality assurance and quality control—Planned and ongoing quality assurance programs or control activities and tasks are an effective method for detecting nonconformity during the realization processes. The quality assurance samples process outputs according to a defined sampling plan and compares them to predefined criteria. When the output does not meet its criteria it may be considered as a nonconformity. The quality assurance process consists of two main quality concepts:
  1. Verification activities—where process outputs are inspected against product output requirements
  2. Validation activities—where process parameters are evaluated against predefined requirements

These quality activities are only reliable when the relevant quality records are appropriately maintained and are accessible.

Reaction to Nonconformity

As soon as nonconformity is detected, the organization must react to it, identify it, segregate it, and ensure, beyond any doubt or possibility, that the nonconforming product will not be used, applied, submitted, to realization processes. Actions of how to react to nonconformity are detailed in chapter 8.7—Control of nonconforming outputs.

Correction of Nonconforming Products

Part of the reaction to nonconformity is its correction. Corrections of nonconforming products may be initiated as reworking or reprocessing. The correction includes the repair, rework, reprocessing, or adjustment of the product in order to eliminate the nonconformity and ensure that the product meets its specifications. Actions of how to correct to the nonconformity are well detailed in chapter 8.7—Control of nonconforming outputs.

Dealing with the Consequences of Nonconformity

Reacting and correcting nonconformity shall put you in a position to understand what its effect is, which will help you deal with the consequences—suggesting an action to repair the damage that the nonconformity caused and ensuring that the product meets its specifications. Dealing with the consequences of nonconformity first requires the understanding of

- The effect of the nonconformity on the process
- The effect of the nonconformity on process outputs

Dealing with consequences may be realized through repair, rework, swap of products, or rejection. In order to deal with the consequences effectively, you must

- Determine the scope of the action—in which area of process or product must be the action applied
- Define the objective of the action—what is to be achieved
- Determine the needed activities—which activities are to be undertaken, what are the relevant tools and equipment, what is the necessary documentation, and what are the necessary qualifications
- Evaluate relevant risks—assessing further consequences of those actions and which controls are to be set
- Allocate responsibilities and resources—who should perform those activities and what resources are required
- Accepting concession—when it is required to concede responsibility for activities that led to the nonconformity

Identifying the Root Cause of the Nonconformity

When a nonconformity was detected and reported, it is required to methodically identify and correct or remove its root causes. The real cause for the nonconformity may not be obvious. The objective is to identify the real reason for the nonconformity and to entirely prevent the problem from reoccurring by analyzing the events and factors that led to the nonconformity and identifying which elements of the process deviated from their requirements, were responsible for the nonconformity, or affected the quality of a process or product.

A nonconformity suggests that a set or processes or their outputs do not satisfy their expectations and their requirements are not met. One may say that you need to understand the sequence of events or combinations of conditions that induced the nonconformity. The source of the nonconformity may occur in various elements throughout the QMS: a product, a process, a machine, an employee, a supplier or an infrastructure. Root cause refers to deviations in the predefined conditions of a process in its intended environment. One or several root causes may lead to nonconformity. These conditions were not available at the time of the realization because of a certain sequence of events. These combinations of events and conditions created deviations that affected the process and its outputs. The cause for the deviation of these conditions must be identified and understood. The identification and correction or removal of those factors or conditions will prevent the reoccurrence of the nonconformity. In other words, fixing the problem ought to prevent the whole sequence of events from reoccurring. The challenge is to detect all root causes. The investigation into the root cause shall produce three main conclusions:

1. The reason for the nonconformance that will need to be removed—which conditions or parameter did not achieve their objectives.
2. The severity of the nonconformance—how many resources must be invested in order to remove the root cause.
3. The impact of the nonconformity—which other products, elements, or other areas may have been affected by this nonconformity.

After realizing these, you may need to exercise the following activities:

- Defining the problem and understanding which objective was not achieved. The characterization of the problem should be quantitative or qualitative of the undesirable outcomes.
- Gathering of data, information, and facts related to the problem.
- Analyzing the process and the activities relevant to the nonconformity.
- Associating the gathered data, information, and facts with the operations and activities of the process.
- Understanding which activity or operation produced the problem.
- Analyzing the conditions and factors that influence the activity, which can be divided into the following main types:
  - Conditions regarding the inputs of the process
  - Conditions regarding the planning of the process
  - Conditions regarding the resources that support the process (competence knowledge, documented information, infrastructures)
  - Conditions regarding the controls of the process
- Understanding the influence of those conditions and factors.
- Identifying the root cause for the nonconformity.

The investigation, its findings, and conclusions will be retained as documented information. Once the reason is identified, all the relevant parties and authorities are to be notified in order to allow follow-up treatment of the nonconformity. You may implement and use different methods for identifying the root cause: histograms, fishbone diagrams, using the five Whys analysis, statistical analysis, correlation diagrams, and so on.

Evaluating the Need for Corrective Action

After identifying the root cause that led to the nonconformity, you must decide whether a corrective action is necessary. The question is, whether the deviation is considered to be a nonconformity or whether it is tolerable. Intuitively, deviation or variation from the specification is considered a nonconformity when process specifications have not been met. There are, however, cases where the deviation is considered and accepted. Not every deviation will prevent the product from meeting its requirements. It is under the responsibility of the organization to define what deviations are acceptable.

I suggest here a review for measuring the severity of the nonconformity, and to decide whether a corrective action is actually required. Basically, the organization shall take corrective actions when quality management goals are not achieved. But there are cases or situations where you may choose to live with the root cause. In case you decide not to initiate a corrective action, provide sufficient justification. You may define criteria, a checklist, or develop an independent review of the matter in order to reach a decision. Possible reasons for not initiating corrective actions:

- The root cause happened only once and the organization cannot in any way control it.
- The nonconformity and its related corrective action are not worth the investment.

The decision will relate to all product or process elements that may have been affected; you may decide that corrective action is required on all elements or aspects of the process or none at all.

Implementing Corrective Action

After the first reaction to the nonconformity, the analysis of the root cause, and deciding that a corrective actions is necessary, it is required to initiate a corrective action—an activity that will eliminate the cause of the nonconformity and will prevent its recurrence. Elimination of the nonconformity means the correction of a product or a process and the assurance that the nonconformity will not happen again. Implementing the corrective action includes planning activities in order to correct identified deviations and variations from the objectives. Your goals should be to

- Eliminate the cause for the detected nonconformity
- Prevent potential nonconformity

The elimination of the nonconformity must be effective. This means that objectives for the corrective actions of elimination are to be determined, and a control is to be implemented to ensure that they will be achieved. In the spirit of the ISO 9001:2015 and the process approach the issue shall be planned and incorporated into the QMS. I suggest here some guidelines for planning and carrying out a corrective action:

- Characterizing and clarifying the nonconformity
  - Identification of nonconformities
  - Analysis of the root cause

- Setting objectives
  - Determining the operative objectives
  - Defining the timeframes
- Planning
  - Promoting a systematic method for the evaluation of root causes and the determination of appropriate solutions
  - Planning actions to be implemented, approved, and tested
  - Defining the required results that will approve that the corrective action is effective and the root cause is eliminated
  - Carrying out the implementation of the actions
  - Defining the controls needed to be implemented in order to test the effectiveness of the corrective action
  - Defining of the expected records that will demonstrate the results
- Defining the resources
  - Determining responsibilities and authorities for the execution and implementation of the corrective action
  - Determining the necessary tools and equipment
  - Identifying the training needs
  - Identifying the relevant documentation
- Evaluation of the corrective actions and its results:
  - Evaluation and review of risks and the consequences of actions that may affect the realization processes or the product
  - Review of the effectiveness of the corrective actions and the achievement of the objectives
  - Defining which documents must be updated

You may design a form that follows all of the above and ensures that all issues were referred to or include these issues in an existing form.

The planned corrective actions shall be appropriate to the effects of the nonconformities encountered. Conformity of goods and services as well as the severity and impact of the root cause play a main role when planning the corrective action. On one hand, the actions must be sufficient to ensure the conformity of goods or services and customer satisfaction but, on the other hand, the efforts must be balanced against the need for a corrective action:

- Effects on conformity of goods and services—It is necessary to measure the impact of the nonconformity or its influence on the ability of QMS elements to achieve or support conformity of goods and services.
- Effects on customer satisfaction—It is necessary to measure the impact of the nonconformity on customer satisfaction.

Consideration of these implications shall determine the extent and degree of the corrective action. What might influence the decision?

- Regulatory requirements
- Effects on product requirements related to the goods and services
- Effects on customer requirements and product specification as determined by the customer
- Effects on processes and activities needed to realize the product

Effectiveness of the Corrective Action

The organization is required to review the effectiveness of the corrective action. The goal is to examine whether the corrective action has reached its objectives. It will be achieved through the evaluation of results of the corrective action. The action shall then receive a status (I suggest here some types): open, in process, or closed. The review shall also consider the results: success, failure, or perhaps more time or other actions are required to examine its effectiveness. The results of the review are to be documented and serve as documented information. When the corrective action was found effective, the organization may close it. A closed and effective corrective action means

- The corrective action was suitable to the nature of the nonconformity.
- The corrective action could be introduced to the necessary elements of the QMS and be properly implemented.
- The objectives of the corrective action have been achieved.
- The corrective action is effective and the nonconformity did not, and will not, reoccur.

The results of the nonconformity shall be evaluated with the objective of ensuring that problems have been resolved, determining whether additional problems have not been introduced, ensuring that adverse effects have been eliminated, and changes have been correctly implemented in the QMS activities. While closing the corrective action, the identification of the approver who concluded the corrective action is to be documented.

When a corrective action is found to be ineffective, it is recommended that you take a new one, because an unsuccessful or ineffective corrective action indicates that the root cause of nonconformity has not been eliminated.

Addressing Risks and Opportunities

Results of analyzing and understanding the root cause of the nonconformity may suggest new risks or opportunities in the QMS that must be addressed. When associating risks with the nonconformity, you must consider the significance of the nonconformity, its frequency, and how well can the users of the QMS identify it. The main goal is to promote a proactive approach that will enable the identification and handling of potential quality problems before the nonconformity occurs. Risks to the QMS are associated with the nonconformity when it creates a situation where

- The QMS will not achieve it quality objectives
- Desirable results of process may not be accepted
- Process or product characteristics may not be achieved
- Adverse effects are possible
- Customer satisfaction may be harmed

When such situations occur, they must be submitted to the planned activities for addressing risks as suggested in clause 6.1—Actions to address risks and opportunities.

In practice, you must plan the interface between the QMS processes—corrective action and risk-based thinking.

Implementing Changes in the QMS

A corrective action may trigger changes or updates to the QMS. Such changes, initiated by, or resulting from the corrective action in QMS elements are to be controlled and implemented. Changes may occur in quality policy, quality objectives, processes, procedures, work instructions, resources, infrastructures, process environment, and documented information. This is why you have to evaluate where such changes may occur when initiating a corrective action. Changes will be submitted to the appropriate controls as suggested in clause 6.3—Planning of changes or clause 8.5.6—Control of changes (depending on the context).

Retaining Documented Information

Documentation of nonconformance and the following corrective action is required by the ISO 9001 Standard. A detailed report shall be defined and prepared to describe the nonconformity and the subsequent actions taken. This report shall support the closed-loop process of planning a correction and evaluating its effectiveness.

The first level of documentation is the primary information gathered about the nonconformity and any information that would help you to investigate the nonconformity later on. The objective of the information is to assist you in mapping the problem and tracking down the root cause. This information was gathered while the nonconformity was identified (see clause 8.7—Control of nonconforming outputs). Bear in mind that this documentation is the first step in a process that will later lead to a corrective action or improvement. The second level of the documentation is the corrective action initiated to remove the root cause and will serve as a follow-up tool to evaluate its effectiveness. Let us review which details, information, and data regarding the nonconformity you should document:

- Statement of the starting date
- Identification and details of the product (product name, model, catalog number, serial number, batch number, date of manufacture, etc.)
- Identification of the individual that detects it (employee, customer, supplier, regulatory authority, etc.)
- Identification of who is responsible for handling the nonconforming product—it may be an individual or a committee
- Type of nonconformity (production error, customer complaint, supplier error, etc.)
- Time and date of receiving or detecting the nonconformity
- Quantities that were identified as nonconforming
- Description of how the nonconforming products were identified or labeled
- Specific areas where the nonconformity was detected or related to the product (department, production hall, machine, etc.)
- Description of the nonconformity (e.g., why it failed to meet requirements—you can provide here a literal description, or even attach other documents such as a statistical analysis)

- Immediate treatment that was applied to the matter
- Reference to documented evidence
- Relevant procedures or work instructions
- Categorization of the nonconformity—this is not required by the ISO 9001 Standard but is highly recommended and will assist you later with a statistical analysis of the nonconformity
- Details and information of the investigation of the root cause
- The evaluation of the need for a corrective action

Details, information, and data regarding the corrective action:

- Statement of the starting date
- Details for reference or identification of the corrective actions
- The team that participates and is responsible for the planning and implementation of the corrective action
- Description of the relevant nonconformity or reference to its documentation
- Description of immediate or short-term corrective action (if undertaken), its results, and reference to the relevant records
- Reference to the results of the root cause analysis investigation and the conclusions
- Description of the solution and planned corrective action to be taken with reference to the relevant factors: departments, products, processes, packages, tools and equipment, monitoring and measuring devices, and controls
- The objectives of the actions taken, including timeframes
- Review and results of the implementation and it effectiveness

The record may appear as a form, a work plan, a project plan, or refer to such dossiers. However, the principle demands a specification of activities or milestones assigned to responsibilities and framed with objectives of time and results. Useful documentation of the plan will also include the results of the implementation. In the case the corrective action was found ineffective, then a following action—an improvement or a change to the corrective action—must be initiated. When designing a form, I recommend designing it logically as well as chronologically. Bear in mind that the form is a tool in the hand of the employee while gathering the information about the nonconformity and initiating an investigation. You may attach records of customer complaints or pictures of nonconforming process outputs on the form.

I suggest here a format for the documentation that I find very effective because it covers all the aspects mentioned in this chapter and follows a rational course for handling nonconformity and applying the corrective actions effectively. You may adapt it to your needs.

1. The first part refers to the relevant interested parties:

Customer/supplier/department:	Date of creation:
	Date of update:
Details of contact person	External Reference No.

2. This part refers to the product or process where the nonconformity was detected:

Identification of product or process	
Identification of relevant complaint	
Identification of involved resources	

3. Definition of the team that will handle the nonconformity (personnel or other interested parties that are involved):

Name	Role (involvement in the nonconformity)

4. Description of the nonconformity—the failure and which objectives were not met:

Description of the nonconformity

5. Description of the immediate measures taken:

	Responsibility	Date	Effectiveness
Immediate measures taken to handle the nonconformity			
Measures taken to ensure that no nonconforming products will not be delivered to the customer			

6. Evaluation of other products affected by the nonconformity:

Are any other products affected by the nonconformity	☐ Yes	☐ No
List of affected products		

7. Analyzing the reasons for the nonconformity and description of the introduced corrective actions on processes and resources:
   a. Actions taken to eliminate the root cause
   b. Actions taken to prevent recurrence of the root cause

Definition of the basic problem and the root causes	Description of the durable corrective action for each root cause detected	Responsibility	Date	Effectiveness

8. Updating the relevant documentation:

Documentation	Relation to the nonconformity	Responsibility	Date

10.3 Continual Improvement

The ISO 9001 Standard strives to introduce improvements in the organization into a method that will control it and will ensure that its objectives will be achieved. Continual improvement is a type of learning where the organization evaluates itself constantly, makes informed judgments and decisions based on the results of such analyses, and initiates actions for improvement. The standard promotes the approach of the PDCA (Plan-Do-Check-Act) cycle, which will be discussed in detail later on. The ISO 9001 Standard requirements:

- The organization shall continually improve the suitability, adequacy, and effectiveness of the QMS.
- The organization shall review and evaluate the following types of results:
  - Results of analysis and evaluation
  - Results of management review
- The objective is to determine if there are opportunities for improvements, which may be considered as inputs for continual improvement.

Improving the QMS

Continual improvement implies that changes that happen frequently and systematically can have a significant effect on the QMS. Controlling and improving those changes made to the QMS leads to process and product optimization. The organization is required to continually improve the suitability, adequacy, and effectiveness of the QMS. The following distinctive characteristics of QMS shall be improved:

- Adequacy—refers to the ability of the QMS, its processes, and products (and services) to meet the needs and expectations of interested parties satisfactorily.
- Suitability—refers to the quality of having planned processes, resources, and controls that may assist in achieving the quality objectives.
- Effectiveness—refers to the quality and capacity of being able to achieve planned objectives.

An effectively implemented method for continual improvement is likely to achieve the following goals:

- Enhancing effectiveness of processes and driving process results—Through the improvement of factors that influence the processes, and interrelations between QMS entities, process performance, and productivity of processes increases.
- Improving the quality of products—Improvements in processes, operations, and work activities improve process outputs.

- Creating value to the organization and reducing quality costs—Through treatment of factors that cause nonconformity.
- Enhancing customer satisfaction—Through the delivery of full conform products.
- Enabling meeting quality objectives and driving alignment of processes with business strategy—The actions for improvements are aimed at quality objectives that were not achieved.

Inputs for Continual Improvement

In order to be able to improve processes or products, you must first know what is needed to be improved and need have the necessary inputs to indicate where improvements are needed. The Standard defines two quality management tools for performance evaluation that may introduce inputs to continual improvement:

- Results of management review—The management review collects many inputs from many QMS activities that refer to their performance. The outputs of the management review shall serve as inputs to the method of continual improvement. The top management has a significant role in introducing the approach of continual improvement in the organization and is committed to promoting it and ensuring effective results. Opportunities for improvements were submitted to the management review as inputs and were reviewed. The outputs of the management review refer to improvements the organization should initiate.
- Results of analysis and evaluation—The results of analysis and evaluation shall be used to evaluate opportunities for improvement. Monitoring, measurement, analysis, and evaluation activities gather data and information that refer to the performance of processes and the quality of its outputs. Process analysis
  - Identifies processes or products that may improve performance and helps in determining the root causes of the inadequacies
  - Assists the organization in understanding how and where improvements should be applied to gain the maximum achievement of objectives

For example, information of KPI—These key process indicators deliver data concerning the performance and effectiveness of their related processes. Those results indicate where objectives were not achieved or where nonconformity may occur or has already occurred. Be aware that too much information may result in "too much" and irrelevant information.

Other inputs that may be submitted to continual improvement are suggested in chapter 10—Improvement under the paragraph Identifying Opportunities for Improvement. Those inputs, recommendations, ideas, needs, suggestions, or propositions will be introduced to the process of continual improvement.

PDCA Cycle

The Plan-Do-Check-Act (PDCA) is a concept (one of many) that drives continual improvement in the organization. The ISO 9001 Standard employs the process approach that incorporates the PDCA cycle. The process approach, promoted by the

ISO 9001 standard, systematically identifies and manages processes that operate the quality system and maintain its interactions. The process approach uses the PDCA cycle to improve itself constantly.

The principle of the PDCA cycle exists in all of our daily business activities. We use it both formally and informally, and the PDCA cycle never ends. Its objective is to maintain continuous improvement. The method combines planning, implementing, controlling, and improving of the different operations of the QMS; the PDCA can be implemented at the core process, at a minor process or even at several processes altogether, a product, or a resource. The PDCA is divided into four stages:

1. Plan—The planning stage is the starting point and will take place after you have located your need for improvement where you already have all the information and inputs. During this stage you may ask yourself the following questions:
   - Why is the improvement necessary? Indicate which quality or process objectives have not been met or which quality problem has occurred.
   - What needs to be done? Set the activities needed to achieve improvement. Break down the goal of the improvement into feasible activities.
   - How much must be done? Set the objectives, goals, and targets that you would like to achieve. Of course they must be derived from the ultimate goal of the improvement.
   - Where must it be done? Set the scope, areas, processes, and process outputs where you would like to achieve improvement.
   - Who should do it and how should they do it? Set the resources and responsibilities required to achieve the results.
   - When must it be done? Set a timeframe in which you would like to achieve the improvement.

   Consider writing the plan with the participants themselves; they may provide appropriate inputs and set effective activities. Try to make the "Do" part (the next stage of the PDCA) easier for them to perform. During the "Plan" stage you can use quality planning tools mentioned in clause 6 of the standard: actions to address risks and opportunities, quality objectives, and planning of changes.

2. Do—This is the second stage where you realize your plan. This stage refers to the operation of your plan. During this stage, you submit the plan (through the relevant participants) for execution on the related areas or scopes in the organization. This is a hard part where you will put words into actions. Therefore, try to make the plan as clear, precise, and detailed as possible. The outputs of the "Do" will be incorporated in your quality activities for the operation of the QMS mentioned in clause 8 of the standard—planning and control, determining and reviewing the requirements for products and services, design and development, control of externally provided processes, products and services, control of production and service provision, etc.

3. Check—In the third stage you assess the effectiveness of the operation of the plan. You will monitor and measure the activities that you have planned and evaluate their progress according to the objectives of the plan. Then, you must report the results. This part is crucial and for every objective set in the "Plan" stage, you must evaluate the results. This stage will verify that the activities were performed as planned and validate that their outputs conform to their objectives. Depending on the results and conclusions of this stage, you would advance to the next stage—"Act." In the

"Check" stage, you may use your quality tools for performance evaluation as suggested in clause 9 of the standard, such as monitoring, measurement, analysis and evaluation, internal audit, and review.

4. Act—In this fourth stage, you are acting for improving the performance of your plan. At any point where you are not satisfied with the results of the plan's progress, review your plan and act to improve it. Examine in which parameters you failed to reach the objectives and try again, or try it in another way. This is how you maintain constant improvement. In the "Act" stage you will need to practice principles of leadership and commitment, as suggested in clause 5 of the standard, such as taking accountability for the effectiveness of the QMS and engaging, directing, and supporting persons to contribute to the effectiveness of the QMS.

Figure 10.1 demonstrates how the PDCA cycle integrates into the QMS.

Support processes play an important role in the PDCA cycle because they transfer significant resources such as competence of personnel, technology, and knowledge to the improvements.

Although documented information is not required, I would document its application. The documentation will help prove that you have implemented the technique of the PDCA in your QMS. The most common method is managing the process with a form that will gather and manage all the information in one place

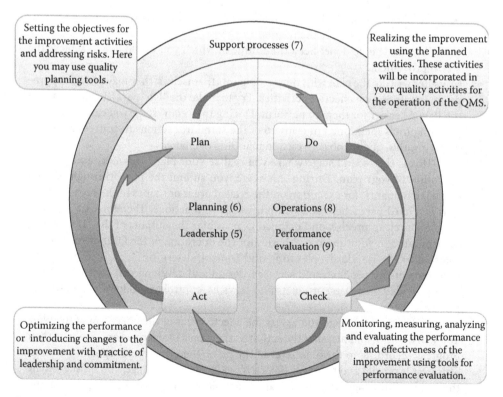

Figure 10.1   Integrating the PDCA cycle in the quality management system.

and will be used as a report for the PDCA activity. The form may be designed in the following format:

- The report will cover all the stages of the PDCA cycle and therefore will be respectively divided.
- To each section of the report you will detail the following:
  - Who are the responsible parties to this stage?
  - What are the required inputs?
  - What are the required resources?
  - What are the expected outputs?
  - Were the activities of this section effective?
  - What is the relevant documentation?
- Details of the improvement—Reference to a process, activity, product, or resource. In this part, you will identify to which element or entity of the QMS the improvement concerns.
- Planning of the improvement (Plan):
  - Description of the case (I tend to be as detailed as possible).
  - Who or how the need for improvement was identified: results of monitoring, measurement, analysis and evaluation, nonconformity, reported by an employee, need for innovation, or output of management review.
  - What is the reason for the improvement; which objective has not been met?
  - Which action will be executed?
  - Who are the responsible parties that will accomplish it?
  - By when should the action be accomplished?
  - What are the expected results and how they will be measured for their effectiveness?
  - How will it be documented?
- Implementing the actions (Do)—Here you need to collect and gather data and information regarding the carrying out of the planned actions. For each action you need to indicate the following:
  - What are the sources for the data and information regarding the operations of the initiated actions?
  - Which reports will be used to deliver the data and information?
  - Who are the responsible parties to deliver the data?
- Controlling the accomplished actions (Check)—Which control measures will be applied in order to ensure the effectiveness of the actions taken. Controls may be management review, audits, or analysis and evaluation. At this stage you will ensure
  - That all actions were undertaken (verification)
  - That the results are satisfying (validation)
  - The level of effectiveness
- Initiating further actions (Act)—According to the results of the control, you must decide whether further corrective actions must be initiated in order to meet the objectives or optimize them:
  - Repetition of the planned actions
  - Change of the already planned actions
  - Initiating new actions

Naturally, this stage brings us back to the first stage—planning.

I illustrated the PDCA cycle with the next form. You may modify it according to your needs:

Details	
Process/Product/Resource/Org. Unit	
Further Details	

Planning					
Description of the case					
Identifier					
Relevancy (to quality objective)					
Actions to be undertaken	Responsibility	Resources	Documentation	Schedule	Expected Results
Action 1					
Action 2					
Action 3					

Operation				
	Type of Data or Information	Source	Report	Responsibility
Action 1				
Action 2				
Action 3				

Control					
Control to Be Applied	Type of Control	Date Received	Date	Effectives	Remarks
Action 1					
Action 2					
Action 3					

Optimization						
	Argument	Responsibility	Resources	Documentation	Schedule	Expected results
Action 1						
Action 2						
Action 3						

Index

Printed in the United States
by Baker & Taylor Publisher Services